Diet, Brain, Behavior

Practical Implications

Diet, Brain, Behavior

Practical Implications

EDITED BY
Robin B. Kanarek and Harris R. Lieberman

CRC Press
Taylor & Francis Group
Boca Raton London New York

CRC Press is an imprint of the
Taylor & Francis Group, an **informa** business

CRC Press
Taylor & Francis Group
6000 Broken Sound Parkway NW, Suite 300
Boca Raton, FL 33487-2742

Printed in the United States of America on acid-free paper
Version Date: 20110815

International Standard Book Number: 978-1-4398-2156-5 (Hardback)

Visit the Taylor & Francis Web site at
http://www.taylorandfrancis.com

and the CRC Press Web site at
http://www.crcpress.com

Contents

Preface vii
Editors ix
Contributors xi

Chapter 1 Mental Energy and Fatigue: Science and the Consumer 1

Harris R. Lieberman

Chapter 2 Hydration and Brain Function 7

Kristen E. D'Anci

Chapter 3 Diet as an Analgesic Modality 19

Alexis M. Codrington, Yoram Shir, and John Pereira

Chapter 4 Breakfast and Adult and Child Behaviors 53

Andrew P. Smith

Chapter 5 Diet, Physical Activity, and Substrate Oxidation: Implications for Appetite Control, Weight Loss, and Body Composition 71

Mark Hopkins, Neil A. King, and John E. Blundell

Chapter 6 The Reward Deficiency Hypothesis: Implications for Obesity and Other Eating Disorders 103

Brenda M. Geiger, Erin N. Umberg, and Emmanuel N. Pothos

Chapter 7 Potential Consequences of Obesity on Cognitive Behavior 133

Nicole A. Jurdak and Robin B. Kanarek

Chapter 8 Dietary Supplements for Weight Loss 153

Igho Onakpoya and Edzard Ernst

Chapter 9 Sweet Taste Preferences and Cravings in Gestational Diabetes Mellitus (GDM): Implications for Diet and Medical Management 169

Beverly J. Tepper, Lisa M. Belzer, John C. Smulian, and Shou-En Lu

Chapter 10 Homocysteine, B Vitamins, and Cognitive Function 189
Joshua W. Miller

Chapter 11 Creatine, Brain Functioning, and Behavior 215
Patricia J. Allen, Kristen E. D'Anci, and Robin B. Kanarek

Chapter 12 Theanine, Mood, and Behavior 237
Jessica E. Smith and Peter J. Rogers

Chapter 13 Caffeine: Practical Implications 271
Andrew P. Smith

Chapter 14 Caffeine Effects on Aggression and Risky Decision Making 293
Caroline R. Mahoney, Tad T. Brunyé, and Grace E. Giles

Index 313

Preface

Nutritional neuroscience is a rapidly growing interdisciplinary field. Interest in the role of nutrition in brain functioning and behavior has gained the attention of both the scientific community and the general population. A clear example of this is the proliferation of books, magazines, newspaper articles, television and radio shows, and Internet sites addressing questions related to the effects of nutrients on behavior. Additionally, during the past decade there has been an explosion in research and publications in this field.

The objective of this book, *Diet, Brain, Behavior: Practical Implications*, is to comprehensively address practical and applied issues in nutritional neuroscience. To represent the broad scope of research in the field of nutritional neuroscience, the topics in this book are quite diverse. Moreover, the subject matter of each chapter was chosen to ensure there is current—or the potential for future—applicability to practical, applied issues. Thus, authors were asked to provide chapters based in part on the practical and applied nature of their own research interests.

Obesity is one of the most important public health issues of the twenty-first century. While it has long been known that obesity is associated with myriad metabolic problems, the effects of excess body weight on brain functioning and behavior have only recently been recognized. Chapters will review research assessing the interaction of obesity and behavior, and discuss strategies for weight loss which could ultimately improve both physical and mental health.

The use and abuse of dietary supplements or food components with known or hypothetical central nervous system effects, such as caffeine, creatine, and theanine, have also emerged as important public health issues. Caffeine, especially when consumed in the form of energy drinks, has been the focus of scientific and lay interest and concern. Creatine, which can enhance physical performance, may also improve cognitive function. The amino acid theanine, found in tea, has been studied as a potential relaxing and calming agent. This monograph includes chapters that focus on caffeine, creatine, theanine, and mental energy.

Chapters are also devoted to day-to-day food-related activities which have the potential to affect mental activities including eating breakfast, maintaining fluid intake, and taking daily vitamins. Additionally, a chapter is dedicated to the important question of the role of diet in pain sensitivity.

This book will be of interest to readers who have backgrounds in a wide range of fields including nutrition, neuroscience, psychology, and exercise physiology, as well as in related clinical fields such as medicine, dietetics, and occupational therapy. Additionally, scientists and administrators working in government agencies responsible for public health and food and dietary regulations, or working for food, dietary supplement, and pharmaceutical companies, will find this book to be extremely useful.

In closing, we would like to thank the authors of each chapter for their dedication and scholarly efforts that made this volume possible.

Editors

Robin B. Kanarek received a BA in biology from Antioch College in Yellow Springs, Ohio, and an MS and a PhD in psychology from Rutgers University in New Brunswick, New Jersey. She is a professor of psychology at Tufts University in Medford, Massachusetts, where she also served as dean of the Graduate School of Arts and Sciences. Her primary research interests are in the area of nutrition and behavior. She has conducted research on the effects of nutritional variables on the development of obesity, the physiological and behavioral factors influencing diet selection in experimental animals and humans, the role of nutrients in determining the consequences of psychoactive drugs, and the importance of nutrition for cognitive behavior in children and adults. She has authored or coauthored more than 100 books, book chapters, and articles, and has presented her research at numerous international and national conferences. Her research has been funded by the National Institutes of Health, as well as by other government agencies and by private companies. Dr. Kanarek has been actively involved in graduate education and teaching throughout her time at Tufts, serving as the mentor for more than fifteen PhD students. In 2000, she was named John Wade Professor and received the Tufts University Senate Professor of the Year award.

Dr. Kanarek's experience includes research fellow, Division of Endocrinology, University of California, Los Angeles (UCLA) School of Medicine, and research fellow in nutrition at Harvard University. She is a member of the editorial boards of *Physiology and Behavior*, *Nutritional Neuroscience*, and the *Tufts Diet and Nutrition Newsletter* and is a past editor-in-chief of *Nutrition and Behavior*. In addition, she regularly reviews articles for peer-reviewed journals including *Science*, *Brain Research Bulletin*, *Pharmacology Biochemistry and Behavior*, *Brain Research*, *Journal of Nutrition*, *American Journal of Clinical Nutrition*, and *Annals of Internal Medicine*. From 1995 to 2001, and again from 2008 to 2011, she was a member of the National Academy of Sciences, Committee on Military Nutrition Research. Dr. Kanarek has also served on review committees for the National Science Foundation, the National Institutes of Health, and USDA Nutrition Research, and as a member of the Program Committee of the Eastern Psychological Association. She is a fellow of the International Society for Behavioral Neuroscience. Her other professional memberships include the Society for the Study of Ingestive Behavior and the Society for Neurosciences.

Harris R. Lieberman is a research psychologist in the Military Nutrition Division of the U.S. Army Research Institute of Environmental Medicine (USARIEM) in Natick, Massachusetts. Dr. Lieberman is an internationally recognized expert in the area of nutrition and behavior and has published more than 100 original full-length papers in scientific journals and edited books. He has been an invited lecturer at numerous national and international conferences, government research laboratories, and universities.

Dr. Lieberman received his PhD in physiological psychology in 1977 from the University of Florida. On completing his graduate training, he was awarded a National Institutes of Health (NIH) fellowship to conduct postdoctoral research at the Department of Psychology and Brain Science at the Massachusetts Institute of Technology (MIT). In 1980, he was appointed to the research staff at MIT and established an interdisciplinary research program in the Department of Brain and Cognitive Sciences to examine the effects of food constituents and drugs on human behavior and brain function. Key accomplishments of the laboratory included the development of methods for assessing the effects of food constituents and environmental factors on human brain function and the determination that specific foods and hormones reliably altered human performance and mood.

In 1990, Dr. Lieberman joined the civilian research staff at USARIEM, where he has continued his work in nutrition, behavior, and stress. From 1994 to 2000, he was chief or deputy chief of the military nutrition program at USARIEM. His current research addresses the effects of various nutritional factors, dietary supplements, diets, and environmental stress on human performance, brain function, and behavior. He holds two patents for novel technologies to assess and enhance cognitive performance.

Contributors

Patricia J. Allen, MS
Psychology Department
Tufts University
Medford, MA
Email: patricia.allen@tufts.edu

Lisa M. Belzer, PhD
Department of Food Science
Rutgers University
New Brunswick, NJ
Email: phdsunflower@gmail.com

John E. Blundell, PhD
Institute of Psychological Sciences
Faculty of Medicine and Health
University of Leeds
Leeds, UK
Email: j.e.blundell@leeds.ac.uk

Tad T. Brunyé, PhD
U.S. Army Natick Soldier Research
Development and Engineering Center
Natick, MA
Email: thaddeus.brunye@us.army.mil

Alexis M. Codrington, PhD
Department of Anesthesia
McGill University Health Centre
Montreal General Hospital
Montreal, Canada
Email: alexis.codrington@mail.mcgill.ca

Kristen E. D'Anci, PhD
Department of Psychology
Nutrition and Neurocognition Laboratory
Tufts University
Medford, MA
Email: kristen.danci@tufts.edu

Edzard Ernst, PhD
Complementary Medicine
Peninsula Medical School
Exeter, Devon, UK
Email: edzard.ernst@pms.ac.uk

Brenda M. Geiger
Pharmacology and Experimental Therapeutics
Tufts University School of Medicine
Boston, MA
Email: brenda.geiger@tufts.edu

Grace E. Giles
U.S. Army Natick Soldier Research
Development and Engineering Center
Natick, MA
Email: grace.emelia@gmail.com

Mark Hopkins, MS
Department of Sport, Health, Leisure & Nutrition
Leeds Trinity University College
Leeds, UK
Email: m.hopkins@leedstrinity.ac.uk

Nicole A. Jurdak, MS
Department of Psychology
Tufts University
Medford, MA
Email: nicole.jurdak@tufts.edu

Robin B. Kanarek, PhD
Department of Psychology
Tufts University
Medford, MA
Email: robin.kanarek@tufts.edu

Neil A. King, PhD
Institute of Health and Biomedical Innovation
Queensland University of Technology
Brisbane, Australia
Email: n.king@qut.edu.au

Harris R. Lieberman, PhD
Military Nutrition Division
U.S. Army Research Institute of Environmental Health
Natick, MA
Email: harris.lieberman@us.army.mil

Shou-En Lu, PhD
University of Medicine and Dentistry of New Jersey
School of Public Health
Piscataway, NJ
Email: lus2@umdnj.edu

Caroline R. Mahoney, PhD
Consumer Research and Cognitive Science
U.S. Army Natick Soldier Research
Development and Engineering Center
Natick, MA
Email: caroline.mahoney@us.army.mil

Joshua W. Miller, PhD
University of California, Davis
School of Medicine
Department of Medical Pathology and Laboratory Medicine
Sacramento, CA
Email: jwmiller@ucdavis.edu

Igho Onakpoya, MS
Complementary Medicine
Peninsula Medical School
Exeter, Devon, UK
Email: igho.onakpoya@pcmd.ac.uk

John Pereira, MD
Faculty of Medicine
University of Calgary
Chronic Pain Centre
Calgary, Canada
Email: john.pereira@albertahealthservices.ca

Emmanuel N. Pothos, PhD
Department of Molecular Physiology and Pharmacology
Tufts University School of Medicine
Boston, MA
Email: emmanuel.pothos@tufts.edu

Peter J. Rogers, PhD
School of Experimental Psychology
University of Bristol
Bristol, UK
Email: peter.rogers@bristol.ac.uk

Yoram Shir, MD
Department of Anesthesia University
Alan Edwards Pain Management Unit
McGill University Health Centre
Montreal, Canada
Email: yoram.shir@muhc.mcgill.ca

Andrew P. Smith, PhD
School of Psychology
Cardiff University
Cardiff, UK
Email: smithap@cardiff.ac.uk

Jessica E. Smith
School of Experimental Psychology
University of Bristol
Bristol, UK
Email: jessica.smith@bristol.ac.uk

John C. Smulian, MD, MPH
Lehigh Valley Hospital
Allentown, PA
Email: john.smulian@lvh.com

Beverly J. Tepper, PhD
Department of Food Science
Rutgers University
New Brunswick, NJ
Email: tepper@aesop.rutgers.edu

Erin N. Umberg
Pharmacology and Experimental Therapeutics
Sackler School of Graduate Biomedical Sciences
Tufts University
Boston, MA
Email: eumberg@gmail.com

1 Mental Energy and Fatigue

Science and the Consumer

Harris R. Lieberman

CONTENTS

Caffeine: A Definitive Example of a Food Component That Increases Mental Energy and Reduces Fatigue 2
Caffeine and Personalized Nutrition 4
Methods to Assess Mental Energy 4
Conclusions 5
Disclaimer 5
References 5

Products intended to enhance "mental energy" are widely available in the United States, and recently their popularity has increased dramatically. Examples include energy drinks such as Monster® and Red Bull®; energy shots such as 5-Hour Energy™; and a wide variety of dietary supplements such as ginseng and ginkgo biloba (Reissig et al. 2009; Gorby et al. 2010). Other beverages, especially coffee and colas, have long been associated with increased mental energy or related behavioral effects, and their popularity undoubtedly is related to their ability to increase alertness. Over-the-counter caffeine, in pill form, has been available for decades as a performance enhancer. It appears that consumers seek energy beverages and shots to increase their levels of self-perceived energy, and the effects they desire are primarily associated with mental state, not physical energy (Childs 2001; Lieberman 2001, 2006, 2007; O'Connor 2006). Physical energy is a relatively straightforward concept which is well defined scientifically. The concept of mental energy is not clearly defined and has only recently been the focus of substantial scientific inquiry (Cook and Davis 2006).

It is clear that consumers frequently report suffering from lack of energy, fatigue, and tiredness, and this interferes with their daily lives (Childs 2001). Such symptoms are associated with behavioral processes including mood states like fatigue and cognitive performance rather than physical energy. Even though the concept of mental energy is not clearly defined, it can be distinguished from physical energy. Furthermore, there is scientific consensus that mental energy can be measured using appropriate standardized methods that were developed to address other issues such as assessment of psychiatric symptoms (Cook and Davis 2006; Lieberman 2007; O'Connor 2006). An individual's perception of his or her own level of mental energy is clearly a mood state, similar to the state of fatigue, and this can certainly be measured with self-report mood questionnaires, such as the Profile of Mood States (POMS) (McNair et al. 1971). However, mental energy can also be considered to be the "ability or willingness to engage in cognitive work" (O'Connor 2006; Lieberman 2007) which can be measured with a wide variety of techniques used by psychologists and others.

Current scientific interest in this concept of mental energy has apparently been the indirect result of the desire of the lay public to optimize this mental state and, consequently, the attempts of food and dietary supplement manufacturers to meet consumer demand with new products. Examples of the emerging industry and academic interest in mental energy include two conferences sponsored by the International Life Sciences Institute (ILSI) in 2000 (Clarkson 2001) and 2004 (Cook and Davis 2006) and a symposium at the Experimental Biology annual meeting in 2010 (Milner and Seligson 2010). Several review papers have summarized and evaluated the literature on specific dietary supplements that potentially alter mental energy. These reviews evaluated caffeine, ginseng, ephedra, gingko biloba, and omega-3 polyunsaturated fatty acids since these are all considered to be the supplements and dietary constituents that are most likely to affect mental energy (Lieberman 2001; Gorby et al. 2010).

As discussed in several publications, caffeine provides the best example of a food constituent that enhances mental energy because it decreases fatigue and increases alertness as measured by self-report questionnaire (Lieberman 2001, 2007). Caffeine's effects on cognitive performance, such as increasing visual vigilance, further indicate it can increase mental energy (Lieberman 2007). Caffeine's mechanism of action, competitive antagonism of central adenosine receptors, is well documented and consistent with its effects on mental energy. There is much less certainty about the effects of the other products on mental energy, although ephedra, which was withdrawn from the market for safety reasons, is a stimulant that appears to have some energy-increasing behavioral effects (Lieberman 2001). There is little convincing evidence that ginseng, ginkgo biloba, and omega fatty acids increase energy, nor is there evidence these compounds substantially modify central nervous system (CNS) receptors associated with alertness.

CAFFEINE: A DEFINITIVE EXAMPLE OF A FOOD COMPONENT THAT INCREASES MENTAL ENERGY AND REDUCES FATIGUE

About 80 percent of the U.S. adult population consumes caffeine on a regular basis, and about 80 percent of caffeine in the American diet is obtained from coffee (Barone and Roberts 1996). Tea, colas, energy drinks, and energy shots also usually contain caffeine, as do many popular dietary supplements marketed as weight loss products or performance enhancers. Levels of caffeine in specific foods vary greatly, with coffee containing the most caffeine (about 65–110 mg per cup), tea an intermediate amount (about 40–60 mg per cup), and cola and some other soft drinks about 40 mg per serving (Table 1.1). Energy beverages and energy shots are gaining in popularity

TABLE 1.1
Estimated Caffeine Content of Selected Beverages, Foods, and Dietary Supplements

Item	Caffeine Content (mg/serving)
Coffee (5 oz)	
Drip method	90–150
Instant	40–108
Decaffeinated	2–5
Tea, loose or bags	
One-minute brew (5 oz)	9–33
Iced tea (12 oz)	22–36
Chocolate products	
Hot cocoa (6 oz)	2–8
Chocolate milk (8 oz)	2–7
Milk chocolate (1 oz)	1–15
Baking chocolate (1 oz)	35
Cola beverages (12 oz)	
Coca-Cola® Classic	35
Diet Coke®	47
Pepsi®	38
Diet Pepsi®	36
Other soft drinks (12 oz)	
Dr. Pepper®	41
Mountain Dew®	55
Pibb Xtra®	41
Barq's® Root Beer	23
Energy drinks	
AMP™ (16 oz)	142
Monster Energy™ (16 oz)	160
Red Bull® (8.3 oz)	80
Rockstar® (16 oz)	160
Energy shots	
5-Hour Energy® Shot (2 oz)	138
DynaPep™ Micro Shot (4 mL)	80
Extreme Energy™ 6-Hour Shot (2 oz)	220
Jolt® Endurance Shot (2 oz)	150
Dietary supplements	
Hydroxycut™ Hardcore X (2 pills)	200
Zantrex® 3 (2 pills)	320
Stacker 2® Ephedra Free (1 pill)	200
Metabolift™ (2 pills)	176
Slenderite™ (2 pills)	75

Source: Modified from Lieberman et al. (2010).

and can contain substantial quantities of caffeine (80 to 220 mg per serving). Unlike coffee, tea, or colas, energy shots are typically consumed quickly, providing a large bolus dose of caffeine. Dietary supplements containing caffeine, which are often marketed as weight loss aids, can contain 75 to 320 mg per dose (Table 1.1). For a review of these and other weight loss supplements, see Chapter 8 in this volume by Onakpoya and Ernst.

Caffeine has specific and limited effects on human cognitive function. When consumed in a range of doses found in foods and dietary supplements, including very modest doses, caffeine increases vigilance, reduces choice reaction time, and alters mood states, in particular increasing vigor and decreasing fatigue (Lieberman et al. 1987, 2002; Fine et al. 1994; Smith et al. 1999, 2005; Kamimori et al. 2005; Childs and de Wit 2006; Hewlett and Smith 2007). Such behavioral parameters, including alertness, fatigue, and vigilance, are clearly appropriate measures of mental energy. In rested individuals, caffeine does not reliably alter higher-order cognitive functions such as learning, memory, and reasoning, but its effects do generalize to a wide variety of cognitive functions in sleep-deprived humans. In several studies, caffeine in moderate doses has been shown to directly alter self-perceived mental energy as assessed by a questionnaire specifically designed to measure caffeine's effects (Leathwood and Pollet 1982–1983; Amendola et al. 1998). For a comprehensive discussion of why caffeine is the best example of a compound that increases mental energy, see Lieberman (2001).

CAFFEINE AND PERSONALIZED NUTRITION

As discussed above, consumers can directly experience the effects of caffeine by awareness of changes in their mood state, specifically increased vigor and decreased fatigue. Accordingly, they can titrate (adjust) the amount of caffeine they consume by the selection of products and the timing of the consumption of these products. This is not accomplished by reading product labels and calculating caffeine intake, but rather by experiencing (feeling) the effects of products and by learning over time what is optimal for them. Furthermore, it appears that genetic variations in sensitivity to caffeine influence their patterns of consumption (Cornelis et al. 2007).

METHODS TO ASSESS MENTAL ENERGY

Although well-established, generally accepted methods for measuring mental energy do not currently exist, there are a variety of accepted methods used to assess closely related functions. Standardized mood questionnaires, such as the POMS, provide the best method for assessing mental energy. Mood questionnaires are simple to administer and provide a reliable and valid measure of mental states such as fatigue and vigor that directly correspond to the concept of mental energy. Such questionnaires are accepted and widely used in many scientific and clinical fields, in addition to psychology (Vollmer-Conna et al. 1997; Mooleenar et al. 1999; Bolmont et al. 2000; Krupp and Elkins 2000; O'Connor 2004; Lieberman 2007).

Certain tests of cognitive performance also appear to assess mental energy, but these are more difficult to administer and standardize from laboratory to laboratory. Some investigators believe such tests are more "objective" than mood questionnaires, but there is fundamentally no inherent advantage to tests of cognitive performance compared to standardized mood questionnaires. Tests of cognitive performance appear to assess factors related to mental energy, including reaction time and vigilance. Tests that assess more complex cognitive functions, such as learning, memory, and logical reasoning, are not sensitive to mental energy-related parameters. Several other methods could also potentially be useful for assessing mental energy, including the electroencephalogram (EEG), functional magnetic resonance imaging (fMRI), and activity monitors (see Lieberman [2007] for further discussion of this topic).

CONCLUSIONS

The concept of mental energy, although not as well defined as other mental states, is emerging as a useful construct for describing fatigue-related mental states. Many consumers are seeking products to increase their mental energy, probably as a result of the extensive demands placed on them by society. In an effort to provide products that address this need, scientists and industry have attempted to define the concept of mental energy and adapt standardized methods to assess it.

DISCLAIMER

Portions of this chapter are based on previous reviews by the author (Lieberman 2001, 2007). This work was supported by the U.S. Army Medical Research and Materiel Command (USAMRMC). The views, opinions, and/or findings in this report are those of the authors, and should not be construed as an official Department of the Army position, policy, or decision, unless so designated by other official documentation. Citation of commercial organization and trade names in this report do not constitute an official Department of the Army endorsement or approval of the products or services of these organizations.

REFERENCES

Amendola, C. A., J. D. E. Gabrieli, and H. R. Lieberman. 1998. Caffeine's effects on performance and mood are independent of age and gender. *Nutr Neurosci* 1:269–80.

Barone, J. J., and H. R. Roberts. 1996. Caffeine consumption. *Food Chem Toxicol* 34:119–29.

Bolmont, B., F. Thullier, and J. H. Abraini. 2000. Relationships between mood states and performances in reaction time, psychomotor ability, and mental efficiency during a 31-day gradual decompression in a hypobaric chamber from sea level to 8848 m equivalent altitude. *Physiol Behav* 71 (5): 469–76.

Childs, E., and H. de Wit. 2006. Subjective, behavioral and physiological effects of acute caffeine in light, nondependent caffeine users. *Psychopharmacology* 185:514–23.

Childs, N. M. 2001. Consumer perceptions of energy. *Nutr Rev* 59:S2–S4.

Clarkson, P. M. 2001. Introduction. *Nutr Rev* 59:S1.

Cook, D. B., and J. M. Davis. 2006. Introduction: mental energy: defining the science. *Nutr Rev* 64:S1.

Cornelis, M. C., A. El-Sohemy, and H. Campos. 2007. Genetic polymorphism of the adenosine A2A receptor is associated with habitual caffeine consumption. *Am J Clin Nutr* 86:240–4.

Fine, B. J., J. L. Kobrick, H. R. Lieberman, B. Marlowe, R. H. Riley, and W. J. Tharion. 1994. Effects of caffeine or diphenhydramine on visual vigilance. *Psychopharmacology* 114:233–8.

Gorby, H. E., A. M. Brownawell, and M. C. Falk. 2010. Do specific dietary constituents and supplements affect mental energy? Review of the evidence. *Nutr Rev* 68:697–718.

Hewlett, P., and A. Smith. 2007. Effects of repeated doses of caffeine on performance and alertness: new data and secondary analyses. *Hum Psychopharmacol* 22:339–50.

Kamimori, G. H., D. Johnson, D. Thorne, and G. Belenky. 2005. Multiple caffeine doses maintain vigilance during early morning operations. *Aviat Space Environ Med* 76:1046–50.

Krupp, L. B., and L. E. Elkins. 2000. Fatigue and declines in cognitive functioning in multiple sclerosis. *Neurology* 55 (7): 934–9.

Leathwood, P., and P. Pollet. 1982–1983. Diet-induced mood changes in normal populations. *J Psychiatr Res* 17:147–54.

Lieberman, H. R. 2001. The effects of ginseng, ephedrine and caffeine on cognitive performance, mood and energy. *Nutr Rev* 59:91–102.

Lieberman, H. R. 2006. Mental energy: assessing the cognitive dimension. *Nutr Rev* 64:S10–S13.

Lieberman, H. R. 2007. Cognitive methods for assessing mental energy. *Nutr Neurosci* 10:229–42.

Lieberman, H. R., C. E. Carvey, and L. A. Thompson. 2010. Caffeine. In *Encyclopedia of Dietary Supplements*, ed. P. M. Coates, 90–100. New York: Informa Health Care.

Lieberman, H. R., W. J. Tharion, B. Shukitt-Hale, K. L. Speckman, and R. Tulley. 2002. Effects of caffeine, sleep loss, and stress on cognitive performance and mood during U.S. Navy SEAL training. *Psychopharmacology (Berl)* 164 (3): 250–61.

Lieberman, H. R., R. J. Wurtman, G. S. Garfield, C. H. Roberts, and I. L. Coviella. 1987. The effects of low doses of caffeine on human performance and mood. *Psychopharmacology* 92:308–12.

McNair, D. M., M. Lorr, and L. F. Droppleman. 1971. *Profile of Mood States Manual*. San Diego, CA: Educational and Industrial Testing Service.

Milner, J., F. H. Seligson (Chairs). 2010. *The State of the Science of Diet and Mental Energy*. Symposium conducted as part of the ILSI North America Special Conference at Experimental Biology 2010, Anaheim, CA.

Mooleenar, M., P. A. Desmond, D. J. Mascord, G. A. Starmer, B. Tattam, and E. R. Volkers. 1999. The effects of ephedrine on the development of fatigue in a prolonged driving-related task. *Hum Psychopharmacol Clin Exp* 14:415–27.

O'Connor, P. J. 2004. Evaluation of four highly cited energy and fatigue mood measures. *J Psychosom Res* 57 (5): 435–41.

O'Connor, P. J. 2006. Mental energy: assessing the mood dimension. *Nutr Rev* 64: S7–S9.

Reissig, C. J., E. J. Strain, and R. R. Griffiths. 2009. Caffeinated energy drinks: a growing problem. *Drug Alcohol Depend* 99:1–10.

Smith, A., W. Sturgess, and J. Gallagher. 1999. Effects of a low dose of caffeine given in different drinks on mood and performance. *Hum Psychopharmacol Clin Exp* 14:473–82.

Smith, A., D. Sutherland, and G. Christopher. 2005. Effects of repeated doses of caffeine on mood and performance of alert and fatigued volunteers. *J Psychopharmacol* 19:620–26.

Vollmer-Conna, U., D. Wakefield, A. Lloyd, et al. 1997. Cognitive deficits in patients suffering from chronic fatigue syndrome, acute infective illness or depression. *Br J Psychiatry* 171:377–81.

2 Hydration and Brain Function

Kristen E. D'Anci

CONTENTS

Introduction 7
Thirst and Water Intake Regulation 8
  Fluid Regulation 8
  Significance of Plasma Sodium 9
Dehydration and the Brain 9
  Dehydration and Mental Performance 10
    Young Adults 10
    Infants and Children 11
    Older Individuals 13
Potential Mechanisms 13
Conclusions 14
References 15

INTRODUCTION

Water is an essential component of the human body, comprising 55–60 percent of total body weight in adults and up to 75 percent of total body weight in children and infants. Water is required for thermoregulation and cellular function, and is critical for life. Mild levels of dehydration (approximately 2 percent loss of body weight) produce alterations in mood and cognitive performance as well as disruptions in physical performance (Popowski et al. 2001). The impact of under-hydration and hydration in physical activity, particularly in athletes and in the military, has been of considerable interest and is well described in the scientific literature (Maughan et al. 2007; Murray 2007; Sawka and Noakes 2007). What remains less well understood is the relationship between hydration and brain function. A number of early studies indicate that water loss in hot climates and under conditions of vigorous physical activity produces significant decrements in cognitive performance (D'Anci 2005 ; Popkin et al. 2010). However, in comparison to earlier reports, results of studies published more recently have shown milder effects of hydration status on cognitive function (Popkin et al. 2010). Even though definitive conclusions have yet to be made, there is evidence that mild levels of dehydration can lead to alterations in cognitive function and disruptions in subjective mood.

The role of water in health is generally characterized in terms of variance from an ideal hydrated, or euhydrated, state. As described in this chapter, the concept of dehydration encompasses both the process of losing body water as well as the physical state of dehydration. In general, research on fluid intake and physical or mental functioning compares individuals in a euhydrated state, usually achieved by provision of water sufficient to overcome water loss, to those in a dehydrated state, which is achieved by withholding of fluids over time and/or during periods of heat stress or high activity. This type of dehydration is referred to in the literature as "voluntary dehydration" to distinguish it from involuntary dehydration. Involuntary dehydration characterizes the fluid loss from illness, especially in children.

THIRST AND WATER INTAKE REGULATION

Fluid Regulation

Fluid regulation is hormonally mediated to maintain optimal sodium levels and blood volume. Several of the sensations associated with thirst, such as dryness of the mouth or throat, induce drinking. However, people also drink in response to a variety of cultural, social, and psychological factors. The type and amount of fluid consumed are dependent not only upon the relative palatability and temperature of the fluid, and meal type and size, but also on water safety and availability. The act of drinking may not be directly involved with a physiological need for water, but can be initiated by habit, ritual, taste, or desire for a warming or cooling effect (Rolls 1991). Similarly, cessation of fluid intake is influenced by both physiological mechanisms and external variables. Even when an urgent physiological need is present, individuals may refrain from drinking as in the case of some religious observances (e.g., during Ramadan), or as an attempt to reduce the need to void the bladder. The latter reason is of significant concern in older individuals who may underconsume fluids to avoid the need to urinate during the night or because of impairments in mobility. Moreover, physiological feedback, such as distension of the stomach, can end drinking before the restoration of fluid balance.

The balance between loss and gain of fluids maintains body water within relatively narrow limits (Andersson 1978). Routes of water loss from the body include the urinary system, the skin, respiratory surfaces, and the gastrointestinal tract. The primary avenues for restoring water balance are fluid and food ingestion, with water produced in the metabolism of food making a minor contribution (Greenleaf 1982).

The primary regulation of thirst is controlled separately by osmotic pressure and body fluid volume, and as such is regulated by the same mechanisms that control central blood pressure, and water and solute reabsorption in the kidneys. Despite large variations in salt and water intake, homeostatic mechanisms work to sustain a normal plasma osmolality of 275–290 mOsm/kg and maintain normal sodium levels between 135 and 145 mEq/L. Increases in plasma osmolality, and activation of osmoreceptors and baroreceptors, stimulate hypothalamic release of argenine vasopressin, which in turn acts on the kidney to decrease urine volume and promote water retention. Sensations of thirst are generated at a higher plasma osmolality than that which stimulates vasopressin release. Thus, the release of vasopressin results first in a concentration of urine and conservation of body water and then a subsequent drive to increase fluid intake.

Based on population studies, fluid intakes generally are considered adequate for maintaining fluid balance in most free-living people (Bellisle et al. 2010), although precise recommendations for fluid intake are still needed (Popkin et al. 2010). Under normal conditions, the sensation of thirst results in an intake of fluid adequate to restore water balance. However, under conditions or high environmental heat or following high levels of physical activity, voluntary fluid intake may be inadequate to offset fluid deficits when individuals are allowed to drink according to thirst (Bar-Or et al. 1980; Nicolaidis 1998). Thus, mild to moderate dehydration can persist for some hours in hot climates or after the conclusion of physical activity. This so-called voluntary dehydration is frequently exploited as the independent variable when examining the role of hydration in cognitive performance.

Significance of Plasma Sodium

Hypernatremia, defined as plasma concentrations of sodium over 145 mEq/L, can result from decreased fluid intake and cause restlessness, altered mental status, confusion, and fatigue. So-called water intoxication, or hyponatremia, occurs when plasma concentrations of sodium drop below 135 mEq/L. Hyponatremia, which can be induced by ingesting large amounts of fluids, results in similar somatic symptoms to dehydration: headache, nausea, fatigue, confusion, and apathy (Overgaard 2005; Stuempfle 2010). Hyponatremia is becoming more understood as a dangerous consequence of overhydration, as a result of a widely publicized case of a woman who died following a radio contest. To win a video-gaming system, contestants drank as much water as they could without urinating (Salzman 2008). The woman, who won the contest, drank approximately 6 L of water over a three-hour period. For reference, the recommended adequate intake of water for women aged 19–30 years is 2.7 L per day (Institute of Medicine 2004). While an extreme case, this example illustrates the importance of understanding the function not only of water but also of sodium balance in hydration status. Exercise-associated hyponatremia is typically seen in elite athletes such as marathoners who ingest large volumes of water without replacing sodium and potassium, and more specific guidelines are being developed for rehydration following strenuous exercise (Stuempfle 2010). Additionally, hyponatremia has been associated with cognitive impairment in schizophrenia and traumatic brain injury (Atchison et al. 1993; Schnur et al. 1993).

DEHYDRATION AND THE BRAIN

Water or its lack influences brain structure and functioning. The very young, the very old, those in hot climates, and those engaging in vigorous exercise may be more at risk for dehydration-related disturbances in brain function. Whatever the cause of loss, if fluids are not replaced, then there is a shrinkage of plasma and extracellular volume which can lead to underperfusion of the brain. In older individuals, urge incontinence is associated with a greater urine loss relative to other forms of incontinence. In this population, urge incontinence is correlated with a decrease in mental performance and with underperfusion of the frontal lobes of the cerebral cortex (Griffiths et al. 1994; Griffiths 1998). Cerebral underperfusion also is associated with confusion, dementia, and lethargy, suggesting that changes in brain hydration levels may be responsible, in part, for the effects of dehydration on cognitive performance.

In hypernatremia, the plasma surrounding cell bodies has a greater concentration of sodium relative to intracellular fluid. Cellular dehydration occurs in hypernatremia as water diffuses across the cell membrane from the intracellular compartment into the extracellular compartment. Such cellular shrinkage is associated with neuronal lesions and brain edema (Finberg et al. 1959). Brain imaging of neonates and children with hypernatremia suggests that increased plasma osmolality is associated with damage to thalamic, cortical, and hippocampal areas of the brain (Korkmaz et al. 2000; Duran et al. 2007; Musapasaoglu et al. 2008).

It is recognized that excessively fast rehydration can cause a rapid influx of water into brain cells resulting in cerebral edema, and that these insults to the brain may be long lasting or even permanent. Potential brain injury resulting from altered hydration status varies with age. Children develop hyponatremic encephalopathy at higher sodium concentration than adults. Furthermore, children have a higher brain-to-skull ratio than adults, leaving less room for brain expansion seen with edema. Evidence suggests that much of the neurological sequelae of dysnatremias are due to damage resulting from rapid changes in fluid balance across cellular concentration gradients, rather than hypo- or hypernatremic states per se.

Dehydration and Mental Performance

Mild dehydration produces alterations in a number of cognitive domains such as concentration, alertness, and short-term memory. These disruptions are of particular concern in children (10–12 years) (Bar-David et al. 2005), young adults (18–25 years) (Gopinathan et al. 1988; Cian et al. 2000, 2001; D'Anci et al. 2009), and the elderly (50–82 years) (Suhr et al. 2004).

Young Adults

In young adults, mild to moderate levels of dehydration, as measured by percent body weight loss, can impair performance on tasks such as short-term memory, perceptual discrimination, arithmetic ability, visuomotor tracking, and psychomotor skills (Gopinathan et al. 1988; Cian et al. 2000, 2001; D'Anci et al. 2009). However, mild dehydration does not appear to alter cognitive functioning in a consistent manner. In some cases, cognitive performance was impaired with reductions of 2–3 percent of body weight, while in others, no differences in behavior were found (Cian et al. 2000, 2001; Szinnai et al. 2005; Adam et al. 2008; D'Anci et al. 2009). Comparing across studies, performance on similar cognitive tests was differently affected by dehydration (Cian et al. 2000; D'Anci et al. 2009). For example, healthy young men who lost 2.8 percent of their body weight either through heat exposure or through treadmill exercise displayed impaired performance on visual perception, short-term memory, and psychomotor tasks (Cian et al. 2000, 2001). In contrast, in a series of studies using exercise in conjunction with water restriction as a means of producing dehydration, only mild decrements in cognitive performance were observed in healthy young men and women athletes (D'Anci et al. 2009). In these latter experiments, the only consistent effect of mild dehydration was significant changes in subjective mood score, including increased fatigue, confusion, anger, and significant decreases in vigor. Finally, twenty-four hours of water deprivation in healthy men and women produced a level of 2.6 percent dehydration from water loss, but no significant alterations in cognitive performance (Szinnai et al. 2005). Taking the results of the previous study together, it can be proposed that heat stress may play a more critical role in the effects of dehydration on cognitive performance than mild fluid loss per se.

Reintroduction of fluids to individuals experiencing mild dehydration could reasonably be expected to reverse dehydration-induced cognitive deficits. However, few studies have examined how fluid reintroduction may alleviate dehydration's negative effects on cognitive performance and mood. One study investigated how water ingestion affected vigor and cognitive performance in young people following twelve hours of water restriction (Neave et al. 2001). While cognitive performance was not affected by either water restriction or water consumption, water ingestion increased self-reported vigor. A similar increase in alertness was observed following water ingestion in participants experiencing either low or high levels of thirst (Rogers et al. 2001). Water ingestion, however, had differing effects on cognitive performance as a function of thirst. Participants experiencing high levels of thirst showed improved performance on a cognitively demanding task when given water to drink, while performance for participants with low levels of thirst declined when given water to drink.

Infants and Children

Involuntary Dehydration

Children are at greater risk for dehydration than adults for several reasons. Relative to adults, young people have a greater surface-to-mass ratio allowing for greater water losses from the skin. During illness, significant water loss can occur through the gastrointestinal tract, and this can be of grave concern in the very young. In developing countries, diarrheal diseases are a leading cause of death in children resulting in approximately 1.5–2.5 million deaths per year (Kosek et al. 2003). Diarrheal illness results not only in a reduction in body water, but also in potentially lethal electrolyte imbalances. Many times mortality in such cases can be prevented with appropriate oral rehydration therapy, in which simple dilute solutions of sugar and salts replace fluid lost by diarrhea. Many consider application of oral rehydration therapy to be one of the hallmark public health developments of the twentieth century (Atia and Buchman 2009).

Children, infants in particular, are dependent upon caregivers for the provision of fluids. In the case of infants, caregivers may not be aware of the extent of water loss and adequate hydration may not be provided (Finberg 1959). As an example, inadequate breastfeeding is becoming more common as a risk factor for dehydration in infants (Laing and Wong 2002). Nursing mothers with insufficient milk may not recognize the signs of progressive dehydration in their infants. While most parents understand what dehydration is, they may not recognize more than one sign of dehydration or at what level of dehydration symptoms are seen (Gittelman et al. 2004).

Voluntary Dehydration

During exercise, children may not understand the need to replace lost fluids (Bar-Or et al. 1980), and both children as well as coaches need specific guidelines for fluid intake (American Academy of Pediatrics 2000). Additionally, children may take longer to acclimate to increases in environmental temperature than adults (Bytomski and Squire 2003; Falk and Dotan 2008). Recent work indicates that children living in hot, arid climates have a high prevalence of voluntary dehydration (Bar-David et al. 2009). It is recommended that young athletes or children in hot climates begin activities in a well-hydrated state and drink fluids over and above the thirst threshold.

Cognitive Performance and Hydration

Relatively few studies have examined the effects of hydration status on cognitive performance in children. Informal observations by school teachers in the United Kingdom indicate that programs encouraging water intake in students could improve student attention and concentration (BBC News 2000). In support of these anecdotal observations, several recent studies have examined the utility of providing water on attentiveness and cognitive functioning in schoolchildren (Benton and Burgess 2009; Edmonds and Burford 2009; Edmonds and Jeffes 2009). In these experiments, children were not fluid restricted prior to cognitive testing, but were allowed to drink as usual. Children were then provided with a drink or no drink 20–45 minutes before the cognitive test sessions. Children in the groups given water showed improvements in visual attention (Edmonds and Burford 2009; Edmonds and Jeffes 2009). However, effects on visual memory were less consistent, with one study showing no effects of drinking water on a visual memory task in 6–7-year-old children (Edmonds and Jeffes 2009) and another showing a significant improvement in a similar task in 7–9-year-old children (Edmonds and Burford 2009). In other research, memory performance was improved by provision of water, while sustained attention was not (Benton and Burgess 2009). Additionally, subjective measures of thirst were reduced in the children given water.

With respect to voluntary dehydration in children, a 2005 study (Bar-David et al. 2005) remains the only one investigating the interaction between measurable dehydration and cognitive performance in children. In this study, Israeli schoolchildren were categorized into euhydrated or dehydrated groups based on urine osmolality, with urine osmolality above 800 mOsm/kg defining dehydration. Cognitive tests were administered at the beginning of the school day and again at lunch time. There were no significant differences on cognitive performance between the groups at the beginning of the school day. At noon, however, students who were initially classified as hydrated tended to perform better on several cognitive tasks than those classified as dehydrated students. Short-term memory scores were significantly higher in hydrated children in comparison to dehydrated children. Additionally, there was a trend for hydrated students to perform better on tasks measuring semantic fluency and semantic flexibility, relative to dehydrated children. These data indicate that mild dehydration is associated with negative effects on cognition throughout the day.

Older Individuals

Aging and Fluid Regulation

With aging, changes in osmoreceptors and baroreceptors, as well as changes in regulatory mechanisms, are related to changes in perception of thirst and fluid intake (Silver and Morley 1992). Following a period of water deprivation, older individuals report less thirst and drink less fluid compared to younger individuals, and drink insufficient amounts to replenish lost body water (hypodipsia) (Phillips et al. 1984; Mack et al. 1994). Furthermore, when offered a highly palatable selection of drinks, older individuals still underconsume fluid relative to younger people (Phillips et al. 1993). Hypodipsia in older individuals can be exacerbated by disease (Miller et al. 1982) and dementia (Albert et al. 1994). In addition, illness and limitations in activities of daily living can lead to further restrictions in fluid intake. Coupled with reduced fluid intake, in aging there is a decrease in total body water stores. Older individuals often have impaired renal fluid conservation mechanisms and impaired responses to heat and cold stress (Vogelaere and Pereira 2005; Thompson-Torgerson et al. 2008). All of these factors contribute to an increased risk of hypohydration and dehydration in the elderly relative to younger individuals.

Aging, Hydration Status, and Brain Function

Dehydration is clinically recognized as a risk factor for delirium and delirium presenting as dementia in the elderly and in the very ill (Lawlor 2002; Culp et al. 2004; Voyer et al. 2009). Recent work shows that dehydration is one of several factors which may predispose elderly residents of long-term care facilities to confusion (Voyer et al. 2009), although in this study daily water intake was used as a proxy measure for dehydration rather than other, more direct clinical assessments such as urine or plasma osmolality.

Little research has examined voluntary dehydration and brain function in older adults. One study compared physical and cognitive performance in young men (24 years) to older men (56 years) during ten days of hill walking in the Scottish highlands (Ainslie et al. 2002). All participants were active and experienced hill walkers who walked an average of 21 km/d and had unlimited access to food and water. Over time, the older men drank less water than the younger men, and showed greater dehydration, as measured using urine osmolality, than younger men. Older men reported lower perceptions of thirst than younger men, and showed significant decrements in psychomotor performance, which were correlated with increased levels of dehydration. These data are consistent with data indicating that thirst sensitivity decreases with age, and suggest that older men may be more susceptible to activity-induced dehydration than younger men.

POTENTIAL MECHANISMS

Dehydration increases circulating levels of stress hormones such as cortisol (Francesconi et al. 1984). In humans, increased levels of cortisol have been associated with decrements in cognitive function (Kirschbaum et al. 1996; Van Londen et al. 1998; Newcomer et al. 1999; Greendale et al. 2000). It is theorized that some of the negative effects of dehydration on mental performance are, therefore, related to the activation of the hypothalamic-pituitary-adrenocortical (HPA) axis and the release of stress hormones. This hypothesis is supported by observations in animals that HPA axis activation induced by stress and/or pharmacological administration of glucocortocoids (stress hormones) can produce dendritic atrophy in hippocampal neurons, and this atrophy is associated with cognitive decrements (Raber 1998). Vasopressin is released when hypothalamic osmoreceptors sense increased extracellular osmolality. Vasopressin release activates the thirst response and thus provokes drinking. Vasopressin may act as a neuromodulator to produce excitatory effects in neural tissue. Elevated levels of vasopressin may enhance cognitive functioning on certain tasks (Van Londen et al. 1998).

In animal studies, long-term dehydration stimulates glutamate and GABA release (Di and Tasker 2004). Chronic dehydration may therefore produce an increase in neuronal activity and enhance the actions of both excitatory (glutamate) and inhibitory (GABA) neurotransmitters. The exact processes that these differing neurotransmitters affect are complex. In brief, inhibitory and excitatory neurotransmitters may have opposing effects on behavior and cognition. However, inhibitory and excitatory neurotransmitters may also have similar effects depending on receptor subtype and localization. As an example, pharmacological *blockade* of GABA-B receptors augments long-term hippocampal-dependent memory (Helm et al. 2005). In contrast, *activation* of GABA-A receptors enhances memory and spatial learning in rats (Maubach 2003). Glutamate receptor antagonists are associated with memory impairments (Parwani et al. 2005), but excitotoxicity produced by high levels of glutamate agonists can also produce damage to hippocampal areas and cognitive decrements (Shikhanov et al. 2005). Dehydration is associated with a decrease in neuronal cell proliferation that is reversed by rehydration (Levine et al. 2002). Although any links between these animal studies and human cognitive performance would be speculative, the existing research suggests that dehydration produces many different physiological effects that can individually or in combination impact cognitive and psychomotor performance.

CONCLUSIONS

Taken together, these studies indicate that low to moderate dehydration can alter cognitive performance. However, rather than indicating that the effects of hydration or water ingestion on cognition are contradictory, many of the studies differ significantly in methodology and in measurement of cognitive behaviors. These variances in methodology underscore the importance of consistency when examining relatively subtle chances in overall cognitive performance. However, in those studies in which significant alterations in cognitive performance were seen as a result of dehydration, most combined heat and exercise. In studies using long-term abstinence from fluids to produce dehydration, fewer, if any, changes in cognitive performance were seen. Thus, in healthy individuals it is difficult to disentangle the effects of dehydration from the effects of heat and exercise on cognitive performance in temperate conditions. Additionally, while the mechanistic effects of more severe dehydration are fairly well described, relatively little is known about the mechanism of mild dehydration's effects on mood and mental performance.

REFERENCES

Adam, G. E., R. Carter III, S. N. Cheuvront, D. J. Merullo, J. W. Castellani, H. R. Lieberman, and M. N. Sawka. 2008. Hydration effects on cognitive performance during military tasks in temperate and cold environments. *Physiol Behav* 93:748–56.

Ainslie, P. N., I. T. Campbell, K. N. Frayn, S. M. Humphreys, D. P. MacLaren, T. Reilly, and K. R. Westerterp. 2002. Energy balance, metabolism, hydration, and performance during strenuous hill walking: the effect of age. *J Appl Physiol* 93:714–23.

Albert, S. G., B. R. Nakra, G. T. Grossberg, and E. R. Caminal. 1994. Drinking behavior and vasopressin responses to hyperosmolality in Alzheimer's disease. *Int Psychogeriatr* 6:79–86.

American Academy of Pediatrics, Committee on Sports Medicine and Fitness. 2000. Climatic heat stress and the exercising child and adolescent. *Pediatrics* 106:158–9.

Andersson, B. 1978. Regulation of water intake. *Physiol Rev* 58:582.

Atchison, J. W., J. Wachendorf, D. Haddock, W. J. Mysiw, M. Gribble, and J. D. Corrigan. 1993. Hyponatremia-associated cognitive impairment in traumatic brain injury. *Brain Inj* 7:347–52.

Atia, A. N., and A. L. Buchman. 2009. Oral rehydration solutions in non-cholera diarrhea: a review. *Am J Gastroenterol* 104:2596–604.

Bar-David, Y., J. Urkin, and E. Kozminsky. 2005. The effect of voluntary dehydration on cognitive functions of elementary school children. *Acta Paediatr* 94:1667–73.

Bar-David, Y., J. Urkin, D. Landau, Z. Bar-David, and D. Pilpel. 2009. Voluntary dehydration among elementary school children residing in a hot arid environment. *J Hum Nutr Diet* 22:455–60.

Bar-Or, O., R. Dotan, O. Inbar, A. Rotshtein, and H. Zonder. 1980. Voluntary hypohydration in 10- to 12-year-old boys. *J Appl Physiol* 48:104–8.

BBC News. 2000. Water improves school test results. Available from http://news.bbc.co.uk/1/hi/education/728017.stm

Bellisle, F., S. N. Thornton, P. Hebel, M. Denizeau, and M. Tahiri. 2010. A study of fluid intake from beverages in a sample of healthy French children, adolescents and adults. *Eur J Clin Nutr* 64:350–5.

Benton, D., and N. Burgess. 2009. The effect of the consumption of water on the memory and attention of children. *Appetite* 53:143–6.

Bytomski, J. R., and D. L. Squire. 2003. Heat illness in children. *Curr Sports Med Rep* 2:320–4.

Cian, C., P. A. Barraud, B. Melin, and C. Raphel. 2001. Effects of fluid ingestion on cognitive function after heat stress or exercise-induced dehydration. *Int J Psychophysiol* 42:243–51.

Cian, C., P. A. Koulmann, P. A. Barraud, C. Raphel, C. Jimenez, and B. Melin. 2000. Influence of variations of body hydration on cognitive performance. *J Psychophysiol* 14:29–36.

Culp, K. R., B. Wakefield, M. J. Dyck, P. Z. Cacchione, S. DeCrane, and S. Decker. 2004. Bioelectrical impedance analysis and other hydration parameters as risk factors for delirium in rural nursing home residents. *J Gerontol A Biol Sci Med Sci* 59 (8): 813–7.

D'Anci, K. E. 2005. Hydration status and cognitive performance in young adults. *Nutr Clin Care* 8:163–6.

D'Anci, K. E., A. Vibhakar, J. H. Kanter, C. R. Mahoney, and H. A. Taylor. 2009. Voluntary dehydration and cognitive performance in trained college athletes. *Percept Mot Skills* 109:251–69.

Di, S., and J. G. Tasker. 2004. Dehydration-induced synaptic plasticity in magnocellular neurons of the hypothalamic supraoptic nucleus. *Endocrinology* 145:5141–9.

Duran, R., N. Aladag, U. Vatansever, O. Temizoz, H. Genchallac, and B. Acunas. 2007. Cranial MR venography findings of severe hypernatremic dehydration in association with cerebral venous thrombosis in the neonatal period. *Pediatr Hematol Oncol* 24:387–91.

Edmonds, C. J., and D. Burford. 2009. Should children drink more water? the effects of drinking water on cognition in children. *Appetite* 52:776–9.

Edmonds, C. J., and B. Jeffes. 2009. Does having a drink help you think? 6–7-year-old children show improvements in cognitive performance from baseline to test after having a drink of water. *Appetite* 53:469–72.

Falk, B., and R. Dotan. 2008. Children's thermoregulation during exercise in the heat: a revisit. *Appl Physiol Nutr Metab* 33:420–7.

Finberg, L. 1959. Pathogenesis of lesions in the nervous system in hypernatremic states. I. Clinical observations of infants. *Pediatrics* 23:40–5.

Finberg, L., C. Luttrell, and H. Redd. 1959. Pathogenesis of lesions in the nervous system in hypernatremic states. II. Experimental studies of gross anatomic changes and alterations of chemical composition of the tissues. *Pediatrics* 23:46–53.

Francesconi, R. P., M. N. Sawka, and K. B. Pandolf. 1984. Hypohydration and acclimation: effects on hormone responses to exercise/heat stress. *Aviat Space Environ Med* 55:365–9.

Gittelman, M. A., E. M. Mahabee-Gittens, and J. Gonzalez-del-Rey. 2004. Common medical terms defined by parents: are we speaking the same language? *Pediatr Emerg Care* 20:754–8.

Gopinathan, P. M., G. Pichan, and V. M. Sharma. 1988. Role of dehydration in heat stress-induced variations in mental performance. *Arch Environ Health* 43:15–7.

Greendale, G. A., D. Kritz-Silverstein, T. Seeman, and E. Barrett-Connor. 2000. Higher basal cortisol predicts verbal memory loss in postmenopausal women: Rancho Bernardo Study. *J Am Geriatr Soc* 48:1655–8.

Greenleaf, J. E. 1982. Dehydration-induced drinking in humans. *Fed Proc* 41:2509–14.

Griffiths, D. 1998. Clinical studies of cerebral and urinary tract function in elderly people with urinary incontinence. *Behav Brain Res* 92:151–5.

Griffiths, D. J., P. N. McCracken, G. M. Harrison, and K. N. Moore. 1994. Urinary incontinence in the elderly: the brain factor. *Scand J Urol Nephrol Suppl* 157:83–8.

Helm, K. A., R. P. Haberman, S. L. Dean, E. C. Hoyt, T. Melcher, P. K. Lund, and M. Gallagher. 2005. GABAB receptor antagonist SGS742 improves spatial memory and reduces protein binding to the cAMP response element (CRE) in the hippocampus. *Neuropharmacology* 48:956–64.

Institute of Medicine. 2004. *Dietary reference intakes for water, potassium, sodium, chloride, and sulfate*. Washington, DC: National Academies Press.

Kirschbaum, C., O. T. Wolf, M. May, W. Wippich, and D. H. Hellhammer. 1996. Stress- and treatment-induced elevations of cortisol levels associated with impaired declarative memory in healthy adults. *Life Sci* 58:1475–83.

Korkmaz, A., S. Yigit, M. Firat, and O. Oran. 2000. Cranial MRI in neonatal hypernatraemic dehydration. *Pediatr Radiol* 30:323–5.

Kosek, M., C. Bern, and R. L. Guerrant. 2003. The global burden of diarrhoeal disease, as estimated from studies published between 1992 and 2000. *Bull World Health Organ* 81:197–204.

Laing, I. A., and C. M. Wong. 2002. Hypernatraemia in the first few days: is the incidence rising? *Arch Dis Child Fetal Neonatal Ed* 87:F158–62.

Lawlor, P. G. 2002. Delirium and dehydration: some fluid for thought? *Support Care Cancer* 10:445–54.

Levine, S., A. Saltzman, B. Katof, A. Meister, and T. B. Cooper. 2002. Proliferation of glial cells induced by lithium in the neural lobe of the rat pituitary is enhanced by dehydration. *Cell Prolif* 35:167–72.

Mack, G. W., C. A. Weseman, G. W. Langhans, H. Scherzer, C. M. Gillen, and E. R. Nadel. 1994. Body fluid balance in dehydrated healthy older men: thirst and renal osmoregulation. *J Appl Physiol* 76:1615–23.

Maubach, K. 2003. GABA(A) receptor subtype selective cognition enhancers. *Curr Drug Targets CNS Neurol Disord* 2:233–9.

Maughan, R. J., S. M. Shirreffs, and P. Watson. 2007. Exercise, heat, hydration and the brain. *J Am Coll Nutr* 26:604S–612S.

Miller, P. D., R. A. Krebs, B. J. Neal, and D. O. McIntyre. 1982. Hypodipsia in geriatric patients. *Am J Med* 73:354–6.

Murray, B. 2007. Hydration and physical performance. *J Am Coll Nutr* 26:542S–548S.

Musapasaoglu, H., A. M. Agildere, M. Teksam, A. Tarcan, and B. Gurakan. 2008. Hypernatraemic dehydration in a neonate: brain MRI findings. *Br J Radiol* 81:e57–60.

Neave, N., A. B. Scholey, J. R. Emmett, M. Moss, D. O. Kennedy, and K. A. Wesnes. 2001. Water ingestion improves subjective alertness, but has no effect on cognitive performance in dehydrated healthy young volunteers. *Appetite* 37:255–6.

Newcomer, J. W., G. Selke, A. K. Melson, T. Hershey, S. Craft, K. Richards, and A. L. Alderson. 1999. Decreased memory performance in healthy humans induced by stress-level cortisol treatment. *Arch Gen Psychiatry* 56:527–33.

Nicolaidis, S. 1998. Physiology of thirst. In *Hydration throughout Life*, ed. M. J. Arnaud. Montrouge: John Libbey Eurotext.

Overgaard, J. 2005. Drink till you drop. *J Exp Biol* 208 (13): vii.

Parwani, A., M. A. Weiler, T. A. Blaxton, D. Warfel, M. Hardin, K. Frey, and A. C. Lahti. 2005. The effects of a subanesthetic dose of ketamine on verbal memory in normal volunteers. *Psychopharmacology* 183:265–74.

Phillips, P. A., C. I. Johnston, and L. Gray. 1993. Disturbed fluid and electrolyte homoeostasis following dehydration in elderly people. *Age Ageing* 22:S26–33.

Phillips, P. A., B. J. Rolls, J. G. Ledingham, M. L. Forsling, J. J. Morton, M. J. Crowe, and L. Wollner. 1984. Reduced thirst after water deprivation in healthy elderly men. *N Engl J Med* 311:753–9.

Popkin, B. M., K. E. D'Anci, and I. H. Rosenberg. 2010. Water, hydration, and health. *Nutr Rev* 68:439–58.

Popowski, L. A., R. A. Oppliger, G. P. Lambert, R. F. Johnson, A. Kim Johnson, and C. V. Gisolf. 2001. Blood and urinary measures of hydration status during progressive acute dehydration. *Med Sci Sports Exerc* 33:747–53.

Raber, J. 1998. Detrimental effects of chronic hypothalamic-pituitary-adrenal axis activation: from obesity to memory deficits. *Mol Neurobiol* 18:1–22.

Rogers, P. J., A. Kainth, and H. J. Smit. 2001. A drink of water can improve or impair mental performance depending on small differences in thirst. *Appetite* 36:57–8.

Rolls, B. J. 1991. Physiological determinants of fluid intake in humans. In *Thirst: Physiological and Psychological Aspects*, ed. D. J. Ramsey and D. A. Booth. London: Springer-Verlag.

Salzman, J. 2008. Is it safe to drink the water? *Duke Environmental Law & Policy Forum* 19:1–42.

Sawka, M. N., and T. D. Noakes. 2007. Does dehydration impair exercise performance? *Med Sci Sports Exerc* 39:1209–17.

Schnur, D. B., E. Wirkowski, R. Reddy, P. Decina, and S. Mukherjee. 1993. Cognitive impairments in schizophrenic patients with hyponatremia. *Biol Psychiatry* 33:836–8.

Shikhanov, N. P., N. M. Ivanov, A. V. Khovryakov, K. Kaspersen, G. M. McCann, P. P. Kruglyakov, and A. A. Sosunov. 2005. Studies of damage to hippocampal neurons in inbred mouse lines in models of epilepsy using kainic acid and pilocarpine. *Neurosci Behav Physiol* 35:623–8.

Silver, A. J., and J. E. Morley. 1992. Role of the opioid system in the hypodipsia associated with aging. *J Am Geriatr Soc* 40:556–60.

Stuempfle, K. J. 2010. Exercise-associated hyponatremia during winter sports. *Phys Sports Med* 38:101–6.

Suhr, J. A., J. Hall, S. M. Patterson, and R. T. Niinisto. 2004. The relation of hydration status to cognitive performance in healthy older adults. *Int J Psychophysiol* 53:121–5.

Szinnai, G., H. Schachinger, M. J. Arnaud, L. Linder, and U. Keller. 2005. Effect of water deprivation on cognitive-motor performance in healthy men and women. *Am J Physiol Regul Integr Comp Physiol* 289:R275–80.

Thompson-Torgerson, C. S., L. A. Holowatz, and W. L. Kenney. 2008. Altered mechanisms of thermoregulatory vasoconstriction in aged human skin. *Exerc Sport Sci Rev* 36:122–7.

Van Londen, L., J. G. Goekoop, A. H. Zwinderman, J. B. Lanser, V. M. Wiegant, and D. De Wied. 1998. Neuropsychological performance and plasma cortisol, arginine vasopressin and oxytocin in patients with major depression. *Psychol Med* 28:275–84.

Vogelaere, P., and C. Pereira. 2005. Thermoregulation and aging. *Rev Port Cardiol* 24:747–61.

Voyer, P., S. Richard, L. Doucet, and P. H. Carmichael. 2009. Predisposing factors associated with delirium among demented long-term care residents. *Clin Nurs Res* 18:153–71.

3 Diet as an Analgesic Modality

Alexis M. Codrington, Yoram Shir, and John Pereira

CONTENTS

Introduction ... 20
Pain ... 20
Acute versus Chronic Pain ... 20
The Analgesic Properties of Dietary Constituents ... 21
Macronutrients ... 21
Dietary Fat ... 21
Dietary Protein ... 24
Carbohydrates ... 25
Micronutrients ... 27
Amino Acids ... 27
Vitamins ... 29
Polyamines ... 31
Magnesium ... 32
Common Foods ... 33
Tart Cherries ... 33
Soy ... 33
Dietary Herbs and Supplements ... 35
Roots and Bulbs ... 35
Devil's Claw (*Harpagophytum procumbens*) ... 35
Ginger (*Zingiber officinale*) ... 36
Turmeric (*Curcuma longa*) ... 36
Tree Bark ... 37
Boswellia (*Boswellia serrata*) ... 37
Leaves ... 37
Sow Thistle (*Sonchus oleraceus*) ... 37
Flowers ... 38
Feverfew (*Tanacetum parthenium*) ... 38
Fruit ... 38
Pineapple (*Ananas comosus*) ... 38
Seeds ... 39
Evening Primrose Oil (*Oenothera biennis*) ... 39
Caloric Restriction ... 39
Future Prospects ... 40
References ... 40

INTRODUCTION

Dietary habits play a crucial role in both the prevention and pathogenesis of multifactorial illnesses such as cancer, heart, and rheumatic diseases. For example, it is well accepted that the consumption of a Mediterranean diet, characteristically low in saturated and high in unsaturated fatty acids, is associated with a lower incidence of coronary artery disease (Roehm 2009). Less explored, but no less plausible, is the idea that diet could also play a significant role in relieving pain. This chapter reviews current knowledge on the effect of diet on acute and chronic pain conditions, and the mechanisms that could mediate this effect. First, however, a short introduction to pain medicine and the unfavorable outlook of current analgesic modalities is mandatory. Only then could the interested reader appreciate the dire need for exploring alternate avenues of therapy such as dietary modification.

PAIN

Pain is defined as an unpleasant sensory and emotional experience associated with actual or potential tissue damage, or is described in terms of such damage. The inability to communicate verbally does not negate the possibility that an individual is experiencing pain and is in need of appropriate pain-relieving treatment. However, the word "pain" can be attributed only to verbally communicating humans. Therefore, in this chapter the terms "pain" and "analgesia" are used when discussing the effect of diet in humans. For data detailing animal experiments, the more scientifically accurate terms used are "nociception" and "antinociception."

ACUTE VERSUS CHRONIC PAIN

The fact that the single word "pain" labels all sensations as soon as they become uncomfortable could lead to the wrong assumption that pain is a uniform pathology mediated through a single neural mechanism. In truth, pain is heterogeneous both in its underlying pathogenesis and clinical presentation. Although a broader review of this topic is beyond the scope of this chapter, it is crucial to discuss the distinction between acute and chronic pain.

Acute pain is usually a short-lived symptom, serving as a warning sign and responding well to a single treatment. Chronic pain, on the other hand, persists longer than the expected time for tissue healing or resolution of the underlying disease process and could last for years. It is often a disease in its own right, is always pathological, and is far less likely to respond even to the most sophisticated treatments. In the majority of people, controlling acute pain is relatively easy if the appropriate therapeutic tools are used. In contrast, chronic pain represents a real therapeutic challenge due to its multiple etiologies and diverse basic mechanisms. Consider, for example, a recent outcome study of more than 600 patients suffering from chronic posttraumatic neuropathic pain for time periods ranging from 1 to 46 years. None of the patients recovered spontaneously, and pain levels were only mildly affected by the various treatments that were tried (Schwartzman et al. 2009). Conservative estimates suggest that approximately 20 percent of the world's population suffers from chronic pain, resulting from injury, musculoskeletal conditions, arthritis, neurological conditions, and cancer. Hence, it has become a leading cause of human suffering, disability, and health care utilization.

This discussion is important since many of the animal experiments studying the effect of diet on pain were done in healthy animals exposed to acute, short-lived noxious stimuli. Animal models of chronic pain are rare and have been used less frequently in dietary experiments. One should, therefore, be careful in trying to interpret some of the data presented in this review as immediately relevant to the human pain scenario.

In part, the fact that approximately 60 million people in the United States alone suffer from chronic pain may be attributed to factors such as deficient education of health practitioners, cultural barriers, legal restrictions, and governmental policies. Nevertheless, a portion of these patients receives the best treatments available. Sadly, these therapies have only a limited effect; most new, patented analgesic medications bring only mild to moderate pain relief (Straube et al. 2010, Arnold et al. 2009) and, many times, are not better than older drugs (Nishishinya et al. 2008). Other therapeutic modalities like spinal injections and nerve blocks are also disappointing, and many times are ineffective in improving the long-term outcome of patients with chronic pain. It is not surprising, therefore, that other types of chronic pain therapy, including diet, are being eagerly explored.

THE ANALGESIC PROPERTIES OF DIETARY CONSTITUENTS

This review pertains mainly to the direct antinociceptive effect of dietary ingredients. However, diet can also affect nociception indirectly by changing the pharmacokinetics and pharmacodynamics of analgesic medications. For example, changes in drug-metabolizing enzyme systems happen frequently by phytochemical-containing foods like fruits, vegetables, herbs, spices, and teas, as well as by dietary macroconstituents. Food can also delay gastric emptying and change the solubilization of drugs and hepatic blood flow, parameters that are crucial for drug transport and metabolism (reviewed in Harris et al. 2003).

MACRONUTRIENTS

Marked by the development of agriculture and major changes in manufacturing techniques, the Neolithic and Industrial eras brought about food staples and food-processing procedures which altered key nutritional factors such as macronutrient composition (Cordain 2007). Consequently, it is believed that the nutritional characteristics of the human diet have been adversely affected and have had far-reaching effects on human health and well-being. All three dietary macronutrients, fat, protein, and carbohydrates, could directly affect nociception and pain in animals and humans.

Dietary Fat

No other topic of dietary analgesia has attracted more attention and was more explored than the effect of dietary fat on nociception. Tests in healthy human volunteers showed that even a single high-fat, low-carbohydrate diet was superior in decreasing experimental acute pain compared to an isoenergetic low-fat, high-carbohydrate diet (Zmarzty et al. 1997). While multiple fatty acids have been implicated as possessing analgesic properties, the most abundant research pertains to polyunsaturated fatty acids (PUFA), especially omega-3 (traditionally regarded as "good fatty acids") and omega-6 (regarded as "bad fatty acids"), and their role in inflammatory pain.

Omega-3 Polyunsaturated Fatty Acids (N-3 PUFA)

In contrast to some other dietary elements, where the link between diet and pain is promising but still somewhat tenuous, the analgesic role of n-3 PUFA is supported by sound principles in biochemistry, physiology, and evidence-based medicine. For better understanding of the anti-inflammatory/analgesic role of n-3 PUFA, a short introduction of their metabolic pathways is warranted. N-3 PUFAs are essential nutrients of which the best studied are eicosapentaenoic acid (EPA) and docosahexaenoic acid (DHA). Both are mainly derived from cold water fish, but could also be converted, in small quantities, from α-linolenic acid, another n-3 PUFA found in plants, animals, and milk. These two n-3 PUFAs are converted to prostaglandins through the cyclooxygenase (COX) enzymatic system. Arachidonic acid, one of the omega-6 PUFAs, competes with EPA and DHA, to be converted into prostaglandins as well. While arachidonic acid is converted into pro-inflammatory prostaglandins, DHA and EPA are converted into anti-inflammatory prostaglandins, PGE3, or less inflammatory leukotrienes. Therefore, increased intake of n-3 PUFA will decrease the production of pro-inflammatory prostaglandins while dietary fat intake that is predominantly n-6 PUFA will lead to more pro-inflammatory precursor production.

The average North American is approximately one-third fat by weight. The type of fatty acids that make up body fat is dietary dependent. It is not surprising, then, that the average Japanese, traditionally consuming a diet rich in fish, has roughly equal amounts of bodily n-6 PUFA and n-3 PUFA. In contrast, North Americans traditionally consume diets rich in n-6 PUFA and have a body composition that is approximately four times higher in n-6 PUFA compared to n-3 PUFA. Consuming a pro-inflammatory diet inevitably leads to the consumption of more anti-inflammatory drugs on a population basis. This might explain why, by some crude estimates, North Americans consume approximately as much anti-inflammatory medication as the rest of the world combined. The use of these medications is directly associated with significant human morbidity and mortality from damage to the gastrointestinal, cardiovascular, and renal systems. Thus, adopting a safer but equally effective anti-inflammatory agent could be extremely beneficial. There are currently sufficient data indicating that n-3 PUFA fulfill these two conditions: they are safe for long-term use; the most common side effects of n3-PUFA are eructation (burping) and a fishy aftertaste, both occurring only in a minority of patients, and possible slight prolongation of bleeding time; and they can efficiently reduce inflammation-related nociception and pain.

In animals, dietary supplementation of n-3 PUFAs was associated with modified serotonin, norepinephrine, and dopamine neurotransmission (Chalon 2006; Su 2008), changes shown to modulate mood and nociception. Furthermore, diet enriched with n-3 PUFA reduced both inflammatory markers and neuropathic-like nociception post nerve injury in a rodent model (Martin and Avendaño 2009). N-3 PUFA could exert this effect through modulation of the capsaicin receptor or transient receptor potential cation channel V1 (TRPV1) (Matta et al. 2007). Although multiple lines of evidence showed that n-3 PUFA are associated with decreased inflammatory pain in rodents, this might not be true for other hypernociceptive conditions. Rats with neuropathic-like hypernociception following partial sciatic nerve ligation (PSL; Seltzer et al. 1990) developed significant hypernociception when fed an n-3 PUFA-rich diet, compared to rats fed an n-6-PUFA-rich diet (Pérez et al. 2005). Interestingly, the significant association between hypernociception and n-3 PUFA was further validated when directly measuring the fatty acid content of the injured nerve; there was a direct correlation between higher neural content of n-3 PUFA and levels of hypernociception (Soleimannejad et al. 2008). These results suggest that in noninflammatory hypernociceptive models, n-3 PUFA could be pronociceptive. Indeed, n-3 PUFA enhanced conduction through slow tetrodotoxin-resistant sodium channels in rat dorsal root ganglion neurons (Hong et al. 2004), resulting in hyperpolarization that could augment nociception.

There are convincing human data supporting the use of n-3 PUFA in a multitude of painful conditions, mainly related to inflammatory processes. For example, a recent non-placebo-controlled study showed that a combination of EPA and DHA allowed 59 percent of patients with neck or back pain to discontinue their prescribed medications for pain (Maroon and Bost 2006). In placebo-controlled studies, omega-3 supplementation reduced the clinical pain of patients with rheumatoid arthritis (Cleland et al. 1988; Kremer et al. 1995), and a recent review noted that there was "high-level evidence" for the benefit of fish oil in rheumatoid arthritis (Proudman et al. 2008). Open labeled studies showed that n-3 PUFA can relieve pain and other symptoms in fibromyalgia patients (Puri 2004), as well as alleviating menstrual pain (Harel et al. 1996). No studies to date examined the dose–response curve of n-3 PUFA in patients with inflammatory diseases. Daily doses of 3–5 g of EPA and DHA combined have been recommended in these patients (Proudman et al. 2008).

Alpha-Lipoic Acid

Second in its popularity as an anti-inflammatory fatty acid is alpha-lipoic acid (α-LA). Although abundant in meats and liver, and to a lesser degree in fruits and vegetables, it is unlikely that appreciable amounts of α-LA are consumed in the typical Western diet; rather, its primary source is dietary supplements. Its distinct anti-inflammatory properties are attributed to it being a reactive oxygen species scavenger (Shay et al. 2009); one study has shown a clinically significant decrease in serum interleukin-6 (IL-6) levels by 15 percent following four weeks of α-LA supplementation (Sola et al. 2005). Its major analgesic effect relates, however, to its interaction with regulatory components of the insulin-signaling cascade; α-LA improved skeletal muscle glucose uptake and decreased insulin resistance in animal models (Jacob et al. 1996). In humans, α-LA improved glucose levels in patients with type 2 diabetes (Jacob et al. 1995). It is not surprising, therefore, that α-LA has the incurable and many times painful disease occurring in approximately 30 percent of diabetes patients; a meta-analysis of four clinical trials showed a significant improvement of pain following α-LA supplementation in patients with diabetic polyneuropathy of the feet and lower limbs (Ziegler et al. 2004). As convincing is a retrospective study testing patients' outcome when switched from pathogenetic treatment with α-LA to more conventional drug therapies for neuropathic pain, such as gabapentin (Ruessmann 2009). In this study a cohort of 443 diabetic patients with chronic painful neuropathy were treated with α-LA for a mean period of five years. After stopping this treatment, 293 patients were switched to gabapentin (600–2400 mg/day), while 150 patients remained untreated after becoming asymptomatic. In the untreated group, 73 percent of patients developed new neuropathic symptoms as soon as two weeks after terminating α-LA treatment. In the gabapentin group, 55 percent were nonresponders requiring alternative treatment. In addition, the frequency of outpatient visits had almost doubled when α-LA therapy was stopped. This study is unique, being one of the few studies comparing dietary therapy with orthodox analgesic measures and clearly showing the advantage of the former.

The Role of Cholinergic Pathways

Less explored, but probably not less important, is the anti-inflammatory role of fat through the cholinergic pathway. The cholinergic anti-inflammatory pathway involves inflammatory factor-mediated activation of the vagus nerve, which serves as a sensor of inflammation or injury. Following this activation, information is relayed to the central nervous system (CNS) via the hypothalamic-pituitary axis. This triggers the release of acetylcholine from efferent vagus nerve endings (Tracey 2005). Stimulation of the vagus nerve prevents the release of pro-inflammatory cytokines, such as tumor necrosis factor alpha (TNF-α) and IL-1 (Tracey 2002; Bernik et al. 2002). This inflammatory reflex provides the brain with the means to regulate the cytokine response in a localized, controlled, and organ-specific manner. Dietary fat can activate this cholinergic anti-inflammatory pathway since the consumption of a high-fat diet reduced circulating levels of cytokines (Luyer et al. 2005). Indeed, when these experiments were repeated in animals subjected to vagotomy, the administration of the high-fat diet no longer prevented increased cytokine activity (Pahl et al. 2003).

The fat/vagal anti-inflammatory circuit is further complicated by another key player, cholecystokinin (CCK), a neuropeptide released from the duodenum after consumption of dietary fat. CCK triggers several digestive functions including the activation of afferent vagus nerve signals that induce satiety. Importantly, the administration of CCK receptor antagonists to rats fed a high-fat diet impaired the fat-induced suppression of cytokine release (Luyer at al. 2005). However, the role of CCK in nociception is far from being resolved; a study in healthy human volunteers, fed either medium- or long-chain fatty acid meals, showed that levels of experimental pain and hyperalgesia were similar although CCK plasma levels differed significantly among participants. Moreover, participants did not differ in their response to alfentanil, a short-acting opioid (Pahl et al. 2003).

Dietary Protein

There is a meager amount of research related to the possible antinociceptive effect of dietary protein. Experiments in rats indicated that changing dietary amounts of protein per se has no effect on nociception (Kanarek et al. 1999). However, supplementation of specific proteins was associated with analgesia. For example, soy protein possessed antinociceptive properties in rat models of chronic nociception, and potentially in humans as well (see the "Soy" section in this chapter). Lactoferrin, a ubiquitous milk protein, is another analgesic protein whose production by neutrophils in the periphery and microglia in the CNS is increased under inflammatory conditions. Interestingly, when given orally or parenterally, milk lactoferrin was an immunomodulator that suppressed pro-inflammatory cytokine production in vitro (Crouch et al. 1992) and in vivo (Zimecki et al. 1998), and increased production of anti-inflammatory cytokines like IL-10 (Togawa et al. 2002). In addition, lactoferrin significantly suppressed inflammatory hypernociception in a rat model of rheumatoid arthritis (Hayashida et al. 2004) and decreased neuropathic-like nociception in nerve-injured rats (Onal et al. 2010). Lactoferrin has also been shown to possess μ-opioid receptor–mediated antinociception properties in a variety of acute pain tests in rats (Hayashida et al. 2003). This milk protein is widely used as an over-the-counter product for two main indications: an anti-inflammatory agent and an immune stimulant. Its analgesic role in humans, however, has not been substantiated by controlled studies.

Carbohydrates

It has been argued that the rise in carbohydrate consumption has caused the epidemic levels of many diseases (Cordain et al. 2005). Paradoxically, both low- and high-carbohydrate diets have been proclaimed to possess substantial health benefits (Bowman and Spence 2002; Delbridge et al. 2009). Accordingly, carbohydrate-based diets were suggested to mitigate or prevent a variety of diseases including diabetes, cardiovascular disease, and cancer. Similarly, multiple anecdotal and scientific reports claim that both low and high carbohydrate consumption is capable of ameliorating pain symptoms. In trying to resolve this paradox, a discussion on the types of carbohydrates consumed and their contrasting effects on physiological processes is mandatory.

There are two common types of carbohydrates in food, simple and complex. Simple carbohydrates, such as refined sugars and grains, are found mainly in processed foods (e.g., cookies, breakfast cereals, and soft drinks) and are high on the glycemic index. Complex carbohydrates are low on the glycemic index and include sources of dietary fiber such as whole grains, whole-wheat breads, brown rice, green vegetables, and fresh fruits. Refined sugars and grains make up 39 percent of the total daily energy consumed in Western diets, compared to whole grains which make up a mere 3.5 percent (Cordain 2007). While the rise in carbohydrate consumption in general has been argued to be the cause of epidemic levels of many diseases, the major culprit may be the steep rise in refined carbohydrate levels in human diets. High-glycemic-index carbohydrates, which the body rapidly converts to sugar, spike blood sugar levels and consequently result in an excess production of insulin. Rapid glucose and insulin responses spur inflammation and inflammatory pain (Kreisberg and Patel 1983) by increasing the production of major inflammatory mediators like arachidonic acid, prostaglandin E2 (Kreisberg and Patel 1983), and C-reactive protein (CRP; Liu et al. 2002; Levitan et al. 2008). In contrast, low-glycemic diets may reduce inflammatory pain by decreasing levels of serum inflammatory markers such as IL-6 (Nilsson et al. 2008) and CRP (King et al. 2003; Ajani et al. 2004). Dietary fiber also increases plasma levels of the hormone adiponectin in healthy men and women (Nilsson et al. 2008). In addition to its ability to regulate the metabolism of lipids and glucose, adiponectin also possesses anti-inflammatory properties (Ouchi and Walsh 2007). In patients with obesity-related inflammatory conditions, such as insulin resistance and type 2 diabetes, pro-inflammatory factors inhibit adiponectin production and thus perpetuate inflammation (Fantuzzi 2008). The effect of adiponectin on nociception is far more complex, however, as increased levels of the hormone have been found in the synovial fluid of patients with rheumatoid arthritis (Schaffler et al. 2003; Fantuzzi 2008).

Not all simple sugars are bad when it comes to pain, as evidenced by the tradition of doctors handing out lollipops to children coming for a checkup. Simple household sugar has been shown to be beneficial in newborn infants and prepubertal children undergoing painful procedures (Stevens et al. 2010). Like sucrose, glucose (Blass and Smith 1992; Gradin and Schollin 2005) and artificial sweeteners (Barr et al. 1999) also reduced acute pain behavior in newborn babies undergoing painful stimuli. Unfortunately, sugar is not capable of preventing the development of increased sensitivity and exaggerated pain responses occurring days or months after repeated painful procedures in newborns (e.g., heel lance, venipuncture, or immunization), even when administered prior to each painful stimuli (Taddio et al. 2009). Interestingly, sucrose analgesia was more effective in older children who had a preference for sweet tastes; those who preferred 24 percent or higher sucrose solutions exhibited improved pain tolerance (Pepino and Mennella 2005). In addition, preexisting dietary habits appear to play a role in the efficacy of sweets to elicit analgesia; among children who preferred higher levels of sucrose, the sugar was effective in those that were of normal weight but not in overweight children or children at risk of becoming overweight (Pepino and Mennella 2005).

Mechanisms of Carbohydrate Analgesia

Carbohydrate analgesia could be related to the hedonic experience of sugar consumption since foods that are palatable and elicit pleasure can trigger analgesia (Pepino and Mennella 2005; Foo and Mason 2009). Indeed, the administration of carbohydrate sweeteners directly to the stomach was not associated with analgesia (Ramenghi et al. 1996; Zmarzty and Read 1999). It is possible, therefore, that afferent signals from the mouth play a role in sweet taste analgesia; if the cognitive and taste components of eating are bypassed, the analgesic effect of food is lost.

Several lines of evidence suggest that sweet taste-induced analgesia is associated with the opioid system. For example, a sweet taste triggered the release of endogenous opioids in the brain, plasma, and cerebrospinal fluid (Blass and Watt 1999; Yamamoto et al. 2000), and sugar consumption modulated morphine-induced analgesia (D'Anci et al. 1996; Kanarek et al. 2001; Davis et al. 2006) and delayed the development of tolerance to opioids (D'Anci 1999). Conversely, other studies have not found an association between sugar and opioid-induced analgesia; the administration of an opioid antagonist to newborns did not diminish the analgesic effect of oral glucose (Gradin and Schollin 2005), and tolerance did not develop in neonates who were given repeated doses of glucose (Eriksson and Finnström 2004).

In addition to opioid-induced mechanisms, the dopaminergic system that is involved in central processes of analgesia (Leknes and Tracey 2008) could also mediate sucrose-induced analgesia (Altier and Stewart 1999). It has been shown that sucrose intake activated the pleasure center in the nucleus accumbens by stimulating dopamine release (Hajnal and Norgren 2001; Hajnal et al. 2004). The lack of sugar-induced analgesia in obese children could be related to a dysregulation of the opioid- and dopamine-mediated analgesic mechanisms, due to higher consumption of carbohydrates and sweet foods; higher than normal β-endorphin plasma levels have been observed in obese children (Bernasconi et al. 1988) and an excess intake of sugar increased opioid and dopamine receptor binding (Colantuoni et al. 2001), modifications that may account for the attenuation of the analgesic properties of sucrose.

Although the association between sweet taste and analgesia is evident during infancy, knowledge of the pain-relieving properties of sugars in older children and adults is limited. A number of variables, including gender, blood pressure, preference for sweets, dietary habits, and weight, could moderate sucrose-induced analgesia in older individuals (Lewkowski et al. 2003; Kanarek and Carrington 2004; Pepino and Mennella 2005). Regardless of their preference for sweet taste, sucrose is likely ineffective in reducing pain in adults (Pepino and Mennella 2005). While some studies reported increased tolerance to pain following the consumption of palatable sweet carbohydrates in adults (Mercer and Holder 1997), others reported no association (Miller et al. 1994). Such differences could be due to a variety of physiological and endocrine differences, and warrants further research aimed at clarifying the role of sucrose as an analgesic in adults.

Micronutrients

Micronutrients are nutrients needed in small amounts for maintaining normal bodily functions. There is a clear paradox in the fact that most modern developed societies are exposed to an excess of macronutrients, while still suffering an epidemic of micronutrient deficiency. Unsurprisingly, reports in the popular media have gone so far as to claim that our body is constantly thirsting for the nutrients it needs by stimulating our hunger; excess eating of macronutrients alone, especially fat and carbohydrates, does not satisfy our hunger. Regardless whether these claims have scientific merit, the market is flooded with micronutrient supplements. There is a huge variety of vitamins, minerals, phytochemicals, fatty acids, and amino acids, to mention just a few. Some of these are marketed specifically for the treatment of painful conditions, mainly inflammatory processes, with no solid scientific background in most instances. However, there are myriad scientific data supporting their antinociceptive and analgesic role.

Amino Acids

Whereas dietary protein as a whole has hardly been researched for its analgesic properties, data on the analgesic properties of specific amino acids are abundant. Amino acids like D-phenylalanine (Walsh et al. 1986), tryptophan (Heyliger et al.

1998), and L-arginine (Harima et al. 1991) have been shown to possess antinociceptive properties. In this chapter, however, only taurine and carnitine are discussed in detail due to the more substantial evidence of their antinociceptive properties.

Taurine

An inhibitory amino acid in the central nervous system (Sakai et al. 1985), taurine is obtained from the diet and is endogenously synthesized from cysteine. While having no effect on the sensitivity of intact rats to acute noxious thermal stimuli (Belfer et al. 1998; Serrano et al. 1990), taurine-rich diets decreased inflammatory hypernociception following the intraperitoneal injection of acetic acid (Serrano et al. 1990). This effect was probably opioid dependent since taurine's antinociceptive properties in this model disappeared in naloxone-treated rats (Serrano et al. 1990). Taurine-enriched diets were also antinociceptive in rodent models of phantom limb pain (Belfer et al. 1998) and neuropathy-like diabetic hypernociception induced by streptozotocin injection (Li et al. 2005). In the latter, taurine was able to not only partially reverse hypernociception but also attenuate deficits of nerve conduction, typical of diabetic peripheral neuropathy. Multiple mechanisms of action were suggested to explain taurine's antinociception, including neuroprotection against excitotoxicity (Wu et al. 2005) and modulation of calcium influx into injured cells (Li et al. 2005). To our knowledge, taurine has never been tested for its analgesic properties in humans.

Carnitine

Carnitine is a long-chain amino acid whose main dietary source is meat products. Carnitine supplements are available as either L-carnitine or acetyl L-carnitine, the esterified form of carnitine. The U.S. Food and Drug Administration approved the supplements for people diagnosed with carnitine deficiency.

In animal studies, carnitine has been shown to be antinociceptive in acute pain tests and in experimental models of neuropathic-like pain, diabetic peripheral neuropathy, neuropathy associated with chemotherapy and HIV therapy drugs, and chronic post–nerve injury hypernociception (Chiechio et al. 2007). The most convincing data on the analgesic effect of carnitine in humans come from studies in patients with painful peripheral neuropathy, mainly due to diabetes. A recent review summarized the results of more than 1,600 patients treated daily with at least 2 g of carnitine (Evans et al. 2008). Carnitine therapy decreased pain scores, improved electrophysiological factors, and enhanced nerve regeneration in these patients. Importantly, carnitine therapy was as safe and effective as conventional therapy for patients with diabetes (Sima 2007). Carnitine was also beneficial in patients with chemotherapy-induced peripheral neuropathy (De Grandis 2007) and sciatic pain caused by a herniated disc (Memeo and Loiero 2008).

Numerous analgesic mechanisms have been suggested to explain the unique analgesic properties of carnitine. Carnitine plays an essential role in mitochondrial energy synthesis as it facilitates long-chain fatty acid transport across the cells' inner mitochondrial membrane. Consequently, carnitine increased energy availability, enhanced oxygenation, and prevented toxic accumulation of fatty acids (Longo et al. 2006). In patients with diabetic neuropathy, carnitine may decrease insulin resistance, thus improving cellular uptake of glucose (Mingrone et al. 1999). As well, carnitine improved healing of damaged neurons (McKay, Hart, et al. 2002), enhanced fast and slow axoplasmic transport within nerve cells (Kano et al. 1999), and upregulated the mGlu2 metabotropic glutamate receptors (Chiechio et al. 2002). Based on the impressive body of evidence and the excellent safety profile of carnitine, a more frequent use of this supplement in patients with neuropathic pain conditions could be expected. Unfortunately, its use is not common practice, and in certain countries, for example Canada, it is not available for purchase over the counter or by prescription.

Vitamins

Few dietary ingredients have attracted more attention and involved more unsubstantiated claims than vitamins. These organic compounds are obtained naturally from plant and animal foods and are required as nutrients in tiny amounts by essentially every multicellular organism. Vitamin supplementation in humans has developed into a multibillion-dollar market, promising a variety of benefits, most of them not scientifically proven. Pain relief, although attributed to some vitamins, is only a minor indication for vitamin research or consumption. Of the thirteen known human vitamins, three most explored for their antinociceptive properties will be discussed here in more detail: vitamins B, C, and D.

Vitamin B

The B vitamins are eight water-soluble vitamins that are important in processes of cell metabolism and for maintaining normal immune and nervous system function. B vitamins are mainly found in unprocessed foods, meat, potatoes, bananas, lentils, chili peppers, and yeast. A number of preclinical studies have shown promising antinociceptive results; pyridoxine (B_6) decreased nociception in acute pain tests (Zimmerman et al. 1990), and supplementation with B_1–B_6–B_{12} complex vitamins decreased thermal and chemical hypernociception in mice (Franca et al. 2001). In contrast, however, B_{12} supplementation had no effect on chemical, electric shock, and heat-evoked nociception in rats (Eschalier et al. 1983). In animal models of persistent or chronic pain, B complex vitamins reduced hypernociception in rats undergoing crush injury to their dorsal root ganglion (Song et al. 2009) and following partial injury to a peripheral nerve (Wang et al. 2005). In an inflammatory pain model, B_6 alone decreased hypernociception through peripheral (Zimmerman et al. 1990) and central mechanisms (Fu et al. 1988).

The analgesic role of vitamin B in humans is still not clear, although a few double-blind, randomized, controlled studies in the last three decades showed promise. For example, a short term, three-day therapy with B_1–B_6–B_{12} complex vitamins, in combination with diclofenac, was superior to diclofenac alone in reducing pain associated with lumbago (Mibielli et al. 2009). A twelve-week treatment with the same B complex decreased pain levels and improved objective parameters of neuropathy in patients with painful diabetic neuropathy (Stracke et al. 1996). B_6 and B_{12}, together with folic acid, were beneficial in patients with rheumatoid arthritis (Yxfeldt et al. 2003), while treatment for twelve days with a multivitamin preparation, including vitamin B_6, brought upon mild pain relief in women with chronic cephalgia and facial pain (Mader et al. 1988). Finally, riboflavin (B_2) has been advocated as a migraine prophylactic agent; in a well-designed, randomized, controlled trial, daily use of 400 mg of riboflavin for three months resulted in a 50 percent reduction in migraine attacks in 59 percent of patients, compared to 15 percent of patients given placebo (Schoenen et al. 1998).

Vitamin C

Ascorbic acid, or vitamin C, is found in fresh fruits, especially citrus fruits, berries, and green vegetables. In addition to being an antioxidant, vitamin C is important in processes involving wound healing and the production of collagen. Anecdotal reports claimed it could decrease bone pain, postexercise muscle soreness, and arthritis. However, the most solid scientific data pertain to its analgesic role in neuropathic pain conditions. Vitamin C prevented the development of complex regional pain syndrome type 1 (CRPS-1, previously referred to as reflex sympathetic dystrophy). CRPS is a chronic neurological pain disorder mainly affecting the limbs, characterized by disabling pain, swelling, vasomotor instability, sudomotor abnormalities, and impairment of motor function. This incapacitating neuropathic pain disorder usually develops after minor trauma or surgery, lacks effective therapy, and if not treated early might never be cured. The prevalence of CRPS-1 is especially high after wrist fracture or surgery, where it could be diagnosed in 5–10 percent of patients. In two double-blind, prospective, multicenter trials. Patients with wrist fractures were given a 500 mg dose of vitamin C from the day of injury and for fifty days thereafter. A year after injury, the prevalence of CRPS was approximately four times lower in the vitamin C group, compared to the placebo group (Zollinger et al. 1999, 2007). Identical results were reported when vitamin C was prophylactically given to patients undergoing foot and ankle surgeries (Besse et al. 2009). No studies to date have tested the role of vitamin C supplementation once CRPS has fully developed.

A single randomized study tested vitamin C for postherpetic neuralgia, one of the most painful and resistant chronic neuropathic pain conditions. Plasma levels of vitamin C were significantly lower in patients with postherpetic neuralgia compared to healthy individuals (Chen et al. 2009). Restoration of vitamin C levels via intravenous infusions of high doses of vitamin C (2.5 g/day for three consecutive days) was associated with a significant decrease in spontaneous pain, but had no effect on tactile allodynia (Chen et al. 2009).

Vitamin D

Vitamin D has garnered significant media attention in recent years since its deficiency has been linked to an increased risk of illnesses such as cancer, diabetes, multiple sclerosis, heart disease, and rheumatoid arthritis. Less explored is its association with chronic pain. Vitamin D is a group of fat-soluble pro-hormones, the two major forms of which are vitamin D_2 (ergocalciferol) and vitamin D_3 (cholecalciferol). Vitamin D, obtained from sun exposure, food, and supplements, is biologically inert and must undergo two hydroxylation reactions to be activated to calcitriol in the body. Vitamin D deficiency is neither rare nor exclusive to patients with chronic pain. A substantial proportion of the world's population lives far from the equator and is deprived of direct exposure to the sun, necessary to produce vitamin D_3 in the skin. For example, 97 percent of the residents in Calgary, the sunniest city in Canada, were found to have low serum levels of vitamin D in the winter (Rucker et al. 2002). Moving south does not change things significantly since 87 percent of those living in Memphis, Tennessee, had inadequate vitamin D levels (Alsafwah et al. 2009). Significant deficiency has even been documented in countries such as India, Turkey, and Saudi Arabia, where factors such as oppressive heat keep people indoors and traditional dress keeps them well covered.

A link between low vitamin D levels and chronic pain may exist, although the exact putative mechanism remains a mystery with vitamin D affecting the expression of hundreds of genes. A study in Minneapolis of 150 patients presenting with persistent, nonspecific musculoskeletal pain revealed that 93 percent were vitamin D deficient (Plotnikoff et al. 2003). Similarly, a study in Egypt of 60 patients presenting with chronic low back pain revealed normal vitamin D levels in 18 percent of patients, compared to healthy controls from the same area who were more than twice as likely to have normal levels (Lotfi et al. 2007). Additionally, a large study of over 3,000 women in the United Kingdom demonstrated that having low vitamin D levels roughly doubled the risk of having fibromyalgia (Atherton et al. 2009).

Low vitamin D may also affect how much pain medication one needs. Patients with chronic pain with low serum vitamin D levels were taking on average twice the amount of morphine compared to patients with higher vitamin D levels (Turner et al. 2008). However, whether vitamin D supplementation could have a direct analgesic effect has not yet been resolved. An open label trial showed that supplementation with vitamin D in quantities ranging from 1000 to 3000 IU reduced pain by almost 50 percent in patients with diabetic neuropathy (Lee and Chen 2008). Furthermore, a small case series of six patients with chronic low back pain and low vitamin D levels showed significant improvement once the deficiency was corrected (Schwalfenberg 2009). On the other hand, supplementing vitamin D_2 to people with diffuse musculoskeletal pain for three months failed to show benefit (Warner and Arnspiger 2008). Although the evidence linking low vitamin D levels with chronic pain is preliminary, the substantial risks of low vitamin D to overall health and existing evidence probably merit a "test and treat" approach.

Polyamines

Polyamines (e.g., spermine, spermidine, and putrescine) are organic polycations, first identified in human semen in the late seventeenth century. Polyamines have been investigated extensively for their central role in a variety of cellular processes (Kusano et al. 2007). The main sources of exogenous dietary polyamines are cheese, fruits, meat, certain vegetables, and human milk, together with intestinal and pancreatic secretions, and catabolism products from intestinal cells and luminal bacteria. Several lines of evidence suggest that polyamines are involved in nociception. In contrast, however, to the other dietary constituents discussed in this chapter, polyamines are hypernociceptive agents. They exert this unique effect by facilitating nociceptive transmission as a result of their positive allosteric modulation of the N-methyl-D-aspartate (NMDA) receptor ion channel, leading to central sensitization and pathological pain processes (Kolhekar et al. 1994; Traynelis et al. 1995). It is conceivable, therefore, that a polyamine-deficient diet could decrease nociception. Indeed, such diets were antinociceptive in rodent models of inflammation, post-surgical hypernociception, arthritic, and neuropathic-like models (Rivat et al. 2008). We are not aware of human experiments testing the effect of polyamine-reduced diets on pain. However, these diets have been tried in men with prostatic cancer; beyond their beneficial effects on cancer parameters, these diets have been associated with decreased cancer-related pain (Cipolla et al. 2003). Devoid of major side effects both in rodents and humans, this therapeutic approach could be considered in other chronic pain conditions.

Magnesium

NMDA receptors play a crucial role in plasticity, the process by which the CNS remodels and changes in response to noxious stimuli, and their activation is one of the culprits for the development of persistent and chronic pathological pain states. It is not surprising, therefore, that NMDA receptor antagonists like ketamine could have a profound antinociceptive effect. Unfortunately, the use of most available NMDA receptor antagonists is associated with significant and often intolerable side effects. Fortunately, our body has a natural modulator of NMDA receptor activity, namely, magnesium. Found in such foods as green vegetables, beans, peas, nuts, and whole grains, magnesium is an essential mineral for human nutrition, playing a role in the production and transport of energy, proper muscle activity, bone and teeth formation, and protein synthesis. Preliminary animal and human data indicate that magnesium could also decrease nociception.

In rodents, magnesium-enriched diets decreased nociception in models of visceral pain (Hasanein et al. 2007) and augmented morphine antinociception in neuropathic pain models (Begon et al. 2002). In humans, magnesium was analgesic in both acute and chronic pain conditions; in a placebo-controlled trial, patients receiving an intravenous magnesium solution during and after surgery reported significantly diminished postoperative pain and improved sleep (Tramer et al. 1996). Others have shown that magnesium supplementation decreased posthysterectomy pain in women (Kara et al. 2002) and decreased the consumption of analgesic medications after knee surgery (Koinig et al. 1998).

Magnesium was also effective in reducing pain in pathological pain conditions. When combined with calcium, magnesium reduced the risk of oxaliplatin chemotherapy-induced peripheral neuropathy (Gamelin et al. 2004). A reduction in average pain levels by approximately 40 percent (Turan et al. 2009) was also observed following magnesium supplementation in patients with fibromyalgia shown to have low levels of magnesium (Sendur et al. 2008). In addition, patients suffering from migraines who were treated with oral magnesium for three months reported decreased attack frequency by 42 percent compared to a decrease of only 16 percent with placebo (Peikert et al. 1996). Finally, possessing a laxative effect, magnesium could be especially beneficial in chronic pain sufferers due to the tendency to develop analgesic drug-induced constipation.

Common Foods

In contrast to the abundance of data on the analgesic effect of certain dietary components, data on possible pain-relieving properties of specific foods are scarce. Two common foods with scientific evidence suggesting their possible analgesic role, tart cherries and soy, are discussed in more detail.

Tart Cherries

In humans, the consumption of cherries for analgesia can be traced back to Roman times. When introduced to the New World, cherries were rapidly adopted by First Nations peoples to treat such maladies as a sore throat and deep organ pain. Cherries contain substantial amounts of anthocyanins, which give this fruit, in addition to its distinct color, strong anti-inflammatory and antioxidant properties.

Animal studies showed that tart cherry extracts reduced inflammation-induced pain in rodent models of inflammation (Tall et al. 2004). This effect was dose dependent, and the efficacy of the highest dose (400 mg/kg) was similar to that of indomethacin (5 mg/kg). Subsequent controlled studies in humans showed identical results. For example, a glass of cherry juice was estimated to block the COX enzyme to the same extent as 300 mg of regular aspirin (Wang et al. 1999); when healthy subjects were given a large bowl (280 g) of bing cherries to eat daily for four weeks, their circulating levels of CRP decreased by 25 percent (Kelley et al. 2006); finally, a single serving of cherries was able to increase peak urinary excretion of urate by 73 percent in healthy women (Jacob et al. 2003), making cherries the only potential therapy of gout arthritis, addressing both the painful symptoms and its pathogenesis.

Beyond their effect in pathological pain, cherries might be beneficial in physiological pain conditions as well; following strenuous muscular activity, some soreness and transient decrease in strength is expected. In a randomized, placebo-controlled trial, consumption of 12 oz of cherry juice was beneficial in reducing both post-exertion muscle pain and temporary loss of muscle strength (Connolly et al. 2006). As well, tart cherry juice was superior to placebo in regaining isometric strength and decreasing IL-6 levels in marathon runners (Howatson et al. 2009). Not surprisingly, these advantages have caught the eye of professional sports teams, with a medical trainer of the New York Rangers hockey team noting that players "feel less sore, sleep better and recover faster" after incorporating cherry juice in their diet (Cornell University 2009). Finally, a good night's sleep is what many people with pain crave. Being one of nature's few sources of melatonin (Burkhardt et al. 2001), cherries have been touted for centuries as a sleep aid. A recent pilot study showed that cherry juice was superior to placebo in maintaining a good night's sleep in healthy subjects (CherryPharm 2009).

Soy

The spread of soybean cultivation, its multiplicity of uses, and its global economic, political, and medical significance set it apart from all other major food plants. Although still perceived by many as primarily an industrial crop or animal feed, soy in reality is an extraordinarily versatile and rich food source, gradually attaining growing importance in the world's food future. One of the advantages of soybeans over other plant foods is that they contain 30–40 percent protein, including all the amino acids essential for human nutrition. The increased interest in soy has stemmed from evidence of its many health benefits such as lower rates of breast and prostate cancer and cardiovascular disease in Asian populations, which traditionally consume large quantities of soy.

In animals, both soy protein and soy fat possess antinociceptive properties. For example, in the PSL model (Seltzer et al. 1990), the consumption of diets consisting of 20 percent soy protein was strikingly antinociceptive (Shir et al. 1998, 2001b), and in a mouse model of bone cancer hypernociception, soy protein–enriched diets had a beneficial effect as well (Zhao et al. 2004). Additionally, soy protein was shown to possess anti-inflammatory properties in two rat models of inflammation, developed by injecting complete Freund's adjuvant into the knee joint and carrageenan into the paw (Borzan et al. 2010).

Dietary soy protein may be a unique modality to prevent the development of hypernociception since the pain-suppressive properties of soy protein were predominantly the result of pre- but not postinjury consumption (Shir et al. 2001a); feeding rats a soy-rich diet up to the time of injury, but not after, was sufficient to prevent the development of chronic hypernociception. Discontinuing the soy diet for just one day before surgery deprived rats of their dietary protection, enabling them to develop a full-blown, neuropathic-like hypernociceptive state (Shir et al. 2001a).

The protein in soy is not a lone contributor to the antinociceptive properties of this food; in the PSL model, soy oil, but not canola or sunflower oils, had significant antinociceptive properties (Pérez et al. 2005). Interestingly, soy protein and soy fat interacted synergistically to decrease nociception in this model (Pérez et al. 2004).

To the best of our knowledge, no population studies in humans have been conducted comparing the prevalence of neuropathic pain in populations consuming soy-rich and soy-deficient diets. Few studies have examined the possible association between soy protein and analgesia; diets enriched with soy milk, taken orally each day for three months (34 g of soy protein/day), had a mild analgesic effect on cyclical menstruation-associated breast pain, where 56 percent of healthy women reported favorable results (McFadyen et al. 2000); daily consumption of 40 g of soy protein for three months was found to be safe and effective in partially relieving pain and discomfort associated with osteoarthritis (Arjmandi et al. 2004). However, daily consumption of beverages containing 20 g of soy protein and 160 mg of isoflavones for six weeks did not improve fibromyalgia symptoms (Wahner-Roedler et al. 2008).

The mechanisms of soy anesthesia are not clear; soy protein by itself contains hundreds of constituents, many of which could be antinociceptive. Phytoestrogens, in particular genistein and daidzein, have received the most attention as potential antinociceptive agents. Phytoestrogens inhibit various types of protein-kinase enzymes (Huang et al. 1992), have anti-oxidative properties (Djuric et al. 2001), and possess immunomodulatory as well as anti-inflammatory qualities (Valsecchi et al. 2008). Partial proof for the antinociceptive properties of phytoestrogens came from experiments in PSL-injured rats (Shir et al. 2002) and a rodent model of sciatic nerve neuritis (Valsecchi et al. 2008). In humans, isoflavones decreased pain in women with mastalgia (Ingram et al. 2002). Other nociceptive candidates of soy include phenolic acids, phytates, and saponins, which also possess anti-inflammatory, antioxidant, and immunostimulatory properties.

The minimal amount of clinical data supporting the use of soy as an analgesic modality necessitates additional controlled studies to determine its efficacy, the optimal amount of soy required to reduce pain, and the minimum time window of exposure required to obtain an analgesic effect, both before and after the pain condition has developed. Finally, comparative population studies are essential for correlating the prevalence of neuropathic pain in societies consuming varying amounts of soy in their diets.

DIETARY HERBS AND SUPPLEMENTS

Herbal medicine is the most ancient form of health care and has been used by most cultures throughout history. Herbs and spices have been used for thousands of years not only to enhance food but also to improve health. Sadly, some 15,000 of 50,000 medicinal species are under threat of extinction because of commercial overharvesting, pollution, competition from invasive species, and habitat destruction (Edwards 2009). As well, the transfer of knowledge to future generations of the multitude of traditional remedies wanes or has already been lost. Modern medicine and the pursuit of evidence-based medicine have largely ignored the therapeutic potential of traditional practices to treat a variety of ailments. Until the last couple of centuries, natural sources, especially from plants, have been one of the major contributors to pain therapy. Today, there is increasing interest in alternatives to chronic pain management, and research is slowly emerging to prove what has been known for centuries on the use of food for medicinal purposes. Unfortunately, a lack of appropriate reliable evidence delays the discovery and acceptance of potential dietary analgesic modalities. Of the hundreds of documented examples, a few are mentioned below, categorized according to their morphological origin.

Roots and Bulbs

Devil's Claw (*Harpagophytum procumbens*)

Originating from the Kalahari and Savannah desert regions of Africa, the tuberous roots of devil's claw have been used for centuries for a variety of pain conditions. Animal studies have demonstrated an antinociceptive effect for devil's claw in a variety of hypernociceptive conditions (Mahomed and Ojewole 2004; Uchida et al. 2008), with equal or greater efficacy for pain and inflammation than conventional nonsteroidal anti-inflammatory drugs (NSAIDs; Lanhers et al. 1992; Baghdikian et al. 1997). Preliminary human data have shown potential benefit for chronic pain conditions such as degenerative rheumatic disorders (Warnock et al. 2007), osteoarthritis (Chantre et al. 2000), low back pain (Gagnier et al. 2007), and nonspecific musculoskeletal pain (Grant et al. 2007). While the analgesic mechanisms of devil's claw are unknown, its anti-inflammatory properties are attributed to iridoid glycosides, in particular two active ingredients, harpagoside and beta sitoserol. Interestingly, although possessing anti-inflammatory properties, devil's claw does not produce the same changes in prostaglandins as standard anti-inflammatory drugs, which act as COX-2 inhibitors (Moussard et al. 1992). Daily doses of 50–100 mg of harpagoside extract were effective for short-term improvement of low back pain (Gagnier et al. 2007). For low back pain or osteoarthritis, a daily dose of 2–9 g of crude extract or 600–1200 mg tablets (standardized to contain 50–100 mg of harpagoside) three times per day has been used (Natural Standard Database 2009).

Ginger (*Zingiber officinale*)

Ginger belongs to the Zingiberaceae plant family, which also includes turmeric and cardamom, and grows in Jamaica, India, Haiti, Hawaii, and Nigeria. For centuries, the Ayurvedic system of traditional medicine native to India has used ginger to treat pain associated with rheumatism. Ginger contains gingeroles, natural anti-inflammatory agents which are partly responsible for its potent odor. Gingerols possess anti-inflammatory properties evidenced by inhibition of prostaglandin and leukotriene formation (Kiuchi etal. 1992). Commercially, HMP-33 is a standardized ginger extract that is rich in gingerol and used to treat arthritic conditions (Bliddal et al. 2000). Moderate evidence of efficacy exists for the treatment of osteoarthritis, rheumatoid arthritis, and musculoskeletal disorders with ginger (Srivastava and Mustafa 1992; Gregory et al. 2008; Chrubasik et al. 2007). Although its efficacy ranked close to that of ibuprofen (Bliddal et al. 2000; Haghighi et al. 2005), its use was not associated with decreased use of analgesic medication (Altman and Marcussen 2001). Ginger was also found to be as effective as ibuprofen in relieving pain in women with primary dysmenorrhea (Ozgoli et al. 2009). More rigorous studies are required, however, to fully establish its use for pain relief.

Turmeric (*Curcuma longa*)

Polyphenols are small molecular compounds of multiple plant sources with potent antioxidant properties that may explain the possible health benefits of red wine, grapes, berries, chocolate, and green tea. Polyphenols are also found in substantial quantities in curcumin, the main active component in turmeric. Turmeric is a yellow spice with a celebrated history. It has been used for thousands of years in India and China, and more recently in Indonesia, Thailand, Japan, and Korea, to treat fever, pain, and skin diseases. Its use continues to this day, with the average Indian consuming approximately 1 g of turmeric daily. In India, products such as turmeric soaps, facial creams, and even band-aids can be found.

Substantial scientific evidence exists for the analgesic properties of curcumin. In a small randomized crossover study, patients with rheumatoid arthritis experienced the same symptom relief with curcumin as with conventional NSAIDs (Deodhar et al. 1980). Additionally, curcumin reduced pain at least as well as NSAID in a randomized control trial of forty patients who were recovering from hernia surgery (Satoskar et al. 1986). Curcumin was also found to reduce relapse rate by 75 percent in patients with ulcerative colitis (Hanai et al. 2006). The analgesic properties of curcumin have been attributed to multiple mechanisms including, but not limited to, COX inhibition, anti-TNF effect, and reduced free radical formation (Aggrawal et al. 2007). The most common side effect of curcumin supplementation is an upset stomach. It may also have a mild anticoagulant effect and possibly reduced blood sugar levels in diabetics. Unfortunately, the bioavailability of curcumin is low as it is not well absorbed and undergoes substantial hepatic inactivation. Its bioavailability could be improved, however, if concomitantly consumed with large amounts of black pepper (Shoba et al. 1998).

Tree Bark

Boswellia (*Boswellia serrata*)

Derived from the gum resin of frankincense tree bark, boswellia has been used for thousands of years to treat a variety of conditions (Ammon 2006). Chinese herbalists use boswellia in powder form and in teas to treat rheumatism and menstrual pain (Basch et al. 2004). Modern herbalists are rediscovering the resin as a treatment for pain, arthritis, and other inflammatory conditions (Etzel 1996; Kimmatkar et al. 2003; Singh and Atal 1986; Basch et al. 2004), although some studies do not support these claims (Sander et al. 1998). In one of the most convincing studies to date, boswellia extract was used in a ninety-day, double-blind, randomized, placebo-controlled study to evaluate its efficacy and safety in seventy-five people with osteoarthritis of the knee. The extract significantly reduced pain and improved physical functioning with no major side effects (Sengupta et al. 2008). Preclinical studies have shown anti-inflammatory properties of boswellia extract similar to those of NSAIDs (Safayhi et al. 1992; Singh and Atal 1986) and antinociceptive properties equal in potency to 4.5 mg/kg of morphine (Menon and Kar 1971). Commercially available boswellia extract is standardized to contain 30–65 percent boswellic acid. For arthritic conditions, an oral dose of 400 mg of standardized boswellia three times per day has been used (Natural Standard Database 2010).

Leaves

Sow Thistle (*Sonchus oleraceus*)

Found worldwide, this very common thorny vegetation is native to Europe and North Africa. Notorious for its problematic tendency to invade fields of many crops, it is seemingly an unlikely candidate to be considered beneficial to humans. Nevertheless, it is edible as a leaf vegetable and is frequently consumed, especially in Mediterranean countries, New Zealand, and Brazil. Beyond its importance as a nutritious dietary ingredient, its leaves have been traditionally used for treating general pain condition, headaches, rheumatism, and toothaches. Although its analgesic properties have never been clinically tested, there are convincing preclinical data on its antinociceptive properties. Given orally, sow thistle extracts significantly reduced chemical and thermal-induced nociception in intact mice. This antinociceptive effect was equal to 10 mg/kg of morphine (Vilela et al. 2009), and superior to 10 mg/kg of indomethacin or 1 mg/kg of dexamethasone (Vilela et al. 2010).

Flowers

Feverfew (*Tanacetum parthenium*)

Native to southeastern Europe, feverfew is widespread throughout Europe, North America, and Australia. The medicinal use of its flowers, leaves, and stems for such ailments as headache, stomach ache, toothache, and arthritis dates back to the seventeenth century (Hanrahan 2001). While shown to be ineffective for rheumatoid arthritis (Pattrick et al. 1989), its beneficial effects for migraine relief are more promising. A few randomized, controlled trials have demonstrated that with frequent prophylactic use, feverfew may reduce the frequency, severity, and duration of chronic migraines (Palevitch et al. 1997; Murphy et al. 1988). The sesquiterpene lactone, parthenolide, is believed to be the active constituent of feverfew. Its use in rodents has been associated with significant antinociceptive and anti-inflammatory effects (Jain and Kulkarni 1999). Possible mechanisms of feverfew-induced analgesia include vasodilatation (Barsby et al. 1992; Hay et al. 1994) and the inhibition of prostaglandin E2 (Collier et al. 1980; Pugh and Sambo 1988) and pro-inflammatory cytokine production (Smolinski and Pestka 2003).

A recent Cochrane review has concluded that there is not enough evidence to support the use of feverfew as an analgesic tool (Pittler and Ernst 2004). Accordingly, although 3–4 fresh leaves or 50–250 mg of dry leaves per day have been recommended for migraine relief (Hanrahan 2001; Pittler and Ernst 2004), there are presently no established evidence-based dosage recommendations.

Fruit

Pineapple (*Ananas comosus*)

For centuries, in Central and South America, pineapple has been used for its medicinal properties. Bromelain, a mixture of sulfhydryl proteolytic enzymes found mostly in the stem and fruit of the pineapple plant, was first isolated in the late 1800s, was introduced as a therapeutic compound in 1957, and has been mostly used as an anti-inflammatory agent (Kapes 2005). Numerous studies have demonstrated the anti-inflammatory effect of bromelain in the treatment of phlebitis, sinusitis, cellulitis, and edema (Kelly 1996), as well as its ability to treat arthritis (Brien et al. 2004; Cohen and Goldman 1964), and reduce healing time and pain after surgery and sports-related injury (Seltzer 1962; Masson 1995; Blonstein 1969). Bromelain has been shown to be an effective alternative to NSAIDs without causing gastrointestinal upset (Heinrich et al. 2004).

Although not completely established, bromelain has been shown to act via four different mechanisms: the kinin–kallekrein system, where prekallikrein and bradykinin levels are decreased (Kelly 1996); the coagulation cascade by reducing intermediate clotting factors such as factor X and prothrombin, thus limiting fibrin formation (Kelly 1996); the cytokine system by decreasing neutrophil migration to sites of acute inflammation (Fitzhugh et al. 2008); and the inflammatory pathway by inhibiting pro-inflammatory prostaglandin formation (Kelly 1996). Bromelain extracts are available in tablet or capsule form and are safe when taken in a dose range of 200–2000 mg (Kelly 1996; Kapes 2005).

Seeds

Evening Primrose Oil (*Oenothera biennis*)

This flowering plant is native to North and South America. The seed oil is rich in the essential fatty acid gamma-linolenic acid and is believed to have analgesic and anti-inflammatory properties, particularly for diabetic neuropathy but also for cyclical mastalgia and premenstrual syndrome in women. Multiple studies have been conducted in the last two decades to substantiate these claims (Blommers et al. 2002). However, in contrast to multiple reports in popular media (Bayles and Usatine 2009), the scientific evidence to support such use is lacking. There are preliminary data, however, suggesting the potential benefits of evening primrose oil in patients with rheumatoid arthritis. In a six-month double-blind study comparing 6 g/day of evening primrose oil with olive oil, disease indexes were improved in both groups; evening primrose oil showed mild improvement by relieving morning stiffness, while olive oil was superior in reducing pain (Brzeski et al. 1991).

CALORIC RESTRICTION

Although this chapter focuses on dietary supplementation, a discussion on the antinociceptive effect of diet cannot be complete without addressing the unexpected finding that both chronic food deprivation (60–70 percent of the diet consumed *ad libitum*) and acute fasting episodes can decrease nociception. Multiple studies tested this hypothesis, mostly in intact animals subjected to acute pain tests, but also in persistent hypernociception conditions that are more relevant to humans. For example, mice subjected to an intermittent fasting diet displayed decreased nociception in models of thermal and visceral pain, compared with mice fed *ad libitum* (de los Santos-Arteaga et al. 2003), and chronic fasting was capable of decreasing nociception in the formalin test and in a chronic, posttraumatic pain model (Hargraves and Hentall 2005). Interestingly, these effects were age dependent and were more robust in adult (46–70 weeks old) than in older mice. Indeed, there are similarities between the effect of fasting on nociception and on aging. These include slowing of the aging process; reducing age-related chronic diseases, especially neurodegenerative diseases; regulating neural synaptic plasticity; and extending life span (Ingram et al. 2004; Masoro 1992).

Multiple putative mechanisms of fasting-induced hyponociception have been suggested. These include blocking the age-dependent loss of NMDA receptor 1 subunits of NMDA-subtype glutamate receptors in the hippocampus (Eckles-Smith et al. 2000); altering brain monoamine concentrations, an effect which may in turn modify the normal rate of aging (Kolta et al. 1989); delaying the enhanced expression of genes associated with inflammation and stress in old age (Lee at al. 2000); and enhancing the endogenous κ-opioid system, as evidenced by increased levels of pro-dynorphin mRNA and κ-opioid receptors in the spinal cord of fasting mice (de los Santos-Arteaga et al. 2003). This last mechanism is supported by the observation that following noxious stimulation, c-Fos expression in the dorsal spinal cord was significantly lower in fasting mice compared to controls, indicating that dynorphin could block nociceptive information at the spinal cord level. Supporting the role of endogenous opioids, the hyponociceptive response produced by food deprivation in intact rats was attenuated by naloxone (McGivern and Berntson. 1980). Finally, fasting hyponociception could be mediated through the endogenous cannabinoid (CB) system as intermittent food deprivation in mice was associated with decreased expression of the CB-1 receptor in the spinal cord (Sáez-Cassanelli et al. 2007). Indeed, sub-chronic inhibition of the CB-1 receptor in mice provokes a κ-opioid receptor-dependent antinociception, similar to that found in mice subjected to intermittent food deprivation (Sáez-Cassanelli et al. 2007). These findings, together with the well-established findings that caloric restriction significantly prolongs life expectancy in rodents (Masoro 1992), clearly indicate that it is not only what we eat but also how much we eat that is crucial for our well-being.

FUTURE PROSPECTS

There are currently good indications, and enough supportive data, to show that dietary ingredients are involved in the nociceptive process. The fact that most physicians neglect the "food as medicine" philosophy of Hippocrates is not surprising, however, and could be due to multiple reasons. First is the dominance of Western medicine in medical education and the disregard of complementary and alternative medicine as an important adjuvant to clinical practice. In addition, there is the meager amount of solid research exploring dietary analgesia in an era when medical practice is dictated by evidence-based medicine. Finally, there could be a practical reason for neglecting diet as an analgesic tool. The traditional analgesic armamentarium currently available to the physician is relatively limited, offering a modest variety of non-opioid and opioid analgesics, the key players in the medical approach to patients with pain. Conversely, hundreds of dietary ingredients could potentially be analgesic, thus the choice for drugs rather than food is clearly easier for a physician.

It is unlikely that adhering to the current traditional approach for relieving chronic pain will change the outcome of chronic pain sufferers. Unveiling new routes of therapy, like dietary modification, could be one way to improve outcomes. To achieve that, however, two conditions must be fulfilled. First, more focused research is needed to isolate the more promising dietary ingredients. This is not an easy task considering the fact that dietary research is significantly less lucrative than other research themes in the field of pain medicine, such as drug development. Second, the medical community and the public alike should be better educated on issues related to dietary analgesia.

REFERENCES

Aggarwal, B. B., C. Sundaram, N. Malani, and H. Ichikawa. 2007. Curcumin: the Indian solid gold. *Adv Exp Med Biol* 595:1–75.

Ajani, U. A., E. S. Ford, and A. H. Mokdad. 2004. Dietary fiber and C-reactive protein: findings from national health and nutrition examination survey data. *J Nutr* 134:1181–5.

Alsafwah, S., S. P. Laguardia, M. D. Nelson, et al. 2009. Hypovitaminosis D in African Americans residing in Memphis, Tennessee with and without heart failure. *Am J Med Sci*

Altier, N., and J. Stewart. 1999. The role of dopamine in the nucleus accumbens in analgesia. *Life Sci* 65:2269–87.

Altman, R. D., and Marcussen, K. C. 2001. Effects of a ginger extract on knee pain in patients with osteoarthritis. *Arthritis Rheum* 44:2531–8.

Ammon, H. P. 2006. Boswellic acids in chronic inflammatory diseases. *Planta Med* 72:1100–16.

Arjmandi, B. H., D. A. Khalil, E. A. Lucas, et al. 2004. Soy protein may alleviate osteoarthritis symptoms. *Phytomedicine* 11:567–75.

Arnold, L. M., D. J. Clauw, M. M. Wohlreich, et al. 2009. Efficacy of duloxetine in patients with fibromyalgia: pooled analysis of 4 placebo-controlled clinical trials. *Prim Care Companion J Clin Psychiatry* 11:237–44.

Atherton, K., D. J. Berry, T. Parsons, et al. 2009. Vitamin D and chronic widespread pain in a white middle-aged British population: evidence from a cross-sectional population survey. *Ann Rheum Dis* 68:817–22.

Baghdikian, B., M. C. Lanhers, J. Fleurentin, et al. 1997. An analytical study, anti-inflammatory and analgesic effects of *Harpagophytum procumbens* and *Harpagophytum zeyheri*. *Planta Med* 63:171–6.

Barr, R. G., M. S. Pantel, S. N. Young, et al. 1999. The response of crying newborns to sucrose: is it a "sweetness" effect? *Physiol Behav* 66:409–17.

Barsby, R. W., U. Salan, D. W. Knight, and J. R. Hoult. 1992. Feverfew extracts and parthenolide irreversibly inhibit vascular responses of the rabbit aorta. *J Pharm Pharmacol* 44:737–40.

Basch, E., H. Boon, T. Davies-Heerema, et al. 2004. Boswellia: an evidence-based systematic review by the natural standard research collaboration. *J Herb Pharmacother* 4:63–83.

Bayles, B., and R. Usatine. 2009. Evening primrose oil. *Am Fam Physician* 80:1405–08.

Begon, S., G. Pickering, A. Eschalier, and C. Dubray. 2002. Magnesium increases morphine analgesic effect in different experimental models of pain. *Anesthesiology* 96:627–32.

Belfer, I., E. Davidson, A. Ratner, et al. 1998. Dietary supplementation with the inhibitory amino acid taurine suppresses autotomy in HA rats. *Neuroreport* 9:3103–7.

Bernasconi, S., F. Petraglia, L. Lughetti, et al. 1988. Impaired beta-endorphin response to human corticotrophin-releasing hormone in obese children. *Acta Endocrinol* 119:7–10.

Bernik, T. R., S. G. Friedman, M. Ochani, et al. 2002. Pharmacological stimulation of the cholinergic antiinflammatory pathway. *J Exp Med* 195:781–8.

Besse, J. L., S. Gadeyne, S. Galand-Desmé, et al. 2009. Effect of vitamin C on prevention of complex regional pain syndrome type I in foot and ankle surgery. *Foot Ankle Surg* 15(4):179–82.

Blass, E. M., and B. A. Smith. 1992. Differential effects of sucrose fructose, glucose and lactose on crying in 1- to 3-day-old human infants: qualitative and quantitative considerations. *Dev Psychol* 28:804–10.

Blass, E. M., and L. B. Watt. 1999. Suckling- and sucrose-induced analgesia in human newborns. *Pain* 83:611–23.

Bliddal, H., A. Rosetzsky, P. Schlichting, et al. 2000. A randomized, placebo-controlled, cross-over study of ginger extracts and ibuprofen in osteoarthritis. *Osteoarthritis Cartilage* 8:9–12.

Blommers, J., E. S. de Lange-De Klerk, D. J. Kuik, et al. 2002. Evening primrose oil and fish oil for severe chronic mastalgia: a randomized, double-blind, controlled trial. *Am J Obstet Gynecol* 187(5):1389–94.

Blonstein, J. L. 1969. Control of swelling in boxing injuries. *Practitioner* 203:206.

Borzan, J., J. M. Tall, C. Zhao, et al. 2010. Effects of soy diet on inflammation-induced primary and secondary hyperalgesia in rat. *Eur J Pain* doi:10.1016.

Bowman, S. A., and J. T. Spence. 2002. A comparison of low-carbohydrate *vs.* high-carbohydrate diets: energy restriction, nutrient quality and correlation to body mass index. *J Am Coll Nutr* 21:268–74.

Brien, S., G. Lewith, and A. Walker. 2004. Bromelain as a treatment for osteoarthritis: a review of clinical studies. *Evid Based Complement Alternat Med* 1:251–7.

Brzeski, M., R. Madhok,and H. A. Capell. 1991. Evening primrose oil in patients with rheumatoid arthritis and side-effects of non-steroidal anti-inflammatory drugs. *Br J Rheumatol* 30(5):370–2.

Burkhardt, S., D. X. Tan, L. C. Manchester, et al. 2001. Detection and quantification of the antioxidant melatonin in montmorency and balaton tart cherries (*Prunus cerasus*). *J Agric Food Chem* 49:4898–902.

Chalon, S. 2006. Omega-3 fatty acids and monoamine neurotransmission. *Prostaglandins Leukot. Essent Fatty Acids* 75:259–69.

Chantre, P., A. Cappelaere, D. Leblan, D. Guedon, J. Vandermander, and B. Fournie. 2000. Efficacy and tolerance of *Harpagophytum procumbens* versus diacerhein in treatment of osteoarthritis. *Phytomedicine* 7:177–83.

Chen, J. Y., C. Y. Chang, P. H. Feng, et al. 2009. Plasma vitamin C is lower in postherpetic neuralgia patients and administration of vitamin C reduces spontaneous pain but not brush-evoked pain. *Clin J Pain* CherryPharm. 2009. Cherry benefits. www.cherrypharm.com/site/backed-by-research.html

Chiechio, S., A. Caricasole, E. Barletta, et al. 2002. L-acetylcarnitine induces analgesia by selectively up-regulating mGlu2 metabotropic glutamate receptors. *Mol Pharmacol* 61:989–96.

Chiechio, S., A. Copani, I. V. Gereau, et al. 2007. Acetyl-L-carnitine in neuropathic pain: experimental data. *CNS Drugs* 21:31–38.

Chrubasik, J. E., B. D. Roufogalis, and S. Chrubasik. 2007. Evidence of effectiveness of herbal anti-inflammatory drugs in the treatment of painful osteoarthritis and chronic low back pain. *Phytother Res* 21:675–83.

Cipolla, B., F. Guillí, and J. P. Moulinoux. 2003. Polyamine-reduced diet in metastatic hormone-refractory prostate cancer (HRPC) patients. *Biochem Soc Trans.*

Cleland, L .G., J. K. French, W. H. Betts, et al. 1988. Clinical and biochemical effects of dietary fish oil supplements in rheumatoid arthritis. *J Rheumatol* 15:1471–5.

Cohen, A., and J. Goldman. 1964. Bromelains therapy in rheumatoid arthritis. *Pennsylvania Med J* 67:27–30.

Colantuoni, C., J. Schwenker, J. McCarthy, et al. 2001. Excessive sugar intake alters binding to dopamine and mu-opioid receptors in the brain. *Neuroreport* 12:3549–52.

Collier, H. O., N. M. Butt, W. J. McDonald-Gibson, and S. A. Saeed. 1980. Extract of feverfew inhibits prostaglandin biosynthesis. *Lancet* 2:922–3.

Connolly, D. A. J., M. P. McHugh, and O. I. Padilla-Zakour. 2006. Efficacy of a tart cherry juice blend in preventing the symptoms of muscle damage. *Br J Sports Med* 40:679–683.

Cordain, L. 2007. Implications of plio-pleistocene hominin diets for modern humans. In *Early Hominin Diets: The Known, the Unknown, and the Unknowable*, ed. P. S. Ungar, 363–83. Oxford: Oxford University Press.

Cordain L., S. B. Eaton, A. Sebastian, et al. 2005. Origins and evolution of the Western diet: health implications for the 21st century. *Am J Clin Nutr* 81:341–54.

Cornell University. 2009. CherryPharm: sour power sports drink. www.cals.cornell.edu/cals/public/businesspartnerships/made/cherrypharm.cfm

Crouch, S. P., K. J. Slater, and J. Fletcher. 1992. Regulation of cytokine release from mononuclear cells by the iron-binding protein lactoferrin. *Blood* 80:235–40.

D'Anci, K. E. 1999. Tolerance to morphine-induced antinociception is decreased by chronic sucrose or polycose intake. *Pharmacol Biochem Behav* 63:1–11.

D'Anci, K. E., R. B. Kanarek, and R. Marks-Kaufman. 1996. Duration of sucrose availability differentially alters morphine-induced analgesia in rats. *Pharmacol Biochem Behav* 54:693–7.

Davis, W. M., T. S. Miya, and L. D. Edwards. 2006. The influence of glucose and insulin pretreatment upon morphine analgesia in the rat. *J Am Pharm Assoc* 45:60–2.

De Grandis, D. 2007. Acetyl-L-carnitine for the treatment of chemotherapy-induced peripheral neuropathy: a short review. *CNS Drugs* 21:39–43.

Delbridge, E. A., L. A. Prendergast, J. E. Pritchard, and J. Proietto. 2009. One-year weight maintenance after significant weight loss in healthy overweight and obese subjects: does diet composition matter? *J Am J Clin Nutr* 90:1203–14.

Deodhar, S. D., R. Sethi, and R. C. Srimal. 1980. Preliminary study on antirheumatic activity of curcumin (diferuloyl methane). *Indian J Med Res* 71:632–4.

Djuric, Z., G. Chen, D. R. Doerge, et al. 2001. Effect of soy isoflavone supplementation on markers of oxidative stress in men and women. *Cancer Lett* 172:1–6.

Eckles-Smith, K., D. Clayton, P. Bickford, et al. 2000. Caloric restriction prevents age-related deficits in LTP and in NMDA receptor expression. *Brain Res Mol Brain Res* 78:154–62.

Edwards, R. 2009. Medicinal plants on verge of extinction. *New Sci*, January 10.

Eriksson, M., and O. Finnström. 2004. Can daily repeated doses of orally administered glucose induce tolerance when given for neonatal pain relief? *Acta Paediatr* 93:246–9.

Eschalier, A., O. Aumaitre, A. Decamps, et al. 1983. A comparison of the effects of vitamin B12 and aspirin in three experimental pain models in rats and mice. *Psychopharmacol (Berl)* 81:228–31.

Etzel, R. 1996. Special extract of *Boswellia serrata* (H 15) in the treatment of rheumatoid arthritis. *Phytomedicine* 3:91–4.

Evans, J. D., T. F. Jacobs, and E. W. Evans. 2008. Role of acetyl-L-carnitine in the treatment of diabetic peripheral neuropathy. *Ann Pharmacother* 42:1686–91.

Fantuzzi, G. 2008. Adiponectin and inflammation: consensus and controversy. *J Allergy Clin Immunol* 121:326–30.

Fitzhugh, D. J., S. Shan, M. W. Dewhirst, et al. 2008. Bromelain treatment decreases neutrophil migration to sites of inflammation. *Clin Immunol* 128:66–74.

Foo, H., and P. Mason. 2009. Analgesic accompanying food consumption requires ingestion of hedonic foods. *J Neurosci* 29:13053–62.

Franca, D. S., A. L. Souza, K. R. Almeida, et al. 2001. B vitamins induce an antinociceptive effect in the acetic acid and formaldehyde models of nociception in mice. *Eur J Pharmacol* 421:157–64.

Fu, Q. G., E. Carstens, B. Stelzer, et al. 1988. B vitamins suppress spinal dorsal horn nociceptive neurons in the cat. *Neurosci Lett* 95:192–7.

Gagnier, J. J., M. W. Van Tulder, B. Berman, and C. Bombardier. 2007. Herbal medicine for low back pain: a Cochrane review. *Spine* 32:82–92.

Gamelin, L., M. Boisdron-Celle, R. Delva, et al. 2004. Prevention of oxaliplatin-related neurotoxicity by calcium and magnesium infusions: a retrospective study of 161 patients receiving oxaliplatin combined with 5-fluorouracil and leucovorin for advanced colorectal cancer. *Clin Cancer Res* 10:4055–61.

Gradin, M., and J. Schollin. 2005. The role of endogenous opioids in mediating pain reduction by orally administered glucose among newborns. *Pediatrics* 115:1004–7.

Grant, L., D. E. McBean, L. Fyfe, and A. M. Warnock. 2007. A review of the biological and potential therapeutic actions of *Harpagophytum procumbens*. *Phytother Res* 21:199–209.

Gregory, P. J., M. Sperry, and A. F. Wilson. 2008. Dietary supplements for osteoarthritis. *Am Fam Physician* 77:177–84.

Hajnal, A., and R. Norgren. 2001. Accumbens dopamine mechanisms in sucrose intake. *Brain Res* 904:76–84.

Hajnal, A., G. P. Smith, and R. Norgren. 2004. Oral sucrose stimulation increases accumbens dopamine in the rat. *Am J Physiol Regul Integr Comp Physiol* 286:R31–R37.

Hanai, H., T. Iida, K. Takeuchi, et al. 2006. Curcumin maintenance therapy for ulcerative colitis: Randomized, multicenter, double-blind, placebo-controlled trial. *Clin Gastroenterol Hepatol* 4:1502–6.

Hanrahan, C. 2001. Feverfew. *Encyclopedia of Alternative Medicine*. http://findarticles.com/p/articles/mi_g2603 /is_ 0003/ai_2603000363

Harel, Z., F. M. Biro, R. K. Kottenhahn, et al. 1996. Supplementation with omega-3 polyunsaturated fatty acids in the management of dysmenorrhea in adolescents. *Am J Obstet Gynecol* 174:1335–8.

Hargraves, W. A., and I. D. Hentall. 2005. Analgesic effects of dietary caloric restriction in adult mice. *Pain* 114:455–61.

Harima, A., H. Shimizu, and H. Takagi. 1991. Analgesic effect of L-arginine in patients with persistent pain. *Eur Neuropsychopharmacol* 1:529–33.

Harris, R. Z., G. R. Jang, and S. Tsunoda. 2003. Dietary effects on drug metabolism and transport. *Clin Pharmacokinet* 42:1071–88.

Hasanein, P., M. Parviz, M. Keshavarz, et al. 2007. Modulation of cholestasis-induced antinociception in rats by two NMDA receptor antagonists: MK-801 and magnesium sulfate. *Eur J Pharmacol* 554(2–3).

Hay, A. J., M. Hamburger, K. Hostettmann, and J. R. Hoult. 1994. Toxic inhibition of smooth muscle contractility by plant-derived sesquiterpenes caused by their chemically reactive alpha- methylenebutyrolactone functions. *Br J Pharmacol* 112:9–12.

Hayashida, K, T. Kaneko, T. Takeuchi, et al. 2004. Oral administration of lactoferrin inhibits inflammation and nociception in rat adjuvant-induced arthritis. *J Vet Med Sci* 66:149–54.

Hayashida, K., T. Takeuchi, H. Shimizu, et al. 2003. Novel function of bovine milk-derived lactoferrin on antinociception mediated by mu-opioid receptor in the rat spinal cord. *Brain Res* 965:239–45.

Heinrich, M., J. Barnes, S. Gibbons, and E. M. Williamson. 2004. *Fundamentals of Pharmacognosy and Phytotherapy*. London: Churchill Livingstone.

Heyliger, S. O., C. B Goodman, J. M. Ngong, and K. F. Soliman. 1998. The analgesic effects of tryptophan and its metabolites in the rat. *Pharmacol Res* 38:243–50.

Hong, M. P., H. I. Kim, Y. K. Shin, et al. 2004. Effects of free fatty acids on sodium currents in rat dorsal root ganglion neurons. *Brain Res* 1008:81–91.

Howatson, G., M. P. McHugh, J. A. Hill, et al. 2009. Influence of tart cherry juice on indices of recovery following marathon running. *Scand J Med Sci Sports*. doi: 10.1111.

Huang, J., M. Nasr, Y. Kim, et al. 1992. Genistein inhibits protein histidine kinase. *J Biol Che* 267:15511–5.

Ingram, D. K., R. M. Anson, R. de Cabo, et al. 2004. Development of calorie restriction mimetics as a prolongevity strategy. *Ann N Y Acad Sci* 1019:412–23.

Ingram, D. M., C. Hickling, L. West, et al. 2002. A double-blind randomized controlled trial of isoflavones in the treatment of cyclical mastalgia. *Breast* 11:170–4.

Jacob, R. A., G. M. Spinozzi, V. A. Simon, et al. 2003. Consumption of cherries lowers plasma urate in healthy women. *J Nutr* 133:1826–9.

Jacob, S., E. J. Henriksen, A. L. Schiemann, et al. 1995. Enhancement of glucose disposal in patients with type 2 diabetes by alpha-lipoic acid. *Arzneimittelforschung* 45:872–4.

Jacob, S., R. S. Streeper, D. L. Fogt, et al. 1996. The antioxidant alpha-lipoic acid enhances insulin-stimulated glucose metabolism in insulin-resistant rat skeletal muscle. *Diabetes* 45:1024–9.

Jain, N. K., and S. K. Kulkarni. 1999. Antinociceptive and anti-inflammatory effects of *Tanacetum parthenium* L. extract in mice and rats. *J Ethnopharmacol* 68:251–9.

Kanarek, R. B., and C. Carrington. 2004. Sucrose consumption enhances the analgesic effects of cigarette smoking in male and female smokers. *Psychopharmacology* 173:57–63.

Kanarek, R. B., K. E. D'Anci, J. M. Przypek, et al. 1999. Altering dietary levels of protein or vitamins and minerals does not modify morphine-induced analgesia in male rats. *Pharmacol Biochem Behav* 62:203–8.

Kanarek, R. B., S. Mandillo, and C. Wiatr. 2001. Chronic sucrose intake augments antinociception induced by injections of mu but not kappa opioid receptor agonists into the periaqueductal gray matter in male and female rats. *Brain Res* 920:97–105.

Kano, M., T. Kawakami, H. Hori, et al. 1999. Effects of ALCAR on the fast axoplasmic transport in cultured sensory neurons of streptozotocin-induced diabetic rats. *Neurosci Res* 33:207–13.

Kapes, B. 2005. Bromelain. S.v. in *Gale Encyclopedia of Alternative Medicine*. www.encyclopedia.com/doc/1G2-3435100126.html

Kara, H., N. Sahin, V. Ulusan, and T. Aydogdu. 2002. Magnesium infusion reduces perioperative pain. *Eur J Anaesthesiol* 19:52–6.

Kelley, D. S., R. Rasooly, R. A. Jacob, et al. 2006. Consumption of bing sweet cherries lowers circulating concentrations of inflammation markers in healthy men and women. *J Nutr* 136:981–6.

Kelly, G. S. 1996. Bromelain: a literature review and discussion of its therapeutic applications. *Alt Med Rev* 1:243–57.

Kimmatkar, N., V. Thawani, L. Hingorani, and R. Khiyani. 2003. Efficacy and tolerability of *Boswellia serrata* extract in treatment of osteoarthritis of knee: a randomized double blind placebo controlled trial. *Phytomedicine* 10:3–7.

King, D. E., B. M. Egan, and M. E. Geesey. 2003. Relation of dietary fat and fiber to elevation of C-reactive protein. *Am J Cardiol* 92:1335–9.

Kiuchi, F., S. Iwakami, M. Shibuya, F. Hanaoka, and U. Sankawa. 1992. Inhibition of prostaglandin and leukotriene biosynthesis by gingerols and diarylheptanoids. *Chem Pharm Bull* 40:387–91.

Koinig, H., T. Wallner, P. Marhofer, et al. 1998. Magnesium sulfate reduces intra- and postoperative analgesic requirements. *Anesth Analg* 87:206–10.

Kolhekar, R., S. T. Meller, and G. F. Gebhart. 1994. N-methyl-D-aspartate receptor-mediated changes in thermal nociception: allosteric modulation at glycine and polyamine recognition sites. *Neuroscience* 63:925–36.

Kolta, M. G., R. Holson, P. Duffy, et al. 1989. Effect of long-term caloric restriction on brain monoamines in aging male and female Fischer 344 rats. *Mech Ageing Dev* 48:191–8.

Kreisberg, J. I., and P. Y. Patel. 1983. The effects of insulin, glucose and diabetes on prostaglandin production by rat kidney glomeruli and cultured glomerular mesangial cells. *Prostaglandins Leukot Med* 11:431–42.

Kremer, J. M., D. A. Lawrence, G. F. Petrillo, et al. 1995. Effects of high-dose fish oil on rheumatoid arthritis after stopping nonsteroidal antiinflammatory drugs: clinical and immune correlates. *Arthritis Rheum* 38:1107–14.

Kusano, T., K. Yamaguchi, T. Berberich, et al. 2007. Advances in polyamine research in 2007. *J Plant Res* 120(3):345–50.

Lanhers, M. C., J. Fleurentin, F. Mortier, A. Vinche, and C. Younos. 1992. Anti-inflammatory and analgesic effects of an aqueous extract of *Harpagophytum procumbens*. *Planta Med* 58:117–23.

Lee, C. K., R. Weindruch, and T. A. Prolla. 2000. Gene-expression profile of the ageing brain in mice. *Nat Genet* 25:294–7.

Lee, P., and R. Chen. 2008. Vitamin D as an analgesic for patients with type 2 diabetes and neuropathic pain. *Arch Intern Med* 168:771–2.

Leknes, S., and I. Tracey. 2008. A common neurobiology for pain and pleasure. *Nat Rev Neurosci* 9:314–20.

Levitan, E. B., N. R. Cook, M. J. Stampfer, et al. 2008. Dietary glycemic index, dietary glycemic load, blood lipids, and C-reactive protein. *Metabolism* 57:437–43.

Lewkowski, M. D., B. Ditto, M. Roussos, and S. N. Young. 2003. Sweet taste and blood pressure-related analgesia. *Pain* 106:181–6.

Li, F., I. G. Obrosova, O. Abatan, et al. 2005. Taurine replacement attenuates hyperalgesia and abnormal calcium signaling in sensory neurons of STZ-D rats. *Am J Physiol Endocrinol* Metab 288:E29–36.

Liu, S., J. E. Manson, J. E. Buring, et al. 2002. Relation between a diet with a high glycemic load and plasma concentrations of high-sensitivity C-reactive protein in middle-aged women. *Am J Clin Nutr* 75:492–8.

Longo, N., C. Amat di San Filippo, M. Pasquali, et al. 2006. Disorders of carnitine transport and the carnitine cycle. *Am J Med Genet C Semin Med Genet* 142:77–85.

Lotfi, A., A. M Abdel-Nasser, A. Hamdy, et al. 2007. Hypovitaminosis D in female patients with chronic low back pain. *Clin Rheumatol* 26:1895–901.

Luyer, M. D., J. Willem, M. Greve, et al. 2005. Nutritional stimulation of cholecystokinin receptors inhibits inflammation via the vagus nerve. *J Exp Med* 202:1023–9.

Mader, R., H. Deutsch, G. K. Siebert, et al. 1988. Vitamin status of inpatients with chronic cephalgia and dysfunction pain syndrome and effects of a vitamin supplementation. *Int J Vit Nutr Res* 58:436–42.

Mahomed, I. M., and J. A. O. Ojewole. 2004. Analgesic, anti-inflammatory and antidiabetic roperties of *Harpagophytum procumbens* DC (Pedaliaceae) secondary root aqueous extract. *Phytother Res* 18:982–9.

Maroon, J. C., and J. W. Bost. 2006. Omega-3 fatty acids (fish oil) as an anti-inflammatory: an alternative to nonsteroidal anti-inflammatory drugs for discogenic pain. *Surg Neurol* 65:326–31.

Martin, Y. B., and C. Avendaño. 2009. Effects of removal of dietary polyunsaturated fatty acids on plasma extravasation and mechanical allodynia in a trigeminal neuropathic pain model. *Mol Pain* 25:5–8.

Masoro, E. J. 1992. Retardation of aging processes by food restriction: an experimental tool. *Am J Clin Nutr* 55 (Suppl 6):1250S–2S.

Masoud, H., A. Khalvat, T. Toliat, and S. Jallaei. 2005. Comparing the effects of ginger (*Zingiber officinale*) extract and ibuprofen on patients with osteoarthritis. *Arch Iran Med* 8:267–71.

Masson, M. 1995. Bromelain in the treatment of blunt injuries to the locomotor system: a study of observed applications in general practice. *Fortschr Med* 113:303–6.

Matta, J. A., R. L. Miyares, and G. P. Ahern. 2007. TRPV1 is a novel target for omega-3 polyunsaturated fatty acids. *J Physiol* 578:397–411.

McFadyen, I. J., U. Chetty, K. D. Setchell, et al. 2000. A randomized double blind-cross over trial of soya protein for the treatment of cyclical breast pain. *Breast* 9:271–6.

McGivern, R. F., and G. G. Berntson. 1980. Mediation of diurnal fluctuations in pain sensitivity in the rat by food intake patterns: reversal by naloxone. *Science* 210:210–1.

McKay, Hart A., M. Wiberg, and G. Terenghi. 2002. Pharmacological enhancement of peripheral nerve regeneration in the rat by systemic acetyl-L-carnitine treatment. *Neurosci Lett* 334:181–5.

Memeo, A., and Loiero, M. 2008. Thioctic acid and acetyl-L-carnitine in the treatment of sciatic pain caused by a herniated disc: a randomized, double-blind, comparative study. *Clin Drug Investig* Menon, M. K., and A. Kar. 1971. Analgesic and psychopharmacological effects of the gum resin of *Boswellia serrata*. *Planta Med* 19:333–41.

Mercer, M. E., and M. D. Holder. 1997. Antinociceptive effects of palatable sweet ingesta on human responsivity to pressure pain. *Physiol Behav* 61:311–8.

Mibielli, M. A., M. Geller, J. C. Cohen, et al. 2009. Diclofenac plus B vitamins versus diclofenac monotherapy in lumbago: the DOLOR study. *Curr Med Res Opin* 25:2589–99.

Miller, A., R. G. Barr, and S. N. Young. 1994. The cold pressor test in children: methodological aspects and the analgesic effect of intraoral sucrose. *Pain* 56:175–83.

Mingrone, G., A. V. Greco, E. Capristo, et al. 1999. L-carnitine improves glucose disposal in type 2 diabetic patients. *J Am Coll Nutr* 18:77–82.

Moussard, C., D. Alber, M. M. Toubin, et al. 1992. A drug used in traditional medicine, *Harpagophytum procumbens*: no evidence for NSAID-like effect on whole blood eicosanoid production in human. *Prostaglandins Leukot Essent Fatty Acids* 46:283–6.

Murphy, J. J., S. Heptinstall, and J. R. Mitchell. 1988. Randomised double-blind placebo-controlled trial of feverfew in migraine prevention. *Lancet* 2:189–92.

Natural Standard Database. 2009. Devil's claw (*Harpagophytum procumbens* DC). www.naturalstandard.com

Natural Standard Database. 2010. Boswellia (*Boswellia serrata* Roxb.). www.naturalstandard.com

Nilsson, A. C., E. M. Ostman, J. J. Holst, and I. M. E. Bjorck. 2008. Including indigestible carbohydrates in the evening meal of healthy subjects improves glucose tolerance, lowers inflammatory markers, and increases satiety after a subsequent standardized breakfast. *J Nutr* 138:732–9.

Nishishinya. B., G. Urrútia, B. Walitt, et al. 2008. Amitriptyline in the treatment of fibromyalgia: a systematic review of its efficacy. *Rheumatology* 47:1741–6.

Onal, A, G. Kayalıoğlu, A. Parlar, et al. 2010. Effect of prolonged administration of bovine lactoferrin in neuropathic pain: involvement of opioid receptors, nitric oxide and TNF-alpha. *Life Sci* 86:251–9.

Ouchi, N., and K. Walsh. 2007. Adiponectin as an anti-inflammatory factor. *Clin Chim Acta* 380:24.

Ozgoli, G., M. Goli, and F. Moattar. 2009. Comparison of effects of ginger, mefenamic acid, and ibuprofen on pain in women with primary dysmenorrhea. *J Altern Complement Med* 15:129–32.

Pahl, I. R., W. Koppert, C. Enk, et al. 2003. Different lipid profiles as constituencies of liquid formula diets do not influence pain perception and the efficacy of opioids in a human model of acute pain and hyperalgesia. *Pain* 104:519–27.

Palevitch, D., G. Earon, and R. Carasso. 1997. Feverfew (*Tanacetum parthenium*) as a prophylactic treatment for migraine: a double-blind controlled study. *Phytother Res* 11:508–11.

Pattrick, M., S. Heptinstall, and M. Doherty. 1989. Feverfew in rheumatoid arthritis: a double-blind, placebo controlled study. *Ann Rheum Dis* 48:547–49.

Peikert, A., C. Wilimzig, and R. Köhne-Volland. 1996. Prophylaxis of migraine with oral magnesium: results from a prospective, multi-center, placebo-controlled and double-blind randomized study. *Cephalalgia* 16:257–63.

Pepino, M. Y., and J. A. Mennella. 2005. Sucrose-induced analgesia is related to sweet preferences in children but not adults. *Pain* 119:210–8.

Pérez, J., M. A. Ware, S. Chevalier, et al. 2004. Dietary fat and protein interact in suppressing neuropathic pain-related disorders following a partial sciatic ligation injury in rats. *Pain* 111:297–305.

Pérez, J., M. A. Ware, S. Chevalier, et al. 2005. Dietary omega-3 fatty acids could be associated with increased neuropathic pain in nerve-injured rats. *Anesth Analg* 101:444–8.

Pittler, M. H., and E. Ernst. 2004. Feverfew for preventing migraine. *Cochrane Database Syst Rev*. 2:CD002286.

Plotnikoff, G. A., and J. M. Quigley. 2003. Prevalence of severe hypovitaminosis D in patients with persistent, nonspecific musculoskeletal pain. *Mayo Clin Proc* 78:1463–70.

Proudman, S. M., L. G. Cleland, and M. J. James. 2008. Dietary omega-3 fats for treatment of inflammatory joint disease: efficacy and utility. *Rheum Dis Clin North Am* 34:469–79.

Pugh, W. J., and K. Sambo. 1988. Prostaglandin synthetase inhibitors in feverfew. *J Pharm Pharmacol* 40:743–5.

Puri, B. K. 2004. The use of eicosapentaenoic acid in the treatment of chronic fatigue syndrome. *Prostaglandins Leukot. Essent Fatty Acids* 70:399–401.

Ramenghi, L. A., G. C. Griffiths, C. M. Wood, and M. I. Levene. 1996. Effect of non-sucrose sweet tasting solution on neonatal heel prick responses. *Arch Dis Child* 74:F129–31.

Rivat, C., P. Richebé, E. Laboureyras, et al. 2008. Polyamine deficient diet to relieve pain hypersensitivity. *Pain* 137(1):125–37.

Roehm, E. 2009. The evidence-based Mediterranean diet reduces coronary heart disease risk, and plant-derived monounsaturated fats may reduce coronary heart disease risk. *Am J Clin Nutr* 90:697–8.

Roehm, E. 2009. The evidence-based Mediterranean diet reduces coronary heart disease risk, and plant-derived monounsaturated fats may reduce coronary heart disease risk. *Am J Clin Nutr* 90:697–8.

Rucker, D., J. A. Allan, G. H. Fick, and D. A. Hanley. 2002. Vitamin D insufficiency in a population of healthy western Canadians. *CMAJ* 166:1517–24.

Ruessmann, H. J. 2009. Switching from pathogenetic treatment with alpha-lipoic acid to gabapentin and other analgesics in painful diabetic neuropathy: a real-world study in outpatients. *J Diabetes Complications* 23:174–7.

Sáez-Cassanelli, J. L., G. H. Fontanella, J. M. Delgado-García, et al. 2007. Functional blockage of the cannabinoid receptor type 1 evokes a kappa-opiate-dependent analgesia. *J Neurochem* 103:2629–36.

Safayhi, H., T. Mack, J. Sabieraj, M. I. Anazodo, L. R. Subramanian, and H. P. T. Ammon. 1992. Boswellic acids: novel, specific, nonredox inhibitors of 5-lipoxygenase. *J Pharmacol Exp Ther* 22:1143–46.

Sakai, Y., K. Okamoto, and H. Kimura. 1985. Pharmacological evidence for taurine as an inhibitory neurotransmitter in the cerebellum. *Prog Clin Biol Res* 179:313–19.

Sander, O., G. Herborn, and R. Rau. 1998. Is H15 (resin extract of *Boswellia serrata*, "incense") a useful supplement to established drug therapy of chronic polyarthritis? Results of a double-blind pilot study. *Z Rheumatol* 57:11–6.

de los Santos-Arteaga, M., S. A. Sierra-Domínguez, G. H. Fontanella, et al. 2003. Analgesia induced by dietary restriction is mediated by the kappa-opioid system. *J Neurosci* 23:11120–6.

Satoskar, R. R., S. J. Shah, and S. G. Shenoy. 1986. Evaluation of anti-inflammatory property of curcumin (diferuloyl methane) in patients with postoperative inflammation. *Int J Clin Pharmacol Ther Toxicol* 24:651–4.

Schaffler, A., A. Ehling, E. Neumann, et al. 2003. Adipocytokines in synovial fluid. *JAMA* 290:1709–10.

Schoenen J., J. Jacquy, and M. Lanaerts. 1998. Effectiveness of high-dose riboflavin in migraine prophylaxis. *Neurology* 50:466–70.

Schwalfenberg, G. 2009. Improvement of chronic back pain or failed back surgery with vitamin D repletion: a case series. *J Am Board Fam Med* 22:69–74.

Schwartzman, R. J., K. L. Erwin, and G. M. Alexander. 2009. The natural history of complex regional pain syndrome. *Clin J Pain* 25:273–80.

Seltzer, A. P. 1962. Minimizing post-operative edema and ecchymoses by the use of an oral enzyme preparation (bromelain). *Eye Ear Nose Throat Mon* 41:813–7.

Seltzer, Z., R. Dubner, and Y. Shir. 1990. A novel behavioral model of neuropathic pain disorders produced in rats by partial sciatic nerve injury. *Pain* 43:205–18.

Sendur, O. F., E. Tastaban, Y. Turan, and C. Ulman. 2008. The relationship between serum trace element levels and clinical parameters in patients with fibromyalgia. *Rheumatol Int* 28:1117–21.

Sengupta, K., K. V. Alluri, A. R. Satish, et al. 2008. A double blind, randomized, placebo controlled study of the efficacy and safety of 5-Loxin for treatment of osteoarthritis of the knee. *Arthritis Res Ther* 10:R85.

Serrano, J. S., M. I. Serrano, M. R. Guerrero, et al. 1990. Antinociceptive effect of taurine and its inhibition by naxolone. *Gen Pharmacol* 21:333–6.

Shay, K. P., R. F. Moreau, E. J. Smith, et al. 2009. Alpha-lipoic acid as a dietary supplement: molecular mechanisms and therapeutic potential. *Biochim Biophys Acta* 1790:1149–60.

Shir, Y., J. N. Campbell, S. N. Raja, et al. 2002. The correlation between dietary soy phytoestrogens and neuropathic pain behavior in rats after partial denervation. *Anesth Analg* 94:421–6.

Shir, Y., S. N. Raja, C. S. Weissman, et al. 2001a. Consumption of soy diet before nerve injury preempts the development of neuropathic pain in rats. *Anesthesiology* 95:1238–44.

Shir, Y., A. Ratner, S. N. Raja, et al. 2008. Neuropathic pain following partial nerve injury in rats is suppressed by dietary soy. *Neurosci Lett* 240:73–6.

Shir, Y., R. Sheth, J. N. Campbell, et al. 2001b. Soy-containing diet suppresses chronic neuropathic sensory disorders in rats. *Anaesth Analg* 92:1029–34.

Shoba, G., D. Joy, T. Joseph, et al. 1998. Influence of piperine on the pharmacokinetics of curcumin in animals and human volunteers. *Planta Med* 64:353–6.

Sima, A. A. F. 2007. Acetyl-L-carnitine in diabetic polyneuropathy: experimental and clinical data. *CNS Drugs* 21:13–23.

Singh, G. B., and C. K. Atal. 1986. Pharmacology of an extract of salai guggal ex-*Boswellia serrata*, a new non-steroidal anti-inflammatory agent. *Agents Actions* 18:407–12.

Smolinski, A. T., and J. J. Pestka. 2003. Modulation of lipopolysaccharide-induced proinflammatory cytokine production in vitro and in vivo by the herbal constituents apigenin (chamomile), ginsenoside Rb(1) (ginseng) and parthenolide (feverfew). *Food Chem Toxicol* 41:1381–90.

Sola, S., M. Q. Mir, F. A. Cheema, et al. 2005. Irbesartan and lipoic acid improve endothelial function and reduce markers of inflammation in the metabolic syndrome: results of the irbesartan and lipoic acid in endothelial dysfunction (ISLAND) study. *Circulation* 111:343–48.

Soleimannejad, E., J. Tremblay-Mercier, S. C. Cunane, et al. 2008. Dietary analgesia in nerve-injured rats is associated with fatty acid changes in the peripheral nerve. *Pain Res Manag* 13:148.

Song, X. S., Z. J. Huang, and X. J. Song. 2009. Thiamine suppresses thermal hyperalgesia, inhibits hyperexcitability, and lessens alterations of sodium currents in injured, dorsal root ganglion neurons in rats. *Anesthesiology* 110:387–400.

Srivastava, K. C., and T. Mustafa. 1992. Ginger (*Zingiber officinale*) in rheumatism and musculoskeletal disorders. *Med Hypotheses* 39:342–8.

Stevens, B., J. Yamada, and A. Ohlsson. 2010. Sucrose for analgesia in newborn infants undergoing painful procedures. *Cochrane Database Syst Rev* 1:CD001069.

Stracke, H., A. Lindemann, and K. Federlin. 1996. A benfotiamine-vitamin B combination in treatment of diabetic polyneuropathy. *Exp Clin Endocrinol Diabetes* 104:311–6.

Straube, S., S. Derry, R. A. Moore, et al. 2010. Pregabalin in fibromyalgia: meta-analysis of efficacy and safety from company clinical trial reports. *Rheumatology* 49:706–15.

Su, K. P. 2008. Mind-body interface: the role of n-3 fatty acids in psychoneuroimmunology, somatic presentation, and medical illness comorbidity of depression. *Asia Pac J Clin Nutr* 17:151–7.

Taddio, A., V. Shah, E. Atenafu, and J. Katz. 2009. Influence of repeated painful procedures and sucrose analgesia on the development of hyperalgesia in newborn infants. *Pain* 144:43–8.

Tall, J. M., N. P. Seeram, C. Zhao, et al. 2004. Tart cherry anthocyanins suppress inflammation-induced pain behavior in rat. *Behav Brain Res* 153:181–8.

Togawa, J., H. Nagase, K. Tanaka, et al. 2002. Oral administration of lactoferrin reduces colitis in rats via modulation of the immune system and correction of cytokine imbalance. *J Gastroenterol Hepatol* 17:1291–98.

Tracey, K. J. 2002. The inflammatory reflex. *Nature* 420:853–9.

Tracey, K. J. 2005. Fat meets the cholinergic anti-inflammatory pathway. *J Exp Med* 202:1017–21.

Tramer, M. R., J. Schneide, R. A. Marti, et al. 1996. Role of magnesium sulfate in postoperative analgesia. *Anesthesiology* 84:340–7.

Traynelis, S. F., M. Hartley, and S. F. Heinemann. 1995. Control of proton sensitivity of the NMDA receptor by RNA splicing and polyamines. *Science* 268:873–6.

Turan, Y., H. Koçyigit, S. Ekinci, et al. 2009. Efficacy of oral magnesium treatment on primary fibromyalgia syndrome. *ADÜ Tıp Fakültesi Dergisi* 10:11–5.

Turner, M. K., W. M. Hooten, J. E. Schmidt, et al. 2008. Prevalence and clinical correlates of vitamin D inadequacy among patients with chronic pain. *Pain Med* 9:979–84.

Uchida, S., K. Hirai, J. Hatanaka, et al. 2008. Antinociceptive effects of St. John's wort, *Harpagophytum procumbens* extract and grape seed proanthocyanidins extract in mice. *Biol Pharm Bull* 31:240–5.

Valsecchi, A. E., S. Franchi, A. E. Panerai, et al. 2008. Genistein, a natural phytoestrogen from soy, relieves neuropathic pain following chronic constriction sciatic nerve injury in mice: anti-inflammatory and antioxidant activity. *J Neurochem* 107:230–40.

Vilela, F. C., A. D. Bitencourt, L. D. Cabral, et al. 2010. Anti-inflammatory and antipyretic effects of *Sonchus oleraceus* in rats. *J Ethnopharmacol* 127:737–41.

Vilela, F. C., M. de Mesquita Padilha, L. Dos Santos-E-Silva, et al. 2009. Evaluation of the antinociceptive activity of extracts of *Sonchus oleraceus* L. in mice. *J Ethnopharmacol* 24:306–10.

Wahner-Roedler, D. L., J. M. Thompson, C. A. Luedtke, et al. 2008. Dietary soy supplement on fibromyalgia symptoms: a randomized, double-blind, placebo-controlled, early phase trial. *Evid Based Complement Alternat Med* doi:10.1093.

Walsh, N. E., S. Ramamurthy, L. Schoenfeld, et al. 1986. Analgesic effectiveness of D-phenylalanine in chronic pain patients. *Arch Phys Med Rehabil* 67:436–9.

Wang, H., M. G. Nair, G. M. Strasburg, et al. 1999. Antioxidant and antiinflammatory activities of anthocyanins and their aglycon, cyanidin, from tart cherries. *J Nat Prod* 62:294–6.

Wang, Z. B., Q. Gan, R. L. Rupert, et al. 2005. Thiamine, pyridoxine, cyanocobalamin and their combination inhibit thermal, but not mechanical hyperalgesia in rats with primary sensory neuron injury. *Pain* 114:266–77.

Warner, A. E., and Arnspiger S. A. 2008. Diffuse musculoskeletal pain is not associated with low vitamin D levels or improved by treatment with vitamin D. *J Clin Rheumatol* 14:12–6.

Warnock, M., D. McBean, A. Suter, et al. 2007. Effectiveness and safety of devil's claw tablets in patients with general rheumatic disorders. *Phytother Res* 21:1228–33.

Wu, H., Y. Jin, J. Wei, et al. 2005. Mode of action of taurine as a neuroprotector. *Brain Res* 1038:123–31.

Yamamoto, T., N. Sako, and S. Maeda. 2000. Effects of taste stimulation on beta-endorphin levels in rat cerebrospinal fluid and plasma. *Physiol Behav* 69:345–50.

Yxfeldt, A., S. Wallberg-Jonsson, J. Hultdin, et al. 2003. Homocysteine in patients with rheumatoid arthritis in relation to inflammation and B-vitamin treatment. *Scand J Rheumatol* 32:205–10.

Zhao, C., P. W. Wacnik, J. M. Tall, et al. 2004. Analgesic effects of a soy-containing diet in three murine bone cancer pain models. *J Pain* 5:104–10.

Ziegler, D., H. Nowak, P. Kempler, et al. 2004. Treatment of symptomatic diabetic polyneuropathy with the antioxidant alpha-lipoic acid: a meta-analysis. *Diabet Med* 21:114–21.

Zimecki, M, R. Miedzybrodzki, and S. Szymaniec. 1998. Oral treatment of rats with bovine lactoferrin inhibits carrageenan-induced inflammation: correlation with decreased cytokine production. *Arch Immunol Ther Exp* 46:361–5.

Zimmerman, M., G. D. Bartoszyk, D. Bonke, et al. 1990. Antinociceptive properties of pyridoxine: neurophysiological and behavioral findings. *Ann N Y Acad Sci* 585:219–30.

Zmarzty, S. A., and N. W. Read. 1999. An examination of the effects of isoenergetic intragastric infusions of pure macronutrients on cold pain perception in healthy human volunteers. *Physiol Behav* 65:643–8.

Zmarzty, S. A., A. S. Wells, and N. W. Read. 1997. The influence of food on pain perception in healthy human volunteers. *Physiol Behav* 62:185–91.

Zollinger, P. E., W. E. Tuinebreijer, R. S. Breederveld, et al. 2007. Can vitamin C prevent complex regional pain syndrome in patients with wrist fractures? A randomized, controlled, multicenter dose-response study. *J Bone Joint Surg Am* 89:1424–31.

Zollinger, P. E., W. E. Tuinebreijer, R. W. Kreis, al. 1999. Effect of vitamin C on frequency of reflex sympathetic dystrophy in wrist fractures: a randomised trial. *Lancet* 354:2025–8.

4 Breakfast and Adult and Child Behaviors

Andrew P. Smith

CONTENTS

Introduction.......53
Breakfast and Nutritional Recommendations.......54
Breakfast and Physical Health.......55
Breakfast and Cognitive Function.......55
Studies of Adults.......55
Studies of Children.......58
Effects of Breakfast on Mood.......59
Breakfast and Well-Being.......59
Underlying Mechanisms.......61
Breakfast Is Better Than Nothing.......61
Glycemic Index and Load.......62
Other Mechanisms.......62
Effects on Real-Life Cognitive Function and Safety.......63
Discussion.......63
References.......64
Appendix: Bibliography of Studies of Breakfast and Cognition in Children and Adolescents.......68
Acute Effects of Interventions.......68
Children Differing in Nutritional Status.......69
School Breakfast Programs.......69
Habitual Breakfast Consumption.......70

INTRODUCTION

The term "breakfast" is used in a number of ways. Studies provide varying definitions of breakfast, including eating after overnight fasting, the first meal of the day, eating before the start of daily activities, eating within two hours of waking, any food or beverage consumed between 05:00 and 09:00, eating before 10:00, and eating a meal in the morning that provides between 20 and 35 percent of daily energy needs. Some studies do not define breakfast at all other than investigating eating occasions that are described as breakfast by the participants. Given this variation in definition, one might expect considerable variation in the observed effects. However, there is generally a good consensus about both health and behavioral effects of consuming breakfast, and these are summarized here.

Epidemiological studies have identified a number of behaviors which influence health. Consumption of breakfast is often considered one of these important health-related behaviors (Berkman and Breslow 1983), and there has been considerable research into its effects. There have been recent concerns that fewer people are now eating breakfast, and this has been confirmed in nationally representative surveys (Haines et al. 1996; Kant and Graubard 2006). For example, in the United States breakfast consumption among adults aged twenty to seventy-four years decreased from 86 percent in 1965 to 75 percent in 1991. It has also been found that breakfast consumption has declined in children, with the steepest drop being observed amongst adolescents aged eleven to eighteen years (Siega-Riz et al. 1998). This effect has been confirmed in data from the National Health and Nutrition Examination Surveys (NHANES), and results from the 2001–2002 survey showed that children are less likely to consume breakfast as they get older. For example, among children aged two to five years about 95 percent eat breakfast. However, in the twelve- to nineteen-year age group, less than 70 percent of the children ate breakfast.

The next sections summarize two areas which show why consumption of breakfast is so important.

Breakfast and Nutritional Recommendations

Regular breakfast consumption is associated with higher intake of key vitamins and minerals (Rampersaud et al. 2005; Ruxton and Kirk 1997). This may increase the likelihood of meeting nutritional requirements. Conversely, breakfast skippers may not make up for missed nutrients at other meals (Morgan et al. 1986). Breakfasts containing ready-to-eat-cereal may also improve the diet due to fortification with micronutrients and low fat levels. Indeed, a review of breakfast and the diet of adults confirms that breakfast eaters consume better quality diets that include more fiber and nutrients and fewer calories than breakfast skippers (Timlin et al. 2007). This has been confirmed in children: a review of 47 studies (Rampersaud et al. 2005) showed that breakfast eaters have higher daily intakes of fiber, calcium, vitamin A, vitamin C, riboflavin, zinc, and iron compared to breakfast skippers. The *2005 Dietary Guidelines for Americans* identify whole grains, fat-free and low-fat milk and milk products, fruits, and vegetables as "foods to encourage." Popular breakfast foods help people meet recommendations for these food groups. Breakfast also contributes to whole-grain intake (over 30 percent of the intake) which is known to reduce the risk of diabetes and coronary heart disease.

Milk is the most commonly consumed breakfast food (consumed by over 50 percent of people who eat breakfast at home), and this, again, helps to meet dietary recommendations for this type of food. Similar results have been reported for fruit intake, with fruit or fruit juice consumption at breakfast being linked with greater total fruit intake over the day (Quan et al. 2000).

Breakfast consumption has also been associated with better weight management. Results from the Seasonal Variation of Blood Cholesterol Study (SEASONS; 1994–1998; Merriam et al. 1999) show that the risk of obesity increases over four times in breakfast skippers compared to breakfast consumers (Ma et al. 2003). Other research shows that breakfast consumption is associated with lower BMIs (Cho et al. 2003) and a reduced risk of weight gain (van der Heijden et al. 2007). These results have been confirmed in studies of children and adolescents (e.g., Utter et al. 2007). Other research has examined which components of breakfast are related to better weight management, and the findings support beneficial effects of cereal (e.g., Bazzano et al. 2005) and milk products (Zemel et al. 2004). Breakfast may also play a role in weight management by affecting satiety (Holt et al. 1995; Blom et al. 2006).

Breakfast and Physical Health

Breakfast consumption is associated with lower levels of cholesterol (Stanton and Keast 1989; Resnicow 1991). Similarly, diets rich in fiber and whole grains are associated with a reduced risk of coronary heart disease (Djoussé and Gaziano 2007). Metabolic syndrome is a cluster of risk factors that are linked with being obese and having an increased risk of diabetes and coronary artery disease. Research shows that diets rich in whole grains and dairy products (key components of many breakfasts) are associated with a reduced risk of metabolic syndrome (Baxter et al. 2006).

Dietary fiber from breakfast cereals may also improve digestion (Smith 2010b). There is also evidence that whole-grain wheat may have a pre-biotic effect (Costabile et al. 2008) and that consumption of wheat bran can reduce levels of harmful bacteria such as *Clostridia* (Deaville et al. in preparation). Eating breakfast also contributes nutrients that are important to bone health, with the main source being milk.

Other research (Smith and Rees 2000; Smith 2002a) has shown that breakfast consumption is associated with reduced susceptibility to the common cold. Smith (2002a) found that breakfast consumption was associated with lower cortisol levels. Cortisol often induces immunosuppression, and this provides a plausible mechanism for some of the health benefits associated with breakfast. Indeed, Li et al. (2007) found that breakfast consumption was associated with significantly higher numbers of NK cells and a significantly lower number of T cells than in volunteers who did not regularly consume breakfast.

The main emphasis of the present chapter is on the behavioral effects of eating breakfast, and these will be reviewed in the following sections.

BREAKFAST AND COGNITIVE FUNCTION

Studies of Adults

There has been considerable interest in the acute effects of meals on human performance and mood (for reviews, see Smith and Kendrick 1992; Mahoney et al. 2005). The overall aim of the present section is to review this evidence to determine whether breakfast influences mood and performance, and to examine whether selective effects are observed depending on the type of meal eaten, task performed, and characteristics of the person eating the meal, and whether caffeinated beverages are consumed as part of breakfast.

It is often thought that consumption of breakfast enhances performance, a suggestion which has arisen largely from a series of studies by Tuttle and colleagues more than forty years ago ("the Iowa Breakfast studies"; Tuttle et al. 1949). The main aim of these studies was to evaluate the effects of varying breakfast regimes on physiological performance, but a number of the studies also included some tests of mental performance. Tuttle et al. (1949), in the first experiment of the series, compared the effects of four breakfast regimes: (1) a heavy breakfast, (2) a light breakfast, (3) no breakfast, and (4) coffee only. Results showed that in the no-breakfast condition, there was a tendency toward slower reaction times. However, this was the only condition in which caffeinated coffee was not given, and the results may reflect this. This was replicated when the same subjects were retested. Five out of six of the females showed a significant increase in simple reaction time in the no-breakfast condition, while three out of six showed a significant increase in choice reaction time in the same condition. Clearly, results from studies with such a small number of subjects must be treated with caution.

Tuttle et al. (1950) carried out a similar experiment comparing breakfast and no-breakfast conditions, with testing taking place three hours after breakfast. Six of the ten subjects showed no change in reaction time in the no-breakfast condition (as compared to breakfast), three showed a significant increase in reaction times, while one subject's reaction time increased significantly during the no-breakfast condition. Again, it is difficult to draw confident conclusions from such a study.

Another study (Tuttle et al. 1952) found no effect of breakfast on reaction times. Three breakfast conditions were compared: (a) a bacon, egg, and milk breakfast; (b) no breakfast; and (c) a cereal and milk breakfast. Subjects (males aged sixty to eighty-three years) received the bacon, egg, and milk breakfast for the first five weeks, followed by four weeks on no breakfast and four weeks on cereal and milk. Seven of the eight subjects showed no change in reaction times during the course of the experiment. Although this experiment has the advantage that it examined the long-term effects of breakfast, the small sample size, poor experimental design, and use of only a few measures of performance limits its value.

These early studies have been criticized for having small numbers of subjects, for producing inconsistent findings, and for the use of subjective assessments (Dickie and Bender 1982). The range of performance measures used was also small, being limited mainly to reaction time tasks. However, impaired performance associated with omitting breakfast was observed in other early studies. One study (King et al. 1945) assessed visual and motor functioning two and three hours after the consumption or omission of breakfast. The results showed that these functions were impaired when breakfast was not eaten compared to when it was.

Richards (1972) compared a standard breakfast with a no-breakfast condition. The volunteers were chosen so that half habitually ate breakfast and half no breakfast. A range of performance measures was employed: a visual search task, a short-term memory task, a vigilance task, and a coding task. Testing was carried out in the late morning. Participants were tested on five occasions: once following their normal breakfast, twice following the standard breakfast, and twice following no breakfast. A modified Latin-square design was used to balance the order of conditions. The consumption or omission of breakfast did not alter performance. Rather, performance was most impaired when subjects changed from their normal meal. This led to the view that "the occasional omission of breakfast is more deleterious than the constant omission."

Benton and Sargent (1992) compared the effects of no breakfast and consumption of a high-protein drink on spatial memory and immediate recall of a word list. Half the subjects were habitual breakfast eaters, and half did not usually eat breakfast. Consumption of the high-protein drink increased the speed with which both memory tasks were completed. Benton and Parker (1998) confirmed that breakfast improves aspects of memory and suggested that this may reflect several different mechanisms.

Other studies have suggested that the size and composition of breakfast influence the postmeal response. Lloyd et al. (1996) compared low-fat/high-carbohydrate, medium-fat/medium-carbohydrate, high-fat/low-carbohydrate, and no-breakfast conditions. No clear differences in performance were observed as a function of type of breakfast, but subjects given the low-fat/high-carbohydrate breakfast (which was most similar to their normal meal) reported improved mood compared to the other conditions. Nabb and Benton (2006) compared breakfasts that contained either high or low levels of carbohydrate, fat, or protein. Better memory was associated with consumption of meals that more slowly released glucose into the blood. This benefit of a low glycemic index breakfast has been confirmed in animal studies (Benton et al. 2003) and in children (Wesnes et al. 2003; Ingwersen et al. 2007).

The next section reports two studies (Smith et al. 1993, 1994) which examined the effects of breakfast on mood and a range of different aspects of performance. The type of breakfast was manipulated, and the influence of caffeinated drinks examined. The experiments also investigated whether personality, eating habits, gender, and previous night's sleep modified any effect of breakfast on behavior. The first experiment examined the effects of two types of breakfast on sustained attention tasks (i.e., tasks which show an effect of lunch), mood, and cardiovascular functioning. Volunteers were given either caffeinated coffee or decaffeinated coffee after the meal (or no meal). This was done to investigate whether caffeine modified any effects of breakfast, and, second, as a positive control to show that the tests used here were sensitive to changes in state produced by caffeine (Lieberman 1992).

In the first study, a between-subject design was used and volunteers were assigned to one of the six conditions formed by combining the three breakfast and two caffeine conditions. Volunteers were assigned to a no-breakfast condition, a cooked breakfast condition, or a cereal and toast breakfast. Details of these are shown below:

1. Cereal and toast breakfast: 1 oz cornflakes; 150 ml skim milk 2 tsp sugar; 1 slice wholemeal toast; 10 g polyunsaturated margarine/butter; 25 g marmalade.
2. Cooked breakfast: 2 eggs scrambled, skim milk; 2 thin slices back bacon; 1 slice wholemeal bread/toast; 10 g polyunsaturated margarine/butter

After breakfast, participants were either given decaffeinated coffee or decaffeinated coffee with 4 mg/kg of caffeine tablets added.

Breakfast had no effects on performance of sustained attention tasks. In contrast, caffeine improved performance of these tasks. No interactions between breakfast conditions and personality were found in any of the analyses. Similar results were found when gender was included as a factor.

Smith et al. (1994) examined effects of breakfast on performance of memory tasks. Consumption of breakfast improved recall and recognition of a list of words but had no beneficial effects on working memory or semantic memory tasks. Again, effects of breakfast were not modified by caffeine or by personality and gender. Breakfast had no effect on free recall in the late morning or after lunch, which suggests that the effects of breakfast on episodic memory are restricted to a few hours after the meal.

Smith et al. (1999) extended the above results by showing that consumption of breakfast may also improve spatial memory. However, the most robust effects of breakfast on memory are found in free recall tasks, and these effects have been observed after consumption of high-carbohydrate cereals (Smith in preparation-a) and cereal bars (Smith and Wilds 2009; Smith and Stamatakis 2010). Similarly, a midmorning cereal bar may also have beneficial effects when consumed after a small breakfast (Smith and Wilds 2009).

There have been a few studies that have examined effects of breakfast in elderly adults. Early studies by Tuttle and colleagues (Tuttle 1953; Tuttle et al. 1952) found little evidence for an effect of breakfast on the cognitive function of elderly people. Recent studies have demonstrated both acute effects of breakfast and effects of the breakfast habit. Kaplan et al. (2001) found that carbohydrate intake was associated with improved performance of a short-term memory task, whereas a protein breakfast was associated with reduced forgetting in a paragraph recall task. Smith (1998) found that elderly adults, aged between sixty and seventy-nine years, who ate breakfast cereal every day performed better on a test measuring intellectual functioning than those who consumed breakfast less frequently. It should be noted that this last result could reflect an effect of intelligence on breakfast consumption rather than a causal effect of breakfast consumption on intelligence. Further intervention studies are needed to assess the effects of breakfast on cognitive function in the elderly.

Studies of Children

There have been a number of reviews of the effect of breakfast on the cognitive performance of adolescents and children (Rampersaud et al. 2005; Mahoney et al. 2005; Hoyland et al. 2009) and the main findings can be summarized as follows. There have been over forty studies published on this topic (see this chapter's appendix) in the last sixty years (for details of the literature, see Hoyland et al. 2009). The results confirm the adult literature showing that breakfast has a beneficial effect on cognition, with the strongest support for improvements in memory. This effect is most readily apparent when nutritional status is compromised. Less is known about the effects of different types and sizes of breakfast, so the role of breakfast size and composition requires further consideration. Wyon et al. (1997) reported that children did better on tests of creativity, physical endurance, and mathematical ability when they consumed a high-energy breakfast compared to consumption of a low-energy breakfast. Michaud et al. (1991) confirmed these results using a short-term memory task. Other studies (Mahoney et al. 2005) report an oatmeal breakfast leads to better performance compared to a ready-to-eat cereal (especially in girls).

Most studies have investigated children rather than adolescents. A recent study of high school students (Widenhorn-Müller et al. 2008) showed that breakfast had no effect on sustained attention but improved visuo-spatial memory in males. Studies of school breakfast programs suggest that such interventions can have positive effects which may reflect an effect of these programs on school attendance. In addition, breakfast consumption in children and adolescents is associated with a superior nutritional profile and better weight management.

EFFECTS OF BREAKFAST ON MOOD

A number of studies have shown that consumption of breakfast is associated with a more positive mood including greater alertness, greater hedonic tone, and a reduction in anxiety in the period shortly after consumption. Maridakis et al. (2009) demonstrated that these effects could be demonstrated using a range of differing measuring instruments. Similarly, they have been found with different types of breakfast (e.g., Smith et al. 1993) and when volunteers were free to select from a range of breakfast cereals (Smith et al. 1999) or cereal bars (Smith and Wilds 2009). These results have been confirmed in studies of adolescents (Widenhorn- Müller et al. 2008) and children (Smith 2010a). Other research has suggested that different macronutrients have selective effects on mood (e.g., simple versus complex carbohydrates; Pasman et al. 2003) although it is often unclear whether such effects reflect factors such as acceptability. Similarly, some research suggests that mood is more negative after a low-energy rather than high-energy breakfast (Lluch et al. 2000). Effects of habitual consumption may also modify the acute effects of breakfast, with deviation from habitual breakfast being associated with a more negative mood (Lloyd et al. 1996). The effects of habitual breakfast consumption patterns on longer term well-being will now be considered.

BREAKFAST AND WELL-BEING

Smith (2003) has discussed the relevance on the concept of well-being in nutrition research. The concept of well-being has become increasingly important since the acknowledgment that there is more to health than the absence of disease. In some areas of research, "well-being" has been replaced by "quality of life" or some other term that relates to the ability to function well (both physically and mentally) and to have a positive mood state. In the area of nutrition, the term "functional food" is widely used, and this refers not only to the beneficial effects related to chronic disease but also to potential improved well-being. Consumption of breakfast has been shown to be associated with various aspects of the multidimensional concept of well-being, and some examples are given in the following section.

Wetzler and Ursano (1988) showed that breakfast consumption was associated with better psychological well-being in a cross-sectional analysis of over 6,000 individuals. Similarly, Tanaka et al. (2008) found that skipping breakfast was associated with an increased prevalence of fatigue in medical students. In the largest study, Huang et al. (2010) examined associations between breakfast skipping and health-related quality of life in a national representative sample (*N* = 15,340) from the 2005 Taiwan National Health Interview Survey. The results showed that breakfast skippers had significantly lower scores (poorer well-being) on five out of the eight domain scores on a quality-of-life questionnaire, the SF-36 (lower general health, reduced vitality, poorer social functioning, poorer emotional roles, and reduced mental health).

Smith (1998) examined the relationship between breakfast consumption and subjective reports of health and health-related behaviors in a general population sample (126 subjects aged between twenty and seventy-nine years). Individuals who consumed a cereal breakfast each day were less depressed, were less emotionally distressed, and had lower levels of perceived stress than those who did not eat breakfast each day. Those who consumed breakfast had a healthier lifestyle than the others as they were less likely to be smokers, drank less alcohol, and had a healthier diet. However, the relationship between cereal breakfast consumption and health was present regardless of differences in the other health-related behaviors.

A subsequent study (Smith 1999) attempted to replicate and extend the above result. The general population sample in this study (262 volunteers aged between twenty-one and eighty-five years, mean age: 60.9 years) was older than the sample in the previous study. Individuals who consumed breakfast cereal everyday reported better mental and physical health than those who consumed it less frequently. This association was still present when demographic factors, indicators of lifestyle such as smoking, or other aspects of diet were covaried.

Smith (2003) continued to study this topic and in the next study considered young adults (189 volunteers, aged between nineteen and twenty-one years, mean age 19.6 years) living at home. The results showed that skipping breakfast was associated with reports of poorer health and that regular breakfast cereal consumption is associated with better reported health. The effects of breakfast could not be explained by other health-related behaviors or other aspects of diet.

In the latest study (Smith 2010a), the sample was 213 children (108 female and 105 male; mean age: 8.11 years, SD 2.04 years), recruited from schools in Cardiff, Wales. Baseline measures of breakfast consumption and different aspects of reported well-being, such as mental health, cognitive functioning, alertness, physical health, and digestive problems, were recorded. Following this, children were allowed to try three cereals and selected the one that they found most acceptable (sixty-three chose Cornflakes, sixty-three Rice Krispies, and fifty-three Rice Krispies Multigrain). The groups consumed these cereals on a daily basis for two weeks. Measures of well-being were recorded on days 7 and 14. The breakfast cereal groups were compared with thirty-four children who consumed no breakfast. The baseline results showed that those who consumed breakfast cereal were perceived as having better well-being, including fewer mental health problems, a more positive mood, higher alertness, and fewer bowel problems than those who did not consume breakfast. This was confirmed in the intervention study, with breakfast cereal consumption being associated with reports of lower depression, emotional distress, and fatigue; greater alertness; fewer cognitive problems; and fewer minor symptoms and bowel problems.

These effects were apparent after both the first and second weeks. They were also observed for all cereals. Overall, the results of this study show that breakfast cereal consumption by children is associated with greater well-being.

There is evidence that breakfast consumption per se improves well-being, and effects appear to be most pronounced with breakfast cereal in combination with dairy products (O'Sullivan et al. 2009). One type of breakfast cereal that has a large effect is high-fiber cereal. Research has also shown that increasing dietary fiber from wheat bran cereals decreases fatigue and increases energy (Smith et al. 2001). Smith (in press-b) conducted secondary analyses of data from this study. Initial analyses examined associations between high fiber intake and well-being (emotional distress, fatigue, cognitive difficulties, and somatic symptoms). The results showed that high fiber intake was associated with increased well-being. Subsequent analyses examined whether the effects of total fiber intake could be accounted for by ingestion of specific sources of fiber, namely, breakfast cereal, fruit, or vegetables. The results showed that it was the breakfast cereal that was largely responsible for the increased well-being. Digestive problems are also associated with reduced well-being, and a second set of analyses examined whether the benefits of fiber were due to a reduction in digestive problems. The results showed that digestive problems reduced well-being, but these effects were independent of the effects of fiber.

The next section considers mechanisms that might underlie the behavioral effects of breakfast.

UNDERLYING MECHANISMS

Breakfast Is Better Than Nothing

Results clearly show that consumption of breakfast improves mood and cognition compared to eating no breakfast. It has been suggested that breakfast removes the negative effects of fasting, and the mechanism has often been conceptualized in terms of providing a supply of energy to the brain (see the "Glycemic Index and Load" section in this chapter). However, recent research suggests that the mechanisms are likely to be more complicated. In a series of studies, Smith (in preparation-b) examined when consumption of breakfast cereal led to an improvement in mood and memory. The first study examined effects observed after a cereal lunch (following consumption of a normal breakfast). The cereal consumption was associated with a more positive mood, but no benefits were seen for the free recall task. This suggests that it is not just the meal that is important. A second study examined effects of consuming cereal in the early evening after a day of fasting. Fasting led to a more negative mood and impaired free recall of a list of words. Consumption of cereal improved mood but did not improve memory. This result suggests that fasting is not the only factor involved in the acute effects of breakfast. Finally, the effects of breakfast on the mood and memory of night workers (who slept during the day and had breakfast in the evening) were investigated. Again, breakfast improved mood but had no effect on memory. These results suggest that consumption of breakfast cereal improves mood whenever it is eaten, but the memory effects depend on it being consumed early in the day.

The effects of breakfast, or rather a high-carbohydrate breakfast, have often been explained in terms of changes in serotonin (Fernstrom and Wurtman 1971). Such a mechanism could explain the association between regular consumption of breakfast and well-being, although it may also be the case that it is well-being that influences food choice (Christensen and Somers 1996). It is likely to be the case that a number of peripheral and central mechanisms underlie the effects of breakfast on behavior, and further research is required to elucidate these mechanisms. The next section considers the importance of the macronutrient content of breakfast.

Glycemic Index and Load

A large number of studies have examined the effects of glucose on behavior. The evidence for positive effects is not consistent, although a number of studies have demonstrated beneficial effects of glucose on verbal memory (Hoyland et al. 2008). Other studies have investigated the effects of meals differing in glycemic index, glycemic load, the ratio of slow:rapid availability of glucose, the proportion of simple to complex carbohydrate, or the amount of rapidly versus slowly digested carbohydrate. Glycemic index (GI) and glycemic load (GL) are the most widely used indices. GI provides a measure of carbohydrate quality, not quantity, whereas GL is a product of the food's GI and the amount of carbohydrate per serving. Gilsenan et al. (2009) have reviewed studies comparing the impact of different GLs. Their conclusion was that there is insufficient evidence to support a consistent effect of GL on cognitive performance. A recent study (Micha et al. 2010) examined the effects of glycemic potency (combinations of GI and GL) on cognitive performance of sixty children aged 11–14 years. A low-GI/high-GL breakfast was associated with faster information processing, whereas a high-GI breakfast was associated with better immediate word recall. Further research is now required to determine whether breakfasts with these macronutrient compositions will have beneficial effects on the academic performance of children. The research to date does not inform on the precise mechanisms through which glucose influences cognition. The possible mechanisms are many and varied. For example, glucose is taken up by astrocytes and converted into lactate, which is then released into extracellular space to be taken up as an energy substrate by neurons. Many of the brain's neurotransmitters are derived from glucose metabolism, which suggests that glucose may influence cognitive function by enhancing neurotransmitter synthesis during periods of neuronal activity. Alternatively, there could be a peripheral effect of glucose on memory due to a neural signal being triggered when glucose is transported into cells. GL may also influence gastrointestinal hormonal response, which in turn may have effects on cognition. Factors such as food acceptability may also be related to levels of circulating glucose, and these variables must be controlled when assessing the impact of different meals.

Other Mechanisms

There are clearly a number of other mechanisms through which consumption of breakfast may influence behavior. These may reflect the macronutrient composition (e.g., effects of high-fiber cereals), the micronutrient composition (e.g., fortification of cereals), or a more general influence on dietary intake and health.

EFFECTS ON REAL-LIFE COGNITIVE FUNCTION AND SAFETY

The major practical implications of breakfast consumption are to be seen in the areas of nutritional intake, weight management, and health. Studies of children suggest that breakfast consumption may improve cognition and school attendance, which leads to better academic achievement. Reviews of breakfast consumption and children's academic achievement (e.g., Ells et al. 2008) have concluded that there are short-term benefits. However, they also point out the methodological problems present in many of the studies: failure to consider the impact of habitual diet; little consistency in the methodology across studies; use of measures with no known validation; failure to distinguish nutritional effects from the social effects of breakfast clubs; and, given that most interventions have been of short duration, the results fail to quantify sustainability and longer term benefits.

Little is known about the real-life behavioral implications of consuming breakfast for adults. For example, a literature search revealed no information on breakfast and accidents and errors at work (or outside of work), on road traffic accidents or driving performance, or on productivity at work. There is a link between performance of laboratory tasks and safety issues in that Morris (2008) found that elevating blood glucose level increased the retention of information from a public safety video.

Chaplin and Smith (2011) examined the effects of breakfast consumption on the health and safety of a sample of 870 nurses. The results showed that accidents, injuries, and cognitive failures at work were greater in those who rarely ate breakfast. In addition, stress at work was greater in the breakfast skippers. Further research is now required to extend these findings to consider real-life activities outside of the workplace. In addition, it is essential to carry out interventions rather than just cross-sectional analyses.

DISCUSSION

The obvious conclusion to be drawn from the literature reviewed here is that breakfast is good for you. This is true when one considers a number of different areas such as nutritional intake, weight management, and health. The same conclusion applies when one considers behavioral outcomes, with breakfast being associated with a more positive mood, improved cognition, and, in the longer term, better well-being. These conclusions generally hold for well-nourished children, children with nutritional deficiencies, and adults (young, middle-aged, and elderly). Given the robust evidence for the beneficial effects of breakfast, it is rather surprising that we have made relatively little progress in understanding the underlying mechanisms (both psychological mechanisms and the CNS changes that underpin these). Furthermore, compared to other aspects of eating and drinking (e.g., consuming caffeine), we know relatively little about the practical benefits of breakfast at work, rest, and play. Future research must extend our current knowledge by conducting translational research that will provide appropriate information for future policy and practice.

REFERENCES

Baxter, A. J., T. Coyne, and C. McClintock. 2006. Dietary patterns and metabolic syndrome: A review of epidemiologic evidence. *Asia Pac J Clin Nutr.* 15:134–42.

Bazzano, L. A., Y. Song, V. Bubes, C. K. Good, J. E. Manson, and S. Liu. 2005. Dietary intake of whole and refined grain breakfast cereals and weight gain in men. *Obes Res* 13:1952–60.

Benton, D., and P. Y. Parker. 1998. Breakfast, blood glucose and cognition. *Am J Clin Nutr* 67 (suppl.):772S–778S.

Benton, D., M. P. Ruffin, T. Lassel, S. Nabb, M. Messaudi, S. Vinoy, D. Desor, and V. Lang. 2003. The delivery rate of dietary carbohydrates affects cognitive performance in both rats and humans. *Psychopharmacology* 166:86–90.

Benton, D., and J. Sargeant. 1992. Breakfast, blood glucose and memory. *Biol Psychol* 33:207–10.

Benton, D., O. Slater, and R. T. Donohoe. 2001. The influence of breakfast and a snack on psychological performance. *Physiology and Behavior* 74:559–71.

Berkman, L. F., and L. Breslow. 1983. *Health and Ways of Living.* New York: Oxford University Press.

Blom, W. A., A. Lluch, A. Stafleu, S. Vinoy, J. J. Holst, G. Schaafsma, and H. F. Hendriks. 2006. Effect of a high-protein breakfast on the postprandial ghrelin response. *Am J Clin Nutr* 3:211–20.

Chaplin, K., and A. P. Smith. In preparation. An investigation of the effects of breakfast cereals on alertness, cognitive function and other aspects of the well-being of adults.

Chaplin, K., and A. P. Smith. Submitted. Breakfast and snacks: Associations with cognitive failures, minor injuries, accidents and stress. *Nutrients.*

Cho, S., M. Dietrich, C. J. Brown, C. A. Clark, and G. Block. 2003. The effect of breakfast type on total daily energy intake and body mass index: results from the Third National Health and Nutrition Examination Survey (NHANES III). *J Am Coll Nutr* 22:296–302.

Christensen, L., and S. Somers. 1996. Comparison of nutrient intake among depressed and non-depressed individuals. *International Journal of Eating Disorders* 20:105.

Costabile, A., A. Klinder, F. Fava, A. Napolitano, V. Fogliano, C. Leonard, G. R. Gibson, and K. M. Tuohy. 2008. Whole-grain wheat breakfast cereal has a prebiotic effect on the human gut microbiota: a double-blind, placebo-controlled, crossover study. *Br J Nutr* 99:110–20.

Deaville, E., G. Gibson, and A. P. Smith. In preparation. Prebiotic effects of high fiber breakfast cereals.

Dickie, N. H., and A. E. Bender. 1982. Breakfast and performance in school children. *Br J Nutr* 48:483–96.

Djoussé, L., and J. M. Gaziano. 2007. Breakfast cereals and risk of heart failure in the physicians' health study I. *Arch Intern Med* 167:2080–5.

Ells, L. J., F. C. Hillier, J. Shucksmith, H. Crawley, L. Harbige, J. Shield, A.Wiggins, and C. D. Summerbell. 2008. A systematic review of the effect of dietary exposure that could be achieved through normal dietary intake on learning and performance of school-aged children of relevance to UK schools. *Br J Nutr* 100:927–36.

Fernstrom, J. D., and R. J. Wurtman. 1971. Brain serotonin content: Increase following ingestion of carbohydrate diet. *Science* 174:1023.

Fischer, K., P. C. Colombani, W. Langhans, and C. Wenk. 2002. Carbohydrate to protein ratio in food and cognitive performance in the morning. *Physiol Behav* 75:411–23.

Gilsenan, M. B., E. A. de Bruin, and L. Dye. 2009. The influence of carbohydrate on cognitive performance: a critical evaluation from the perspective of glycemic load. *Brit J Nut* 101:941–9.

Gomez-Pinilla, F. 2008. Brain foods: the effects of nutrients on brain function. *Nature Reviews Neuroscience* 9:568–78.

Haines, P. S., D. K. Guilkey, and Popkin, B. M. 1996. Trends in breakfast consumption of U.S. adults between 1965 and 1991. *J Am Diet Assoc* 96:64–470.

Holt, S. H. A., H. J. Delargy, C. L. Lawton, and J. Blundell. 1999. The effects of high carbohydrate versus high fat breakfasts on feelings of fullness and subsequent food intake. *Int J Food Sci Nutr* 50:13–28.

Holt, S. H., J. C. Miller, P. Petocz, and E. Farmakalidis. 1995. A satiety index of common foods. *Eur J Clin Nutr* 49:675–90.

Hoyland, A., C. L. Lawton, and L. Dye. 2008. Acute effects of macronutrient manipulations on cognitive test performance in healthy young adults: a systematic research review. *Neurosci Biobehav R* 32:72–85.

Hoyland, A., L. Dye, and C. L. Lawton. 2009. A systematic review of the effect of breakfast on cognitive performance of children and adolescents. *Nutr Res Rev* 22:220–43.

Huang, C. J., H. T. Hu, Y. C. Fan, Y. M. Liao, and P. S. Tsai. 2010. Associations of breakfast skipping with obesity and health-related quality of life: evidence from a national survey in Taiwan. *Int J Obes* 34:720–5.

Ingwersen, J., M. A. Defeyter, D. O. Kennedy, K. A. Wesnes, and A. B. Scholey. 2007. A low glycemic index breakfast cereal preferentially prevents children's cognitive performance from declining throughout the morning. *Appetite* 49:240–4.

Kant, A. K., and B. I. Graubard. 2006. Secular trends in patterns of self-reported food consumption of adult Americans: NHANES 1971–1975 to NHANES 1999–2002. *Am J Clin Nutr* 84:1215–23.

Kaplan, R. J., C. E. Greenwood, G. Winocur, and T. M. S. Wolever. 2001. Dietary protein, carbohydrate and fat enhance memory performance in the healthy elderly. *Am J Clin Nut* 74:687–93.

Kelly, T. H., R. W. Foltin, B .J. Rolls, and M. W. Fischman. 1994. Effect of meal macronutrient and energy content on human performance. *Appetite* 23:97–111.

King, C. G., H. A. Bickerman, W. Bouvet, C. J. Harrer, J. R. Oyler, and C .P. Seitz. 1945. Effects of pre-flight and in-flight meals of varying composition with respect to carbohydrate, protein or fat. *Journal of Aviation Medicine* 16:69–84.

Lieberman, H. R. 1992. Caffeine. In *Handbook of Human Performance*, vol. 2, ed. A. P. Smith and D. M. Jones, 49–72. London: Academic Press.

Li, Q., K. Morimoto, A. Nakadai, T. Qu, H. Matsushimaq, M. Katsumata, T. Shimizu, H. Inagaki, Y. Hirata, K. Hirata, T. Kawada, Y. Lu, K. Nakayama, and A. M. Krensky. 2007. Healthy lifestyles are associated with higher levels of perforin, granulysin and granzymes A/B-expressing cells in peripheral blood lymphocytes. *Preventive Medicine* 44:117–23.

Lloyd, H. M., P. J. Rogers, and D. I. Hedderley. 1996. Acute effects on mood and cognitive performance of breakfast differing in fat and carbohydrate content. *Appetite* 27:151–64.

Lluch, A., P. Hubert, N. A. King, and J. E. Blundell. 2000. Selective effects of acute exercise and breakfast interventions on mood and motivation to eat. *Physiol Behav* 68:515–20.

Ma, Y., E. R. Bertone, E. J. Stanek III, G. W. Reed, J. R. Hebert, N. L. Cohen, P. A. Merriam, and I. S. Ockene. 2003. Association between eating patterns and obesity in a free-living U.S. adult population. *Am J Epidemiol* 158:85–92.

Mahoney, C. R., H. A. Taylor, and R. B. Kanarek. 2007. The acute effects of meals on cognitive performance. In *Nutritional Neuroscience*, ed. H. Lieberman, R. Kanarek, and C. Prasad, 73–92. Taylor and Francis.

Mahoney, C. R., H. A. Taylor, R. B. Kanarek and P. Samuel. 2005.Effect of breakfast composition on cognitive processes in elementary school children. *Physiol Behav* 85:635–45.

Maridakis, V., M. P. Herring, and P. J. O'Connor. 2009. Sensitivity to change in cognitive performance and mood measures of energy and fatigue in response to differing doses of caffeine and breakfast. *International Journal of Neuroscience* 119:975–94.

Merriam, P. A., I. S. Ockene, J. R. Hebert, M. C. Rosal, and C. E. Matthews. 1999. Seasonal variation of blood cholesterol levels: study methodology. *J Biol Rhythms* 14:330–39.

Micha, R., P. J. Rogers, and M. Nelson. 2010. The glycemic potency of breakfast and cognitive function in school children. *Eur J Clin Nut* 64:948–57.

Michaud, C., N. Musse, J. P. Nicolas, and L. Mejean. 1991. Effects of breakfast-size on short-term memory, concentration, mood and blood glucose. *J Adolesc Health* 12:53–57.

Morgan, K. J., M. E. Zabik, and G. L. Stampley. 1986. The role of breakfast in diet adequacy of the U.S. adult population. *J Am Coll Nutr* 5:551–63.

Morris, N. 2008. Elevating blood glucose level increases the retention of information from a public safety video. *Biol Psychol* 78:188–90.

Nabb, S., and D. Benton. 2006. The influence on cognition of the interaction between the macro-nutrient content of breakfast and glucose tolerance. *Physiology and Behavior* 87:16–23.

O'Sullivan, T. A., M. Robinson, G. E. Kendall, M. Miller, P. Jacoby, S. R. Silburn, and W. H. Oddy. 2009. A good-quality breakfast is associated with better mental health in adolescence. *Public Health Nutrition* 12:249–58.

Pasman, W. J., V. M. Blokdijk, F. M. Bertina, W. P. Hopman, and H. F. J. Hendriks. 2003. Effect of two breakfasts, differing in carbohydrate composition, on hunger and satiety and mood in healthy mean. *Int J Obes* 27:663–8.

Quan, T., J. Salomon, S.Nitzke, and M. Reicks. 2000. Behaviors of low-income mothers related to fruit and vegetable consumption. *J Am Diet Assoc* 100:567–9.

Rampersaud, G. C., M. A. Pereira, B. L. Girard, et al. 2005. Breakfast habits, nutritional status, body weight, and academic performance in children and adolescents. *Jam Diet Assoc* 105:743–60.

Resnicow, K. 1991. The relationship between breakfast habits and plasma cholesterol levels in schoolchildren. *J Sch Health* 61:81–85.

Rice, J. E., and D. F. Duncan. 1985. Health practices and mental health: an exploratory study. *Psychological Reports* 57:1110.

Richards, M. M. K. 1972. Studies of breakfast and mental performance. *Nutrition* 26:219–23.

Ruxton, C. H. S., and T. R. Kirk. 1997. Breakfast: a review of associations with measures of dietary intake, physiology and biochemistry. *Br J Nutr* 78:199–213.

Siega-Riz, A. M., B. M. Popkin, and T. Carson. 1998. Trends in breakfast consumption for children in the United States from 1965–1991. *Am J Clin Nutr* 67 (suppl.):748S–756S.

Smith, A. P. 1998. Breakfast and mental health. *International Journal of Food Sciences and Nutrition* 49:397–402.

Smith, A. P. 1999. Breakfast cereal consumption and subjective reports of health. *International Journal of Food Sciences and Nutrition* 50:445–9.

Smith, A. P. 2002a. Stress, breakfast cereal consumption and cortisol. *Nutritional Neuroscience* 5:141–4.

Smith, A. P. 2002b. Stress, breakfast cereal consumption and objective signs of upper respiratory tract illnesses. *Nutritional Neuroscience* 5:145–8.

Smith, A. P. 2003. Breakfast cereal consumption and subjective reports of health by young adults. *Nutritional Neuroscience* 6:59–61.

Smith, A. P. 2005. The concept of well-being: relevance to nutritional research. *British Journal of Nutrition* 93 (1): S1–S5.

Smith, A. P. 2010a. An investigation of the effects of breakfast cereals on alertness, cognitive function and other aspects of the well-being of children. *Nutritional Neuroscience* 13:230–6.

.Smith, A. P. 2010b. Breakfast cereal, fibre, digestive problems and well-being. *Current topics in Nutraceutical Research* 8:117–26.

Smith, A. P. 2011. Breakfast cereal, digestive problems and well-being. *Stress and Health.*

Smith, A. P. In preparation-a. Effects of a high breakfast cereal on memory, mood and well-being.

Smith, A. P. In preparation-b. Effects of fasting, breakfast cereal, caffeine withdrawal and caffeine on mood and cognitive function.

Smith, A. P., C. Bazzoni, J. Beale, J. Elliott-Smith, and M. Tiley. 2001. High fibre breakfast cereals reduce fatigue. *Appetite* 37:249–50.

Smith, A. P., R. Clark, and J. Gallagher. 1999. Breakfast cereal and caffeinated coffee: effects on working memory, attention, mood, and cardiovascular function. *Physiol Behav* 67 (1):9–17.

Smith, A. P., and A. Kendrick. 1992. Meals and performance. In *Handbook of human performance,* vol. 2, ed. A. P. Smith and D. M. Jones, 1–23. Health and Performance. London: Academic Press.

Smith, A. P., A. M. Kendrick, and A. L. Maben. 1993. Effects of breakfast and caffeine on performance and mood in the late morning and after lunch. *Neuropsychobiology* 26:198–204.

Smith, A. P., A. M. Kendrick, A. L. Maben, and J. Salmon. 1994. Effects of breakfast and caffeine on performance, mood and cardiovascular functioning. *Appetite* 22:39–55.

Smith, A. P., and G. Rees. 2000. Stress, breakfast cereal consumption and susceptibility to upper respiratory tract infections. *Nutritional Neuroscience* 3:339–43.

Smith, A. P., and C. Stamatakis. 2010. Cereal bars, mood and memory. *Current Topics in Nutraceutical Research* 8:169–72.

Smith, A. P., and A. Wilds. 2009. Effects of cereal bars for breakfast and mid-morning snacks on mood and memory. *International Journal of Food Sciences and Nutrition* 60 (s4): 63–9.

Stanton, J. L., Jr., and D. R. Keast. 1989. Serum cholesterol, fat intake, and breakfast consumption in the United States adult population. *J Am Coll Nutr* 8:567–72.

Tanaka, M., K. Mizuno, S. Fukuda, Y. Shigihara, and Y. Watanabe. 2008. Relationships between dietary habits and the prevalence of fatigue in medical students. *Nutrition* 24:985–9.

Timlin, M. T., M. A. Pereira, M. Story, and D. Neumark-Sztainer. 2008. Breakfast eating and weight change in a 5-year prospective analysis of adolescents: Project EAT (Eating Among Teens). *Pediatrics* 121:e638–e645.

Tuttle, W. 1953. Physiologic response to size and content of breakfast by men over 60. *Journal of Abnormal Child Psychology* 29:34–40.

Tuttle, W., K. Daum, C. J. Imig, B. Randall, and M. T. Schumacher. 1952. Effect of omitting breakfast on the physiologic effect of the aged. *Journal of the American Dietary Association* 26:332–5.

Tuttle, W., M. Wilson, and K. Daum. 1949. Effect of altered breakfast habits on physiologic response. *Journal of Applied Physiology* 1:545.

Utter, J., R. Scragg, C. N. Mhurchu, and D. Schaaf. 2007. At-home breakfast consumption among New Zealand children: associations with body mass index and related nutrition behaviors. *J Am Diet Assoc* 107:570–6.

Van der Heijden, A. A., F. B. Hu, E. B. Rimm, R. M. van Dam. 2007. A prospective study of breakfast consumption and weight gain among U.S. men. *Obesity* 15:2436–69.

Wesnes, K. A., C. Pincock, D. Richardson, et al. 2003. Breakfast reduced declines in attention and memory over the morning in schoolchildren. *Appetite* 41:329–31.

Wetzler, H. P., and R. J. Ursano. 1988. A positive association between physical health practices and psychological well-being. *J Nerv Ment Dis* 176:280–3.

Widenhorn-Müller, K., K. Hille, J. Klenk, et al. 2008. Influence of having breakfast on cognitive performance and mood in 13- to 20-year-old high school students: results of a crossover trial. *Pediatrics* 122:279–84.

Wyon, D. P., L. Abrahamsson, M. Jartelius, et al. 1997. An experimental study on the effects of energy intake at breakfast on the test performance of 10-year-old children in school. *Int J Food Sci Nutr* 48:5–12.

Zemel, M. B., W. Thompson, A. Milstead, K. Morris, and P. Campbell. 2004. Calcium and dairy acceleration of weight and fat loss during energy restriction in obese adults. *Obes Res* 12:582–90.

APPENDIX: BIBLIOGRAPHY OF STUDIES OF BREAKFAST AND COGNITION IN CHILDREN AND ADOLESCENTS

Acute Effects of Interventions

Benton, D., A. Maconie, and C. Williams. 2007. The influence of the glycaemic load of breakfast on the behaviour of children in school. *Physiol Behav* 92:717–24.

Busch, C. R., H. A. Taylor, R. B. Kanarek, et al. 2002. The effects of a confectionery snack on attention in young boys. *Physiol Behav* 77:333–40.

Connors, C. K., and A. G. Blouin 1983. Nutritional effects on behaviour of children. *J Psychiatr Res* 17:198–201.

Cromer, B. A., K. J. Tarnowski, A. M. Stein, et al. 1990. The school breakfast program and cognition in adolescents. *J Dev Beh Pediatr* 11:295–300.

Dickie, N. H., and A. E. Bender. 1982. Breakfast and performance in schoolchildren. *Br J Nutr* 48:483–96.

Ingwersen, J., M. A. Defeyter, D. O. Kennedy, et al. 2007. A low glycaemic index breakfast cereal preferentially prevents children's cognitive performance from declining throughout the morning. *Appetite* 49:240–4.

Ma, G., X. Hu, S. Gao, et al. 1999. Effect of energy intake at breakfast on school performance. *Wei Sheng Yan Jiu* 28:286–8.

Mahoney, C. R., H. A. Taylor, R. B. Kanarek, et al. 2005. Effect of breakfast composition on cognitive processes in elementary school children. *Physiol Behav* 85:635–45.

Marquez Acosta, M., R. Sutil de Naranjo, C. E. Rivas de Yepez, et al. 2001. Influence of breakfast on cognitive functions of children from an urban area in Valencia, Venezuela. *Arch Latinoam Nutr* 51:57–63.

Micha, R., P. J. Rogers, and M. Nelson. 2010. The glycaemic potency of breakfast and cognitive function in school children. *Eur J Clin Nutr* 64:948–57.

Michaud, C., N. Musse, J. P. Nicolas, et al. 1991. Effects of breakfast-size on short-term memory, concentration, mood and blood glucose. *J Adolesc Health* 12:53–7.

Morrell, G., and D. R. Atkinson. 1977. Effects of breakfast program on school performance and attendance of elementary school children. *Education* 98:111–16.

Morris, N., and P. Sarll. 2001. Drinking glucose improves listening span in students who miss breakfast. *Educ Res* 43:201–7.

Pollitt, E., S. Cueto, and E. R. Jacoby. 1998. Fasting and cognition in well- and undernourished schoolchildren: a review of three experimental studies. *Am J Clin Nutr* 67 (suppl.): 779–84.

Pollitt, E., R. L. Leibel, and D. Greenfield. 1981. Brief fasting, stress and cognition in children. *Am J Clin Nutr* 34:1526–33.

Pollitt, E., N. L. Lewis, C. Garza,. et al. 1982–1983. Fasting and cognitive function. *J Psychiatr Res* 17:169–74.

Smith, M. A., and J. K. Foster. 2008. The impact of a high versus a low glycemic index breakfast cereal meal on verbal episodic memory in healthy adolescents. *Nutr Neurosci* 11:219–27.

Vaisman, N., H. Voet, A. Akivis, et al. 1996. Effect of breakfast timing on the cognitive functions of elementary school students. *Arch Pediatr Adolesc Med* 150:1089–92.

Wesnes, K. A., C. Pincock, D, Richardson, et al. 2003. Breakfast reduced declines in attention and memory over the morning in schoolchildren. *Appetite* 41:329–31.

Widenhorn-Müller, K., K. Hille, J. Klenk, et al. 2008. Influence of having breakfast on cognitive performance and mood in 13- to 20-year-old high school students: results of a crossover trial. *Pediatrics* 122:279–84.

Wyon, D. P., L. Abrahamsson, M. Jartelius, et al. 1997. An experimental study on the effects of energy intake at breakfast on the test performance of 10-year-old children in school. *Int J Food Sci Nutr* 48:5–12.

Children Differing in Nutritional Status

Cueto, S., E. Jacoby, and E. Pollitt. 1998. Breakfast prevents delays of attention and memory functions among nutritionally at-risk boys. *J Appl Dev Psychol* 19:219–34.

Grantham-McGregor, S. M., S. Chang, and S. P. Walker. 1998. Evaluation of school feeding programs: some Jamaican examples. *Am J Clin Nutr* 67 (suppl.):785–9.

Lopez, I., I. de Andraca, C. G. Perales, et al. 1993. Breakfast omission and cognitive performance of normal, wasted and stunted schoolchildren. *Eur J Clin Nutr* 47:533–42.

Muthayya, S., T. Thomas, K. Srinivasan, et al. 2007. Consumption of a mid-morning snack improves memory but not attention in school children. *Physiol Behav* 90:142–50.

Noriega, J. A. N. 2000. Method and theory in school breakfast program evaluation. *Estudios de Psicologia* 5:33–48.

Pollitt, E., S. Cueto, and E. R. Jacoby. 1998. Fasting and cognition in well- and undernourished schoolchildren: a review of three experimental studies. *Am J Clin Nutr* 67 (suppl.): 779–84.

Pollitt, E., E. Jacoby, and S. Cueto. 1996. School breakfast and cognition among nutritionally at-risk children in the Peruvian Andes. *Nutr Rev* 54:22–6.

Simeon, D. T. 1998. School feeding in Jamaica: a review of its evaluation. *Am J Clin Nutr* 67 (suppl.):790–4.

Simeon, D., and S. Grantham-McGregor, 1987. Cognitive function, under nutrition and missed breakfast. *Lancet* 2:737–8.

Simeon, D. T., and S. Grantham-McGregor. 1989. Effects of missing breakfast on cognitive functions of schoolchildren of differing nutritional status. *Am J Clin Nutr* 49:646–53.

School Breakfast Programs

Cueto, S., and M. Chinen. 2008. Educational impact of a school breakfast program in rural Peru. *Int J Educ Dev* 28:132–48.

Kleinman, R. E., S. Hall, H. Green, et al. 2002. Diet, breakfast and academic performance in children. *Ann Nutr Metab* 46 (1):24–30.

Lieberman, H. M., I. F. Hunt, A. H. Coulson, et al. 1976. Evaluation of a ghetto school breakfast program. *J Am Diet Assoc* 69:132–8.

Meyers, A. F., A. E. Sampson, M. Weitzman, et al. 1989. School breakfast program and school performance. *Am J Dis Child* 143:1234–9.

Murphy, J. M., M. Pagano, J. Nachmani, et al. 1998. The relationship of school breakfast to psychosocial and academic functioning: cross-sectional and longitudinal observations in an inner-city school sample. *Arch Pediatr Adolesc Med* 152:899–907.

Pollitt, E., E. Jacoby, and S. Cueto. 1996. School breakfast and cognition among nutritionally at-risk children in the Peruvian Andes. *Nutr Rev* 54:22–6.

Richter, L. M., C. Rose, and R. D. Griesel. 1997. Cognitive and behavioral effects of a school breakfast. *S Afr Med J* 87 (1):93–100.

Shemilt, I., I. Harvey, L. Shepstone, et al. 2004. A national evaluation of school breakfast clubs: evidence from a cluster randomized controlled trial and an observational analysis. *Child Care Health Dev* 30:413–27.

Simeon, D. T. 1998. School feeding in Jamaica: a review of its evaluation. *Am J Clin Nutr* 67 (suppl.):790–4.

Vera Noriega, J. A., S. E. Dominguez Ibanez, M. O. Pena Ramos, et al. 2000. Evaluation of the effects of a school breakfast program on attention and memory. *Arch Latinoam Nutr* 50:35–41.

Wahlstrom, K. L., and M. Begalle. 1999. More than test scores: results of the Universal School breakfast pilot in Minnesota. *Topics Clin Nutr* 15:7–29.

Worobey, J., and H. S. Worobey. 1999. The impact of a two-year school breakfast program for preschool-aged children on their nutrient intake and pre-academic performance. *Child Stud J* 29:113–31.

Habitual Breakfast Consumption

Ahmadi, A., Z. Sohrabi, and M. H. Eftekhari. 2009. Evaluating the relationship between breakfast pattern and short-term memory in junior high school girls. *Pak J Biol Sci* 12:742–5.

Gajre, N. S., S. Fernandez, N. Balakrishna, and S. Vazir. 2008. Breakfast eating habit and its influence on attention-concentration, immediate memory and school achievement. *Indian Pediatr* 45:824–8.

Herrero Lozano, R., and J. C. F. Fillat Ballesteros. 2006. A study on breakfast and school performance in a group of adolescents. *Nutr Hosp* 21:346–52.

Lien, L. 2007. Is breakfast consumption related to mental distress and academic performance in adolescents? *Public Health Nutr* 10:422–8.

Lopez-Sobaler, A. M., R. M. Ortega, M. E. Quintas, et al. 2003. Relationship between habitual breakfast and intellectual performance (logical reasoning) in well-nourished schoolchildren of Madrid (Spain). *Eur J Clin Nutr* 57 (1):49–53.

Morales, I. F., M. V. A. Villas, C. J. M. Vega, et al. 2008. Relation between the breakfast quality and the academic performance in adolescents of Guadalajara (Castilla-La Mancha). *Nutr Hosp* 23:383–7.

5 Diet, Physical Activity, and Substrate Oxidation

Implications for Appetite Control, Weight Loss, and Body Composition

Mark Hopkins, Neil A. King, and John E. Blundell

CONTENTS

Dietary Restriction, Physical Activity, and the Propensity for Weight Loss............72
Introduction........72
The Role of Diet and Physical Activity in Weight Gain........72
Dietary Restriction, Physical Activity, and Weight Loss........73
Dietary Restriction and Weight Loss........73
Physical Activity and Weight Loss........74
Diet-Induced and Physical Activity-Induced Weight Loss........74
The Impact of Dietary Restriction and Physical Activity on Energy Balance: Implications for the Efficacy of Weight Loss........75
Interindividual Variability in Biological and Behavioral Responses........77
Biological and Behavioral Compensation to Dietary Restriction and Physical Activity........78
Dietary Restriction and Metabolic Compensation........78
Physical Activity and Metabolic Compensation........79
The Effect of Physical Activity on Energy Intake........80
Nonexercise Activity Thermogenesis........82
Dietary Restriction and Nonexercise Activity Thermogenesis........83
Physical Activity and Nonexercise Activity Thermogenesis........83
The Temporal Profile and Stability of Compensation: Is the Magnitude of Compensation Dose Dependent?........84
Conclusions........85
Substrate Metabolism, Weight Gain, and Appetite Control........86
Introduction........86
Postprandial Trafficking of Dietary Fat........87
Substrate Metabolism as a Physiological Driver of Eating Behavior........89
Conclusions........93
References........94

DIETARY RESTRICTION, PHYSICAL ACTIVITY, AND THE PROPENSITY FOR WEIGHT LOSS

INTRODUCTION

In many countries, governments and health agencies are strongly promoting physical activity (PA) as a means to prevent the accumulation of fatness that leads to weight gain and obesity. However, there is often a resistance to respond to health promotion initiatives. For example, in the UK, the chief medical officer has recently reported that 71 percent of women and 61 percent of men fail to carry out even the minimal amount of physical activity recommended in the government's guidelines. Similarly, the Food Safety Agency has promoted reductions in the intake of fat, sugar, and salt but with very little impact on the pattern of consumption. Why is it that recommendations to improve health are so difficult to implement and produce the desired outcome?

A major reason is that both physical activity and dietary intake are forms of behavior. These are not simple acts but can be regarded as integrated behavioral sequences that are held in place by various environmental contingencies together with attributions and cognitions (self-explanations) that maintain existing sequences (sedentariness, low amounts of leisure time physical activity, and high intakes of fat and sugar) as part of a daily repertoire. This repertoire is made up of habits (habitual sequences of integrated behavior) that are extremely resistant to change. No single statement or policy for change is likely to be applicable to all.

Further, there is considerable individual variability in the impact of physical activity and diet on the body. People should not be expected to respond in the same way. For example, not all people who are exposed to a high-fat diet gain weight; some remain lean,[1] and this leads to the concept of susceptible and resistant phenotypes.[2] Similarly not all individuals who undertake a compulsory program of physical activity lose weight;[3] some actually gain weight and appear to demonstrate a "resistance" to weight loss.[3,4] These individual differences not only are a genuine biological fact (a fact of life) but also can be used to throw light on the mechanisms that link physical activity and diet to fatness. The disclosure of such mechanisms, and the behavioral sequences with which they are associated, can be used to devise programs to motivate and empower people to change behavior. Recognition that individual differences exist and that no single model fits all human beings (not even the majority of those who are overweight, overconsuming, unfit, and living in Western societies) can help to promote an understanding of an extremely complex situation. First, it is important to be aware of some basic relationships among physical activity, diet, and fatness.

THE ROLE OF DIET AND PHYSICAL ACTIVITY IN WEIGHT GAIN

Using data from national surveys, Hill et al.[5] estimated that the average increase in body weight for an adult American over the last decade was approximately 1 kg per year. It was suggested that this increase in weight could be explained by an "energy gap" of less than 50 $kcal\cdot day^{-1}$, which is equivalent to that reported in middle-aged

Australian women.[6] While it is accepted that a prolonged positive energy balance will result in weight gain, whether the current rates of obesity are primarily driven by imbalances created by increased energy intake (EI) or decreased energy expenditure has been debated.[7–9] Recently, it has been suggested that an increased level of EI at the population level is the key driver of the current obesity epidemic, as levels of PA have remained unchanged since the 1980s.[10,11] For example, Westerterp and Speakman[10] found no evidence that total energy expenditure or that expended on PA (measured using doubly labeled water) has declined in North Europe over the last two decades. However, it has been suggested that increases in EI could adequately account for the weight gain seen in the U.S, population.[11,12] Therefore, it is important to understand how PA and diet can lead to weight loss and the issues that can influence the efficacy of both weight loss strategies.

Dietary Restriction, Physical Activity, and Weight Loss

Dietary Restriction and Weight Loss

Dietary restriction is an obvious method for the promotion of weight loss, as EI can be readily reduced, at least in the short term, to create a negative energy balance. While attention has been given to the macronutrient content of the diet, the energy deficit created by dietary restriction appears to have the greatest influence over weight loss.[13] For example, when isoenergetic deficits are created by either high- or low-fat diets, similar changes in body composition are observed.[14] However, under *ad libitum* feeding conditions where the total calorie intake is not fixed, the macronutrient composition of a diet may influence the hunger and satiety response, and therefore weight loss.[15] Current recommendations suggest that any dietary restriction should be moderate in nature, and create an energy deficit of 500–1000 $kcal \cdot day^{-1}$ below that required for body weight maintenance.[16] For an individual consuming 2500 $kcal \cdot day^{-1}$, this translates to a reduction of 20-40 percent of daily EI, and may result in a weekly loss in body weight of approximately 1 kg.[17]

Research has focused on the efficacy of a number of different types of dietary restriction strategies, including low-fat or -calorie diets, very-low-calorie diets, low-carbohydrate CHO diets, or diets with a low glycemic index.[18] Traditional dietary approaches have emphasized low dietary fat intake (< 30 percent of daily energy intake), and there is good evidence that this can produce meaningful reductions in body weight under *ad libitum* feeding conditions.[19] There has been a recent focus on the effectiveness of low-CHO diets, and there is some evidence to suggest that a low-CHO diet is more effective in the short term (six months) for inducing weight loss than a calorie-restricted low-fat diet.[20–24] However, it remains unclear whether differences in weight loss efficacy exist between specific diets in the long term.[25,26]

Recent research has started to explore the influence of the energy density of a diet on weight loss. Energy density represents the energy content of a specific food relative to its weight (i.e., calories per gram of food).[27] Manipulation of the energy density of a diet through the addition of water-rich foods (i.e., lowering the energy density through the addition of fruit, vegetables, and soup) without calorie restriction has been shown to result in clinically meaningful reductions in body weight under free-living conditions.[28,29] While acute and short-term studies have shown that energy density can mediate satiety[30] and satiation,[31] further long-term studies are needed that address the mechanisms behind any long-term effects of energy density on body weight.[27]

Although clinically meaningful weight loss can be achieved with a variety of dietary approaches, long-term maintenance of this new body weight remains a serious issue.[32] Based on findings from a systematic review, Curioni and Lourenco[33] reported weight regain at one year to be 50 percent of that initially lost through dietary restriction. Indeed, it has been suggested that only 20 percent of individuals on a dietary treatment will maintain a weight loss of greater than 10 percent of initial body weight after three years.[18]

Physical Activity and Weight Loss

Physical activity is also a commonly prescribed means of promoting body weight loss. This approach to weight management stems from the fact that PA is a modifiable component of energy balance, accounting for approximately 30 percent of total daily energy expenditure in sedentary individuals.[34] As such, manipulating the amount of PA performed can create either an acute negative energy deficit (total daily energy expenditure > daily energy intake) or an acute positive energy balance (daily energy intake > total daily energy expenditure). Prolonged exposure to such imbalances will be reflected in changes in body weight, with a chronic negative balance leading to a reduction in body weight and fat mass.[35]

In 2001, the American College of Sports Medicine (ACSM) released a position statement recommending that overweight/obese individuals perform a minimum of 150 min/wk^{-1} of moderate-intensity PA to improve health.[36] However, for weight loss it was suggested that 200–300 min/wk^{-1} of moderate-intensity PA was needed to achieve long-term reductions in body weight. Importantly, these recommendations made the distinction between the level of PA needed to bring about improvements in general health and to induce weight loss. It is now becoming clear that differing levels of PA are needed to (1) prevent initial weight gain, (2) elicit weight loss, and (3) prevent subsequent weight regain. In a recent update,[35] the ACSM recommend that 150–250 min/wk^{-1} (equivalent 1200 to 2000 kcal/wk^{-1}) of moderate-intensity PA is needed to prevent weight gain (greater than 3 percent of body weight) in adults. This dose of exercise may also be associated with modest weight loss (1–3 kg). To achieve clinically significant weight loss (≥ 5 percent of body weight), it is suggested that moderate-intensity PA should be performed for at least 225 min·wk^{-1}, with a clear dose-dependent relationship existing whereby higher levels of PA facilitate greater weight loss. Following weight loss, the ACSM suggests that 60 minutes or more per day of moderate intensity PA is needed to prevent weight regain.[35] As such, it is apparent that the volume of exercise required to prevent weight regain following periods of intentional weight loss is much greater than that initially required to prevent weight gain.

Diet-Induced and Physical Activity-Induced Weight Loss

The combined efficacy of dietary restriction and increased PA has been recently examined.[26,37,38] These studies have concluded that a small but significantly greater weight loss can be achieved through a combined approach rather than when dietary restriction or exercise is performed alone. For example, Wu, Chen, and Dam[26] examined data from eighteen randomized control trials lasting six months or longer that compare the effects of diet alone versus diet and exercise in overweight individuals. The pooled weight loss for the combined treatments was 1.14 kg greater than that seen with diet alone, which is consistent with the additional weight loss of 1.1 kg reported by Shaw et al.[37] It has been proposed that diet restriction plays the dominant role in weight loss following a combined treatment, with diet contributing to approximately 80 percent of the weight.[39] However, the exact contribution is unknown and may be moderated by the degree of calorie restriction imposed, with exercise only providing additional weight loss when combined with moderate dietary restriction.[35]

It is also important to note that the addition of exercise to dietary restriction will bring about improvements in cardiovascular fitness[40] and metabolic factors such as glucose tolerance[41] and lipoprotein profiles[42] not seen with diet restriction alone. For example, Larson-Meyer et al.[43] examined the cardiometabolic responses in thirty-six overweight participants during six months of isoenergetic calorie restriction (a 25 percent reduction in EI; $n = 12$), calorie restriction and exercise (a 12.5 percent reduction in EI and a 12.5 percent increase in ExEE; $n = 12$), or a weight maintenance diet ($n = 12$). While there was no difference in weight loss between the calorie restriction and combined groups, maximal aerobic capacity (VO_{2max}) improved only in the combined treatment group (22 ± 5 percent; $P < 0.001$). Furthermore, the combined treatment group experienced greater improvements in insulin sensitivity ($P < 0.05$), LDL cholesterol ($P < 0.05$), and diastolic blood pressure ($P < 0.05$) than the diet-only group.

The Impact of Dietary Restriction and Physical Activity on Energy Balance: Implications for the Efficacy of Weight Loss

Given that dietary restriction and increased PA are promoted for weight loss due to their theoretical ability to create energy deficits, it is important to understand how the homeostatic regulatory system responds to perturbations in energy balance. The classic depiction of weight loss occurring by simply increasing energy expenditure (via PA) or decreasing EI (via dietary restriction) to produce a negative energy balance is simplistic in nature.[44] Such a view assumes that the remaining components of energy balance remain static following the manipulation of EI or PA. However, the regulation of energy balance is a dynamic process in which perturbations to one component can produce responses in other areas.[45] These responses may be compensatory in nature and act to minimize disruptions to homeostasis.[46] For example, dietary restriction (alone) is often associated with concurrent reductions in resting metabolic rate,[47] which may attenuate the energy deficit created and subsequent weight loss.[46] Similarly, it is assumed that increased levels of PA or exercise are met with an increased level of EI, again acting to undermine any exercise-induced energy deficit.[48] This notion of a dynamic energy balance system that adjusts to perturbations is not new, with Jean Mayer suggesting fifty years ago[49] that exercise induces compensatory increases in EI to restore energy balance.

The efficacy of dietary restriction and PA must therefore be viewed in the context of this dynamic model, where perturbation to energy balance may elicit behavioral or biological compensatory responses that minimize the energy deficit created.[45,46] While acute periods of exercise-induced energy surfeits are met with a weak regulatory response to restore energy balance, energy deficit appears to trigger a number of potent signals designed to attenuate any imbalance and resist weight loss.[45,46] This asymmetry is noted by Westerterp,[48] who reports based on studies employing doubly labeled water that while overfeeding does not impact PA levels, underfeeding is met with a reduction in energy expenditure (via decreases in resting metabolic rate [RMR], diet-induced, or activity-induced energy expenditure). Therefore, the appetite regulation responses appear to be more sensitive to dietary manipulations compared with exercise interventions.

This "resistance" to weight loss is apparent in studies examining the efficacy of dietary restriction or PA as a means of reducing body weight. The effectiveness of exercise to reduce body weight in overweight and obese individuals has been extensively reviewed.[37–39,50–54] These studies consistently conclude that the degree of weight loss associated with PA or exercise interventions is typically 1.5–3.0 kg. Importantly, the observed weight loss seen is often below that theoretically expected based on objective measures of exercise-induced energy expenditure (ExEE). Indeed, Borer[53] suggests that when performed without dietary restriction, a daily exercise-induced energy expenditure of 400 kcal produces losses in body fat equivalent to approximately one-third of that predicted. Differences between the measured and predicted weight loss have also been reported following dietary restriction. Goele et al.[55] reported that the mean weight loss in forty-eight overweight and obese females (31.5 ± 6.1 years; BMI 35.4 ± 4.4 $kg{\cdot}m^2$) following fourteen weeks of dietary restriction (1,000 $kcal{\cdot}day^{-1}$) was only 44 percent of the predicted value.

Until recently, there has been little attempt to understand this disparity between predicted and actual weight loss. Differences in the adherence to a prescribed dietary or exercise intervention will undoubtedly introduce variability in any biological outcome measured. This will be particularly evident under free-living conditions, even if adherence to the prescribe intervention is monitored.[56–59] For example, Colley et al.[59] monitored adherence to a 16-week free-living exercise intervention in twenty-nine obese women, by comparing the prescribed exercise dose (1500 $kcal{\cdot}wk^{-1}$) to that objectively measured using heart rate monitors. When the prescribed physical activity was unsupervised, participants on average only achieved approximately half of the prescribed weekly exercise dose (768 ± 516 $kcal{\cdot}wk^{-1}$).

While poor adherence can clearly undermine the outcome of an intervention, it cannot fully explain the discrepancies in predicted and actual weight loss. While the mean weight loss was only 44 percent of that predicted based on the prescribed energy deficit in the study by Goele et al.,[55] poor compliance explained only 50 percent of the difference between measured and predicted weight loss. Indeed, recent longer term exercise interventions, which have imposed a tighter control (i.e., monitored) of the ExEE, still report large variability in weight loss.[3,4,60–63] As such, biological or behavioral responses should be expected to interact during periods of energy deficit to modify the outcomes of an intervention. However, despite being a common means of weight control, a sound understanding of how dietary restriction or increased PA effects total energy expenditure within a dynamic and flexible regulatory system is lacking.

Interindividual Variability in Biological and Behavioral Responses

Large interindividual variability in the biological and behavioral responses to PA has been reported, independent of the ExEE.[64,65] These included changes in VO_{2max},[66,67] insulin sensitivity,[68] resting substrate metabolism,[61] and body composition.[3,4,60–63] Bouchard[67] reported a mean increase of 50 percent in aerobic capacity following twenty weeks of endurance training. However, improvements ranged from 16 percent to 97 percent, despite the volume of exercise being consistent between individuals. However, few studies have attempted to characterize this variability and have focused on the group response to an exercise stimulus. Hence, the phenomenon of individual variability has not been exploited. Overlooking the variability makes the incorrect assumption that all individuals will respond similarly and that any subsequent recommendations concerning the use of PA will be appropriate for all individuals. This is not the case.

Clear evidence exists that the body weight response to long-term exercise will vary between individuals, independently of ExEE (see Figure 5.1). King et al.[3] reported highly divergent changes in body weight and fat mass following twelve weeks of supervised aerobic exercise in thirty-five overweight and obese individuals. Mean weight loss over the intervention was 3.7 ± 3.6 kg, which was concordant with the predicted weight loss based on the total ExEE. However, inspection of the individual responses revealed that changes in body weight and fat mass ranged from −14.7 kg to +1.7 kg and −9.5 kg to +2.6 kg, respectively. Indeed, a clear dichotomy in response could be seen when participants were retrospectively classified as either

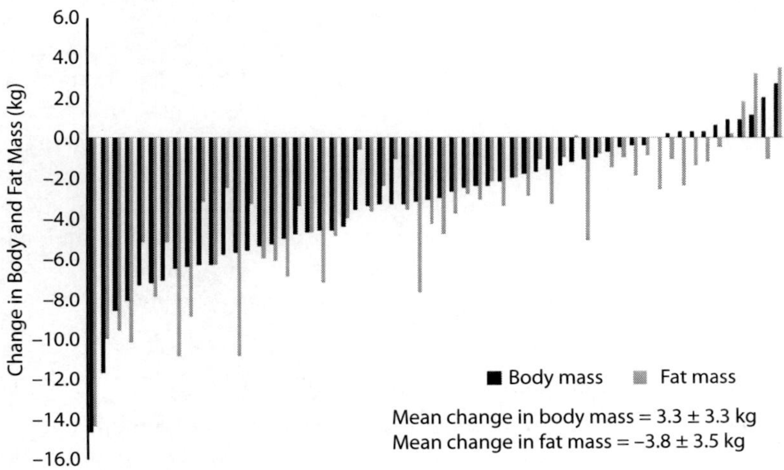

FIGURE 5.1 Individual changes in body mass and fat mass following twelve weeks of supervised aerobic exercise (2500 kcal·wk^{-1}) in fifty-eight overweight and obese individuals. Data are modified from King et al.[3,4]

noncompensators (NCs; $n = 17$) or compensators (Cs; $n = 18$), based on the relationship between actual and predicted weight loss. Compensators lost only -1.5 ± 2.5 kg (approximately half the predicted weight loss), while NCs lost -6.3 ± 3.2 kg. This was independent of total net exercise-induced energy expenditure, which did not differ between groups ($C = 2393 \pm 547$ kcal·wk^{-1}; $NC = 2272 \pm 542$ kcal·wk^{-1}; $P > 0.05$).

Similarly, Barwell et al.[61] reported that following seven weeks of aerobic exercise, large variability in the change in fat mass was evident in fifty-five sedentary women. While the mean loss of fat mass was only -0.97 ± 1.5 kg, examination of the individual responses indicated a range of –5.3 to +2.1 kg. In this instance, the total net ExEE accounted for 36 percent of the variance in fat mass. However, when differences in ExEE were accounted for, wide variation in the residual change in fat mass was still evident (+2.5 to –2.9 kg). These two studies demonstrate the importance of examining the individual responses to exercise, as group or mean data can mask important information regarding the efficacy of an exercise intervention. While the mean weight loss in these studies was consistent with that previously reported (i.e., 1.5–3.0 kg),[37–39,50–54] it is clear that individuals will not respond the same to the same exercise stimuli even under conditions of high compliance.

Biological and Behavioral Compensation to Dietary Restriction and Physical Activity

The phenomenon of individual variability also points to marked differences in the susceptibility (and indeed resistance) to weight loss following periods of dietary- or exercise-induced energy deficits. However, the mechanisms behind this propensity for weight change are ill defined. This may in part been hampered by methodological difficulties in the accurate measurement of EI and energy expenditure,[69] but characterisation of responsive and nonresponsive individuals is imperative if effective strategies for weight control are to be developed.

Recent attempts have been made to identify biological and behavioral factors that may contribute to this variability in weight loss during dietary restriction or increased PA. While studies have identified individual mechanisms relating to metabolic adaptations[61] and behavioral adjustments in EI[3,4,62] and activity energy expenditure nonexercise,[58,60] it is unlikely that such factors will act in isolation and be the sole cause of any resistance to weight loss. Rather, it is likely that individuals experience a number of behavioral and/or biological responses that will act in tandem to influence the response in body weight. Furthermore, this compensatory "profile" will differ between individuals, with certain mechanisms playing a more dominant role in some individuals than others.

Dietary Restriction and Metabolic Compensation

Dietary-induced weight loss has been associated with reductions in total energy expenditure (TEE), associated with reductions in resting metabolic rate,[70,71] the thermic effect of food,[72] and the energy cost of PA after weight loss.[73] Based on the OPEN study,[74] Schoeller[44] suggests that for every 1 kg of body weight lost during dietary restriction, energy expenditure will decrease by 21 kcal·day^{-1} and 15 kcal·day^{-1} in men and women, respectively. However, it has been suggested that this reduction in TEE is often greater than that theoretically expected based on losses of metabolically active tissue (i.e., fat mass and fat-free mass).[75–78] This metabolic response to dietary restriction has been termed "adaptive thermogenesis"[79] and it has been suggested this response may act as a physiological mechanism through which large reductions in weight are defended against during dietary-induced weight loss.[80] Furthermore, the greater than expected reduction in energy expenditure has been shown to persist in weight-reduced individuals who have maintained a lower (stable) body weight for one year, promoting weight regain during this period.[81] However, it should be noted that some have doubted its clinical significance in compensating for the prescribed energy deficit during dietary weight loss.[82]

Adaptive thermogenesis in response to dietary-induced weight loss again appears to be characterized by a high degree of interindividual variability.[80] For example, Weyer et al.[83] reported that, on average, metabolic "overcompensation" in response to long-term spontaneous weight loss in 102 Pima Indians was small. It was estimated that a 15 kg weight change was accompanied by change in 24-hour energy expenditure of 244 $kcal \cdot day^{-1}$, which was 33 $kcal \cdot day^{-1}$ greater than that predicted based on changes in body composition. However, a large standard deviation (±192 $kcal \cdot day^{-1}$) indicated that some individuals experienced large "overcompensatory" responses that would have acted to resist weight loss and encourage weight regain.

The effect of combining exercise and dietary restriction on the adaptive response in metabolism is unclear. For example, Martin et al.[77] and Heilbronn et al.[75] examined the effects of six months of calorie restriction with exercise (12.5 percent calorie restriction plus 12.5 percent increase in energy expenditure via structured exercise) or without exercise (25 percent calorie restriction of baseline energy requirements) on metabolic adaptations in participants from the Pennington CALERIE study. In both cases, the metabolic adaptation to calorie restriction was greater (approximately 6 percent) than that predicted based on changes in body composition, and this addition of exercise to calorie restriction did not affect this response. However, Redman et al.[40] measured metabolic and behavioral compensation in free-living conditions in 48 overweight participants during six months of calorie restriction and exercise (12.5 percent calorie restriction plus 12.5 percent increase in energy expenditure via structured exercise) or calorie restriction alone (25 percent calorie restriction of baseline energy requirements). Using a combination of doubly labeled water and indirect calorimetry, it was observed that TEE (adjusted for sedentary energy expenditure) was significantly less than predicted at six months following calorie restriction (–209 ± 114 $kcal \cdot day^{-1}$). However, no metabolic adaptation was observed following calorie restriction and exercise. This may have important implications for weight regain, as metabolic compensatory adjustments may contribute to the high incidence of weight regain following weight loss.[84]

Physical Activity and Metabolic Compensation

Long-term exercise training produces changes in substrate metabolism at rest and during exercise[85–87] which favor the use of fat as an energy substrate, and individual differences in this variability may mediate the body weight response seen. Following seven weeks of aerobic exercise, Barwell et al.[61] reported large variability in the change in fat mass was reported in fifty-five sedentary women. While the mean fat loss was only −0.97 ± 1.5 kg, examination of the individual responses indicated a range of −5.3 to +2.1 kg. Changes in fasted RQ explained 7 percent of the variance in the fat mass change. As the postintervention RQ measurements were taken 15–24 hours after the final exercise bout, the variability in weight loss may in part reflect differences in postexercise substrate oxidation as well as resting metabolism, as transient increases in fat oxidation can still be seen 24–48 hours after a bout of exercise.[88]

The mediating effect of fasting RQ on body weight during chronic exercise is consistent with the idea that impaired substrate metabolism is a causal factor in the development of obesity and susceptibility to weight regain. For example, a high fasting nonprotein respiratory quotient (RQ) has been found to be predictive of weight gain,[89–91] while formerly obese individuals (who have a higher RQ than never-obese individuals) experience greater weight regain following weight loss.[92–94] Furthermore, obese and formerly obese individuals display an inability to increase fat oxidation in response to increased dietary fat intake (see the "Postprandial Trafficking of Dietary Fat" section of this chapter).

The Effect of Physical Activity on Energy Intake

Given that Barwell et al.[61] found variations in net ExEE accounted for 36 percent of the variance in fat mass, while changes in the fasted RQ explained an additional 7 percent of the variance, this still leaves a large proportion of the variance in fat mass unaccounted for. Intuitively, it could be suggested that changes in EI may also act to mediate the body weight response to exercise and account for some of this unknown variance. In this instance, the change in EI was not a significant predictor of the change in fat mass. However, the food diaries used to assess EI in this study may have lacked sufficient sensitivity to detect small changes in daily EI that could have influenced energy balance and body composition.

Where more sensitive measures of food intake have been used, EI has been shown to be a key determinant of the body weight response to long-term exercise.[3,4] For example, in addition to body composition King et al.[3] also measured subjective rating of appetite and EI during their twelve-week exercise intervention. Significant differences were noted in the direction and magnitude of response between C (who lost −1.5 ± 2.5 kg) and NC (who lost −6.3 ± 3.2 kg) groups. Energy intake and average daily hunger increased by +268 ± 455 $kcal{\cdot}day^{-1}$ and 6.9 ± 11.4 $mm{\cdot}day^{-1}$ in C. However, EI decreased by −130 ± 485 $kcal{\cdot}day^{-1}$ in the NC, while daily hunger remained constant.

A compensatory increase in EI following long-term exercise contradicts the majority of studies examining the acute coupling between a single bout of exercise and EI.[95–98] However, given that the homeostatic regulatory system may be insensitive to acute perturbations to energy balance,[99] but strongly defends against more sustained challenges,[48] acute or short-term studies may fail to capture any compensation that occurs to track increases in ExEE.

Seminal studies by Stubbs demonstrated partial compensation in EI has been observed when exercise is extended over short-term periods.[100,101] Stubbs et al.[100] measured EI in six lean women (BMI = 21.4 ± 1.0 $kg.m^2$) over seven-day periods of no-exercise (0 MJ.day^{-1}), moderate-exercise (1.9 MJ.day^{-1}), and high-exercise (3.4 MJ.day^{-1}) doses. A significant but partial increase in EI was observed, with EI equal to 8.9, 9.2, and 10.0 MJ·day^{-1} for the no-, moderate-, and high-exercise groups, respectively. Approximately 30 percent of the additional energy expenditure caused by the high-exercise dose was compensated for by an increase in EI. While Stubbs et al.[102] failed to observe any compensation in lean males following a similar seven-day period of imposed exercise, Whybrow et al.[101] has again presented evidence of partial compensation in EI to increased ExEE in both males and females. Six lean men and women performed fourteen days of no- (0 MJ·day^{-1}), moderate- (1.5–2.0 MJ·day^{-1}), and high- (3.0–4.0 MJ·day^{-1}) exercise doses. While there was no increase in subjective appetite sensations, partial compensation in EI was observed, again equated to approximately 30 percent of the ExEE.

Taken together, these data appear to capture the initial stages of compensation, whereby EI attempted to track the imposed increases in total energy expenditure. While the degree of compensation was relatively modest and did not fully negate the imposed negative energy balance, if the exercise had been continued for a long period, compensation in EI may have more closely match the elevated levels of energy expenditure. Indeed, Stubbs[103] speculated that it would take several weeks of exercise-induced increments of energy expenditure before an increase in EI occurred. The mechanisms behind this compensatory drive are unclear. As eating behavior is a psychobiological variable,[104] EI can in itself be seen as a behavioral outcome that is driven by a number of physiological and psychological factors. As such, it is important to determine the mechanisms that underlie compensatory eating to long-term exercise, and identify factors that confer susceptibility and resistance to exercise-induced weight loss.

In light of this, King et al.[4] examined the effects of twelve weeks of supervised aerobic exercise on fasting and daily hunger in fifty-eight overweight and obese individuals (BMI = 31.8 ± 4.5 kg/m^2). Large variability in the body weight response to exercise was again observed, with individuals classified as responders or nonresponders based on changes in body composition relative to exercise-induced energy expenditure. Nonresponders ($n = 26$), who only lost 1.0 percent of initial body weight, exhibited a significant increase in fasting ($P < 0.001$) and daily hunger ($P < 0.005$) after the intervention. In contrast, the responders (who lost 5.7 percent of their initial body weight) did not show an increase in daily hunger, despite an increase in fasting hunger ($P < 0.0001$). The satiety quotient[105] associated with a fixed energy meal was measured during the intervention. This approach revealed two processes that acted concurrently to mediate the effect of exercise on appetite regulation (Figure 5.2). While exercise increased the orexigenic drive in the fasted state, postprandial satiety signaling was improved, which was independent of weight loss. This dual process is important as it reflects both changes in homeostatic energy status (orexigenic drive) and the interaction between the homeostatic system and the physiologic action of food on satiety signaling.[4]

These findings are supported by Martins et al.,[62] who measured fasting and postprandial levels of orexigenic (total and acylated ghrelin) and anorexigenic (peptide YY, glucagon-like-peptide-1) peptides in fifteen overweight or obese individuals during a twelve-week supervised exercise program. In the fasted state, significant

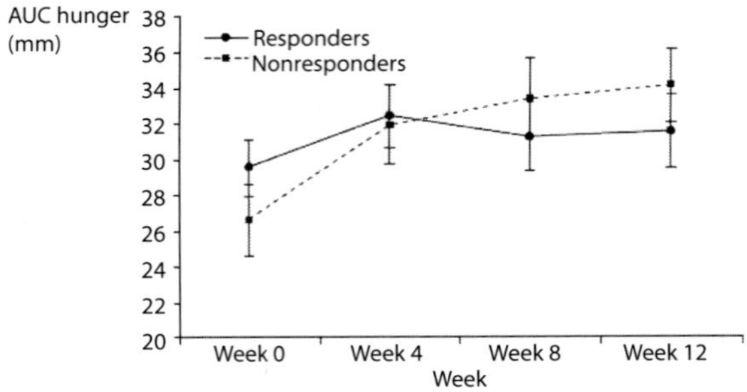

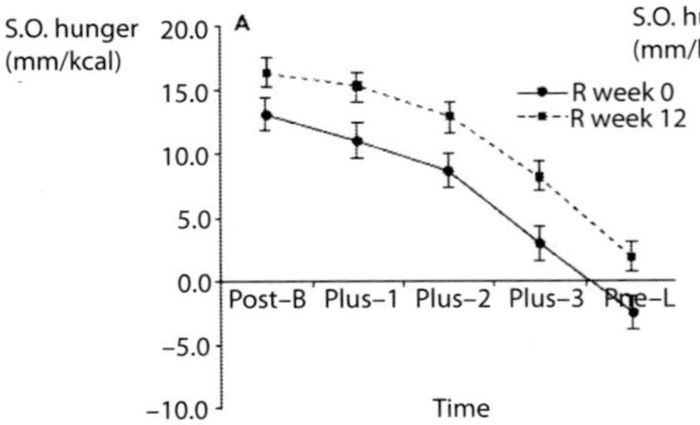

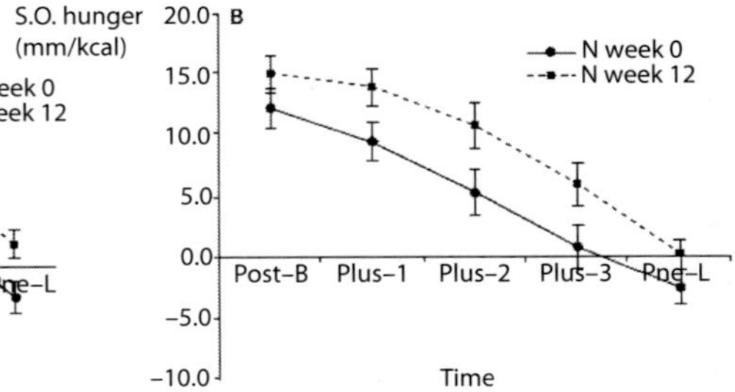

FIGURE 5.2 Top panel: Mean (±SEM) change in daily area under the curve (AUC) for hunger over twelve weeks in the responders and the nonresponders. Bottom panels: Mean (±SEM) immediate and subsequent satiety quotients (SQs) in response to the fixed breakfast preloads at weeks 0 and 12 in responders (R; n = 32; A) and nonresponders (NR; n = 26; B). Data from King et al.[4]

increases in subjective hunger ($P < 0.01$) and desire to eat ($P < 0.05$) were reported at the end of the intervention, while feeling of fullness significantly decreased ($P < 0.01$). It was suggested that these changes in subjective appetite sensations were mediated by changes in orexigenic peptides, as fasting levels of acylated ghrelin increased significantly ($P = 0.011$) post intervention. However, this increase in exercise-induced fasting orexigenic drive again appeared to be offset by an improved satiety response to a fixed meal. There was a significant increase in the suppression of acylated ghrelin following a fixed meal and a tendency toward an increase in the late postprandial release (90–180 mins) of glucagon-like-peptide-1. These hormonal responses would have acted to augment satiety during the postprandial period. It should be noted, though, that as the participants lost weight during the intervention (–3.5 kg), the effects of exercise per se cannot be discerned from the effects of weight loss.

Nonexercise Activity Thermogenesis

Given the dynamic nature of the components of energy balance, behavioral compensation to dietary restriction or increased PA may also be expressed through other areas aside from eating behavior. One such area is the energy cost associated with nonexercise activity thermogenesis (NEAT). NEAT includes the energy cost of all activities (outside of purposeful exercise) and includes the energy expenditure of work and leisure activities, maintenance of posture, fidgeting, etc.[106,107] NEAT is the predominant component of activity thermogenesis, and substantial inter- and intravariability exists in NEAT.[108] As such, it has been suggested that differences in habitual levels of NEAT may play a role in the susceptibility to weight gain,[109] potentially acting as a behavioral compensatory mechanism in response to perturbations to energy balance.[110]

Dietary Restriction and Nonexercise Activity Thermogenesis

Evidence suggests that dietary restriction results in a reduction in activity-induced energy expenditure, particularly when large energy deficits are created.[48] For example, during the Minnesota semistarvation study,[111] thirty-two normal-weight young men voluntarily undertook severe energy restriction for 24 weeks (6.6 $MJ \cdot d^{-1}$ from a maintenance requirement of 14.6 $MJ \cdot d^{-1}$). By the end of this period, participants had achieved a new state of energy balance, with 57 percent of the reduction in total energy expenditure coming from activity-induced energy expenditure. Of this, 40 percent was due the reduced body weight, and 60 percent was due to reductions in free-living PA levels. As such, it appears that reduction in activity-induced energy expenditure may act as a potent auto-regulatory response to help reestablish energy balance during periods of energy restriction and protect against weight loss.[69] While it should be noted that the energy restriction in this study was large, reductions in free-living PA have also been reported in studies with more moderate calorie restrictions.[40,112,113] For example, using doubly labeled water, Redman et al.[40] reported that part of the greater than predicted reduction in total energy expenditure (based on changes in fat mass and fat-free mass) following six months of calorie restriction (25 percent of baseline energy requirements) was due to a significant reduction in free-living PA levels.

Physical Activity and Nonexercise Activity Thermogenesis

Until recently, scientific evidence to support a reduction in NEAT following exercise training has been lacking. However, Manthou et al.[60] reported differences in the change in NEAT between those who lost more or less than the predicted loss in fat mass following eight weeks of supervised exercise. Based on a net ExEE of 30.2 ± 12.6 MJ over the intervention, predicted loss in fat mass should have been 0.8 ± 0.2 kg. However, no mean loss in fat mass was observed (0.0 ± 0.2 kg). The individual responses in fat mass ranged from −3.2 kg to +2.6 kg, with 11 participants losing more than or equal to their predicted fat mass (responders; 5.6 percent reduction in initial body fat) and 23 participants losing less than the expected weight loss (nonresponders; 1.9 percent reduction in initial body fat). As compliance to the exercise intervention was 100 percent, compensation must have occurred to undermine losses in fat mass. Indeed, while there was no difference in total ExEE between responders and nonresponders, activity energy expenditure (energy expenditure of all activity excluding ExEE) increased in the responders (+0.79 ± 0.50 $MJ \cdot d^{-1}$) but decreased in the nonresponders (−0.62 ± 0.39 $MJ \cdot d^{-1}$). As such, it would appear that the nonresponders were compensating for the increased energy expenditure induced by the exercise by becoming less active in other areas of their daily routine.

These findings have been recently supported by Colley et al.,[58] who used doubly labeled water and measured total energy expenditure in seven obese women during an eight-week walking intervention. The prescribed volume of exercise (1500 kcal·wk^{-1}) should have increased total daily energy expenditure by 215 kcal·day^{-1}. However, no change in total daily energy expenditure was actually observed, despite a mean ExEE of 1434 ± 237 kcal·wk^{-1}. This was in part accounted for by a significant reduction in NEAT, equal to 175 kcal·day^{-1} (–22 percent).

The Temporal Profile and Stability of Compensation: Is the Magnitude of Compensation Dose Dependent?

It is unknown whether behavioral or biological compensation is evident at the start of a dietary or exercise intervention, or whether compensation occurs when a specific threshold in weight loss is exceeded. There is some evidence to suggest that the magnitude of compensation seen in response to exercise training may relate to the dose (total ExEE) employed. Church et al.[63] examined the effect of exercise dose on compensatory responses by comparing actual to predicted weight loss across different levels of ExEE. Female participants performed six months of supervised exercise in which they were randomly assigned to one of three groups that expended 4 (72.2 ± 12.4 min·wk^{-1}; n = 139), 8 (136.3 ± 19.4 min·wk^{-1}; n = 85), or 12 (193.7 ± 31.0 min·wk^{-1}; n = 93) kcal·kg·wk^{-1}. Adherence to exercise in each group was > 99 percent, and in the 4 and 8 kcal·kg·wk^{-1} groups, actual weight loss closely mirrored that predicted (Figure 5.3). However, in the 12 kcal·kg·wk^{-1} group, actual weight loss was significantly lower than the predicted weight loss (–1.5 kg vs. –2.7 kg, respectively;

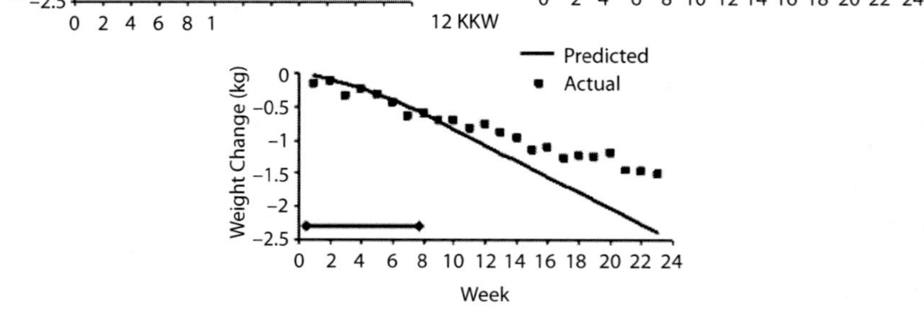

FIGURE 5.3 Predicted and actual weight loss observed in response to six months of exercise at 4, 8, or 12 kcal·kg·wk^{-1}. A similar pattern of weight loss is observed between groups up until week 10. At week 12, the twelve kcal·kg·wk^{-1} group display marked compensation to exercise (as demonstrated by predicted weight loss exceeding actual). Data from Church et al.[63]

$P < 0.05$). Consequently, weight loss was higher in the 8 kcal·kg·wk^{-1} group (–2.1 kg) than the 12 kcal·kg·wk^{-1} group (–1.5 kg), despite a lower total ExEE.

While this suggests that higher exercise doses may elicit greater compensation, negating the effects of a higher ExEE on energy balance, the mechanisms behind the compensation seen in this study were not explored. Indeed, it is worth noting that even within each exercise dose group, compensation between individuals varied widely. While the prevalence of compensation was 72.6 percent in the 12 kcal·kg·wk^{-1} group, the prevalence rate in the 4 and 8 kcal·kg·wk^{-1} groups was still 54.3 percent and 52.8 percent, respectively. There was little separation in weight loss between the three groups during the first ten weeks. At week 10, weight loss in participants in the 12 kcal·kg·wk^{-1} group started to plateau, suggesting that compensation of some form occurred (see Figure 5.3). This may suggest that there is a critical threshold in terms of the degree to which energy balance is perturbed, and once this level has been exceeded, the homeostatic regulatory system triggers compensatory or corrective adjustments to minimize the extent of further perturbations.

The temporal pattern and threshold of compensation in relation to the start of an intervention have not been well studied. Senechal et al.[114] examined the timing and stability of compensatory change in RMR that occurs with dietary-induced weight loss in twenty obese women. While there was no mean change in RMR at the end of the fifteen-week intervention, individual changes in RMR ranged from –320 to +330 kcal·day^{-1}. Interesting, at week 5, there was variability in changes in RMR. Eight individuals experienced a > 5 percent increase in baseline RMR, while eight experienced a > 5 percent decrease in baseline RMR; RMR remained stable in four. After week 5, the direction of change in RMR remained stable until the end of the intervention. Theoretically, an increase in RMR during the early phase of an exercise and/or dietary program would be beneficial for weight loss (and subsequent weight maintenance), providing that such metabolic adaptations were not compensated for in other areas contributing to energy balance (e.g., EI, NEAT, and physical activity levels). While Senechal et al.[114] reported that the direction of response was related to baseline RMR and fat mass, and sympathetic nervous system activity during the first five weeks, further research is needed to address such variability and the impact it has on weight loss and subsequent weight maintenance.

Conclusions

Dietary restriction and increased physical activity are common methods of promoting body weight loss. However, despite this, knowledge concerning the impact of dietary restriction or increased PA on the regulatory control of energy balance is limited. Rather than being a static model, perturbations to energy balance caused by either dietary restriction of increased PA can result in corrective responses in other components of the energy balance equations.[45,46] These compensatory responses can be seen as an auto-regulatory response designed to reestablish energy balance and protect against sustained losses in body weight.[69]

Compensation can be biological and/or behavioral, with evidence suggesting that adaptations can occur in physiological processes such as resting metabolic rate[40,61,76] or the energy cost of exercise,[73] and behavioral outcomes such as EI[3,4,62] or NEAT.[58,60]

Furthermore, the compensatory response to dietary restriction or exercise is highly variable,[3,4,55,60,61,63] and these responses interact to determine the propensity for weight change. As such, it is clear that not all individuals will respond similarly to the same dietary or exercise intervention. While intuitive, this notion has not been reflected in the standard approach to the scientific study or prescription of diet and exercise for weight control. As differences in adherence to an intervention cannot always explain this variability, further research is needed to better understand the factors that drive compensation via EI or other biological and behavioral pathways. Greater knowledge of the mechanisms that confer susceptible and resistances to exercise-induced weight loss will provide more effective dietary and exercise programs.

SUBSTRATE METABOLISM, WEIGHT GAIN, AND APPETITE CONTROL

Introduction

In the previous section, the roles of various biological and behavioral compensatory mechanisms were discussed in relation to the efficacy of PA and diet to induce weight loss. While it is clear that the PA and dietary restriction will not provoke the same physiological or behavioral response in individuals, the nature of compensation that underlines this variability is likely to be multifaceted. Resistance to weight loss will be expressed through a number of compensatory pathways, and consistent with the notion of individual responsiveness, these pathways will likely differ between individuals. For example, some may display metabolic resistance to weight loss, expressed through reductions in RMR[40,75–77] or blunted adaptations in substrate metabolism following dietary restriction or exercise.[61] Others may be particularly susceptible to compensatory adjustments in eating behavior, mediated through homeostatic or nonhomeostatic drivers following long-term exercise.[3,4,62]

Notwithstanding this, certain compensatory responses may have a more potent effect on energy balance, due to their capacity to undermine any energy deficit created. For example, while exercise-induced changes in RMR, substrate metabolism, and diet-induced thermogenesis have the potential to influence total daily energy expenditure, the impact on energy balance is small in comparison to the degree of energy expended on a daily basis. However, changes in EI have the capacity to automatically alter energy balance, as it is possible to create large disparities in energy balance through the consumption or abstinence from energy-dense foods. Indeed, it has been suggested that increased EI at the population level is the key driver of the current obesity epidemic as levels of PA have remained unchanged since the 1980s.[10,11] As such, a better understanding of the regulation of appetite and EI is warranted.

A major tenet of eating behavior is that the control of appetite is a psychobiological process, in which physiological mediators act as drivers of behavior.[104] Appetite regulation is commonly described using the Satiety Cascade (see Figure 5.4).[115] This is particularly important when considering the effects of PA or diet on the regulation body weight. Exercise and diet both provide strong metabolic and hormonal stimuli for behavior change.[104] However, dietary-induced stimuli are more potent. While PA and diet elicit changes in substrate metabolism, the impact of substrate

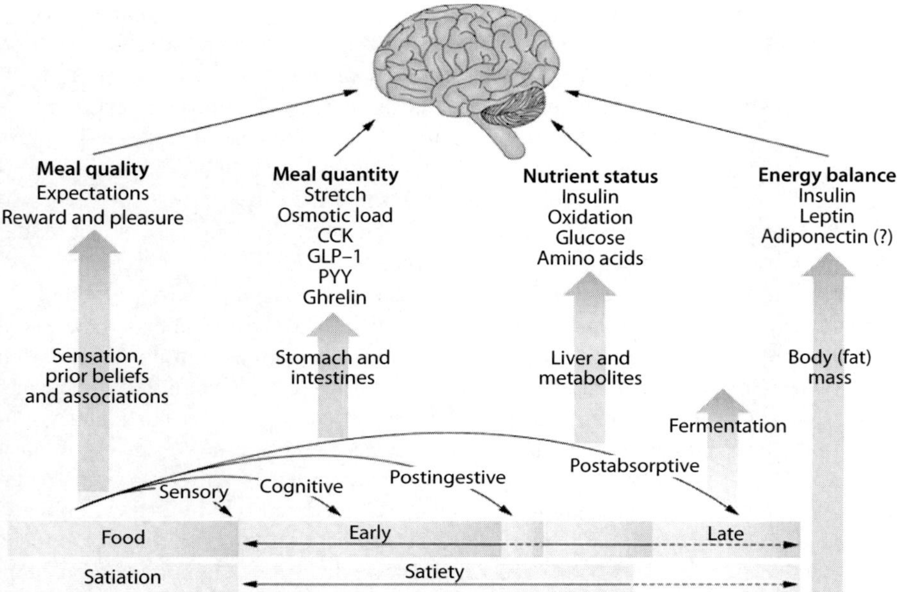

FIGURE 5.4 (*See color insert.*) Behavioral and physiological events that occur following food intake. Satiety cascade from J. Blundell modified by D. Mela. Blundell et al.[115]

oxidation on eating behavior has received less attention. Indeed, the glycogenostatic model[116–118] suggests that glycogen availability is central in regulating food intake, with eating behavior designed to maintain CHO stores (see "Substrate Metabolism as a Physiological Driver of Eating Behavior" in this section). Therefore, manipulations in substrate oxidation could serve as a mediator between exercise- and dietary-induced changes in eating behavior. Furthermore, the metabolic response during the postprandial period following food intake has been implicated in the susceptibility to weight gain.[119,120] As such, the interaction between PA, diet, and substrate metabolism may have important implications for obesity.

Postprandial Trafficking of Dietary Fat

Individuals spend much of their time in a postprandial rather than fasted state. Consequently, the metabolic response to dietary food intake may play an important role in determining an individual's susceptibility to overconsumption and weight gain.[119] Recent attention has been given to the role of nutrient trafficking, and in particular the metabolic fate of dietary fat, in the development of obesity.[120] "Nutrient trafficking" refers to the delivery and redistribution of dietary nutrients to specific tissues and indicates whether a particular dietary nutrient is being oxidized or stored.[119] A propensity for the partitioning of dietary fat for storage rather than oxidation would promote weight gain via positive fat and energy balances.

Bessesen et al.[119] also suggest that nutrient trafficking may also influence the capacity of the homeostatic regulatory system to accurately sense perturbations to fat balance. While there is tight coupling between dietary intake and CHO and protein oxidation, Bessesen et al.[119] suggest that the signal generated by dietary fat ingestion is weak. Unlike for CHO and protein, large disparities exist between the ratio of ingested to stored fat. Consequently, acute fat intake causes minimal perturbations to whole-body adipose levels, thereby eliciting a weak biological signal to increase oxidation to match intake. However, preferential delivery of dietary fat to metabolically active tissue (liver and skeletal muscle), where the disparity between ingested to stored fat is much lower, will enhance the regulatory drive to increase fat oxidation and improved coupling between oxidation and intake.[119]

Obese and formally obese individuals appear to be metabolically vulnerable to high-fat feeding, due to an inability to increase fat oxidation in response to increased fat intake.[121–126] For example, Blaak et al.[124] found that increases in postprandial fat oxidation following a high-fat meal (95 percent fat) were attenuated with increasing BMI. This inability to upregulate postprandial fat oxidation was particularly evident in those displaying low fasting fat oxidation. Furthermore, Westerterp et al.[120] used deuterated palmitic acid delivered in a breakfast to track dietary fat oxidation over twelve hours. The amount of dietary fat oxidized was highly variable and ranged from 4 percent to 28 percent of the fat provided in the breakfast. This inter-individual variability was in part accounted for by differences in body composition, with dietary fat oxidation being negatively correlated with BMI ($r = -0.58$).

Differences in nutrient trafficking also persist in weight-reduced individuals. Astrup et al.[121] studied the effects of isocaloric low- (20 percent), medium- (30 percent), and high- (50 percent) fat diets (three days) on twenty-four-hour energy expenditure and nutrient balances in nine postobese and nine weight-matched lean women. Although no group differences existed following the low-fat diet, the post-obese women failed to increase fat oxidation to a level that matched fat intake on the high-fat diet. While the control group remained in fat balance, the postobese exhibited preferential fat storage (at a rate of 11 g·day^{-1}), lower twenty-four-hour energy expenditure ($P = 0.02$), and a negative CHO balance (postobese −41.8 g·day^{-1}). It was postulated that not only would the increased drive for fat storage negatively impact on long-term weight gain, but also a negative CHO balance would augment glycogen depletion.

Individual differences in the postprandial handling of dietary nutrients may also act as a risk factor, promoting weight gain.[125,127] Giacco et al.[127] found that lean individuals (n = 8) with a family history of obesity (both parents having a BMI ≥25.0 kg/m^2) exhibited attenuated levels of fat oxidation in response to a high-fat meal (51 percent fat, 34 percent CHO, and 15 percent protein) compared to matched lean individuals (n = 8) with no history of family obesity. It was suggested that the postprandial metabolic response in those with a family history of obesity would confer susceptibility to weight gain or obesity due to the augmented storage of adipose when exposed to a habitual diet high in fat.

The role that PA may play in mediating the trafficking of dietary nutrients, and in particular dietary fat, has received little attention to date. Long-term exercise training has been shown to increase fat oxidation at rest[128] and during exercise.[85,87] However, whether regular exercise alters the postprandial trafficking of fat is unknown. Given the upregulation in fat oxidation following exercise training, it is feasible to suggest that regular exercise may help encourage the oxidation of fat and attenuate storage. It has been shown that imposed periods of physical inactivity influences nutrient trafficking. Bergouignan et al.[129] examined dietary oleate and palmitate trafficking in eight lean women undertaking two months of imposed bed rest. While bed rest had no effect on oleate oxidation (a monounsaturated fat), physical inactivity encouraged the partitioning of palmitate (a saturated fat) toward storage, promoting its accumulation in skeletal muscle.

As such, it appears that the metabolic fate of dietary fat may play an important role in determining the propensity for weight gain. A preference toward storage of dietary fat, as has been observed in obese and formally obese individuals, would encourage the accumulation of adipose tissue and intramuscular triglycerides. While the intraindividual variability in nutrient trafficking has not been reported, given the inherent variability associated with markers of resting[89,130] and exercise substrate metabolism,[131–133] it is likely that this also demonstrates a high degree of variability. Again, this will act to confer susceptibility or resistance to weight gain and weight loss. Theoretically, it can be suggested that exercise might have a positive effect on nutrient trafficking, with the upregulation of resting and exercising fat oxidation observed following exercise training extending to the postprandial period and augmenting the oxidation rather than storage of fat.

Substrate Metabolism as a Physiological Driver of Eating Behavior

The role of nutrient trafficking in determining weight gain highlights the potential importance of substrate metabolism in the complex interaction between PA, diet, and obesity. It is known that substrate oxidation during fasted, postprandial, and exercise conditions is influenced by nutritional status.[134] However, the idea that substrate metabolism may stimulate behavioral changes in EI, satiety, and food preference has received less attention. Indeed, rather than being simply a response to exercise or food intake, substrate metabolism may also act as a biological driver of eating behavior.

It is intuitive that the metabolic fate of dietary macronutrients will confer properties on behavior in the interests of regulating energy balance and fuel stores. This forms part of a psychobiological approach to appetite control in which physiological mediators act as drivers of behavior. Substrate metabolism has long been implicated in the energostatic control of food intake, in which increased fatty acid oxidation (FAO) is thought to reduce EI via the maintenance of postmeal satiety.[135] While the exact mechanisms are poorly understood, changes in hepatic energy status (hepatocellular ATP:ADP ratio) resulting from altered FAO may influence EI via the stimulation of vagal afferent nerve activity.[136] Although hepatic FAO has been suggested to be pivotal in this control, Langhans[137,138] has recently suggested that FAO in intestinal enterocytes may also be important (although this has been challenged).[139] Scharrer and Langhans[140] initially demonstrated that mercaptoacetate, a CoA-dehydrogenase inhibitor that suppresses FAO, stimulated EI in rats fed a high-fat diet. Increased EI following the pharmacological inhibition of FAO has been consistently replicated using other substances such as methyl-palmoxirate and etomoxir.[135] Recently, Gatta et al.[141] demonstrated that the administration of (-)-hydroxycitrate (HCA) in time-blinded males enhanced postmeal satiety and delayed subsequent meal requests. The authors attributed this effect to an HCA-induced increase in non-estrified fatty acid (NEFA) oxidation, although as substrate oxidation was not directly measured, the exact mechanisms remain unclear. Furthermore, efforts to suppress EI by the stimulation of FAO have failed to consistently show an effect on eating behavior, potentially as whole-body changes in FAO are buffered by endogenous adipose pools, thereby failing to elicit any appetite regulating signal.[135,137]

However, peripheral changes in nutrient availability may still influence EI. The control of EI has often been described using a negative feedback model, whereby perturbations to energy balance trigger corrective responses (EI) to restore homeostasis (energy availability or body weight).[142] Due to its limited storage capacity, carbohydrate (CHO) balance is tightly regulated and its restoration is a high metabolic priority following depletion.[119] Indeed, daily oxidation rates are equivalent to that consumed via daily dietary intake.[143] As such, large perturbations to CHO stores can be readily achieved through dietary restriction or aerobic exercise. For example, whole-body CHO storage is typically 400–800 g,[143] while ninety minutes of high-intensity exercise could induce a total CHO oxidation of a similar order.[144] This may have implications for the peripheral control of short-term EI, as dynamic changes in blood glucose (glucostatic theory)[49] and glycogen availability (glycogenostatic theory)[117] have been linked to satiety, meal initiation, and EI.

The glycogenostatic theory[116–118] suggests that glycogen availability is critical in determining EI. As whole-body CHO storage is approximately equal to that oxidized daily, the drive to select food to maintain these stores is a major determinant of daily EI. Flatt[116–118] suggested that eating behavior is designed to maintain these CHO stores at a specific set point, with any challenges to availability strongly defended. Carbohydrate depletion (via diet or exercise) acts as an internal biological cue to trigger a strong negative feedback response that elicits compensatory eating to restore availability.[117] This theory was based around a negative relationship between CHO balance on one day and *ad libitum* EI on the following day in mice.[117] Since the theory was proposed, there have been numerous attempts to replicate these findings in humans. These studies have manipulated glycogen availability over short-term periods (1–3 days) using exercise, diet, or both. However, attempts have often produced equivocal findings.

Due to the increased energy demand, a bout of aerobic exercise will attenuate muscle and liver glycogen availability. Almeras et al.[145] reported that immediate postexercise compensatory eating was related to substrate metabolism during preceding exercise bout. In line with similar studies,[95,96,98] no mean increase in EI was reported in lean men (n = 11) following ninety minutes of cycling (60 percent VO_{2max}). However, when participants were divided into "high" or "low" fat oxidizers based on their exercise RQ, postexercise EI was significantly lower ($P < 0.05$) in the high fat oxidizers. Exercise induced a –1.7 MJ net energy deficit in the high fat oxidizers, but a positive energy balance of a similar order in the low fat oxidizers. It was suggested that the postexercise orexigenic drive was lower in the high fat oxidizers, as a lower exercise RQ would have attenuated any exercise-induced glycogen depletion. However, as muscle glycogen levels were not measured, this can only be inferred. As the mean difference in RQ between the two groups was small (0.02), differences in the exercise-induced perturbation to muscle glycogen would likely also be small. Indeed, as habitual dietary intake was not controlled, it cannot be ruled out that the higher RQ in the high-fat oxidizers was actually driven by lower initial levels of glycogen. Despite this, this study does suggest that substrate availability may act as a driver of EI.

In a series of studies, Stubbs et al.[146] manipulated CHO balance using one-day isoenergetic depletion (3 percent CHO) or control (47 percent CHO) diets, with *ad libitum* EI being assessed over the subsequent day. Similarly, Shetty et al.[147] manipulated glycogen availability using two-day isoenergetic diets composed of 79 percent, 48 percent, or 9 percent CHO, with *ad libitum* EI measured over the subsequent two days. In both cases, the dietary-induced manipulation of glycogen availability (which was not measured) had no impact on EI. Instead, dietary CHO changes were compensated for by alterations in substrate oxidization that reestablished CHO balance. However, in a third study, Stubbs et al.[148] manipulated glycogen availability using *ad libitum* 20 percent, 40 percent, and 60 percent fat diets over a seven-day period. A negative relationship ($P = 0.0082$) between CHO balance on one day and EI on the following day was observed, accounting for 5–10 percent of the variance in EI. While again inferences are made concerning glycogen availability based on short-term CHO balance, this does suggest that glycogen availability provides a modest feedback mechanism through which EI is regulated. However, any increase in orexigenic drive arising from energy availability may take time before it is reflected in eating behavior. This is consistent with the notion that the homeostatic regulatory system is not sensitive to immediate changes in energy balance,[99] with external or nonhomeostatic factors influencing whether such perturbations are expressed behaviorally.

Snitker et al.[149] has also reported a relationship between CHO balance and EI. Exhaustive exercise and three-day high- (75 percent) or low- (10 percent) CHO diets were used to manipulate glycogen stores ($n = 8$), with *ad libitum* EI assessed for two subsequent days. Despite a 46 ± 21 percent difference in muscle glycogen, no treatment differences in EI were observed. However, EI on the second day of feeding was negatively correlated ($P = 0.03$) with CHO balance on the first day, accounting for 9 percent of the variance in EI. This suggests that CHO balance may affect the short-term regulation of EI, but there is a delay before this is manifested in food intake. The partitioning of CHO for either storage or oxidation would influence glycogen availability, with habitual oxidation maintaining CHO stores at a lower level. In turn, this may elicit compensation to augment these stores. Furthermore, Sparti et al.[150] examined the effects of acute energy and CHO deficit on subsequent *ad libitum* feeding in healthy men. Following a day of CHO deprivation (and fat and protein intake at one-third of baseline intake), participants consumed foods of either high- or low-CHO composition in a cross-over design. Ad libitum intake on these days indicated that food intake was regulated to restore CHO balance rather than energy balance.

Recently, Pannacciulli et al.[151] also reported that twenty-four-hour CHO oxidation predicted *ad libitum* EI, independent of twenty-four-hour energy balance. Energy and nutrient balances were measured in energy stable individuals (sixty-seven men and forty-five women; mean body fat = 31 ± 8 percent), with *ad libitum* EI measured for three days. It was found that twenty-four-hour RQ ($r = 0.28$; $P < 0.005$), twenty-four-hour CHO oxidation ($r = 0.40$; $P < 0.0001$), and twenty-four-hour CHO balance

($r = -0.34$; $P < 0.001$) all predicted subsequent EI. Furthermore, weight gain over the feeding period (1.0 ± 1.1 kg; range = −1.2 to +4.9 kg) was positively correlated with twenty-four-hour CHO oxidation ($r = 0.23$; $P = 0.01$) and negatively correlated with twenty-four-hour CHO balance ($r = -0.20$; $P = 0.03$). Galgani et al.[152] also found that CHO balance was related to *ad libitum* EI following an isoenergetic switch from a high-CHO diet to a high-fat diet. Galgani et al.[152] measured nutrient balance in a respiratory chamber for four days in fifty-eight individuals. On the first day, participants consumed a diet consisting of 49 percent CHO, which was then followed by a three-day high-fat diet (50 percent fat). Ad libitum food intake was then measured over a four-hour period following the high-fat diet. As would be expected, switching to the high-fat diet resulted in a significant reduction in twenty-four-hour RQ (0.87 ± 0.02 to 0.83 ± 0.02; $P < 0.0001$). It was also found that CHO balance over the four-day period was a significant predictor of subsequent *ad libitum* EI ($R^2 = 0.10$; $P = 0.01$).

While the findings are not always in agreement, it appears that the regulation of CHO may play a role in the regulation of food intake. The findings that a lower or negative CHO balance predicts increased *ad libitum* EI following periods of dietary intervention are consistent with the glycogenostatic theory of appetite control. Those who display or respond to a dietary-induced reduction (i.e., negative) CHO balance exhibit higher CHO oxidation relative to CHO intake. As such, this will act to attenuate storage and potentially trigger increased EI. It should be noted however, that while the findings of Stubbs et al.,[148] Snitker et al.,[149] Pannacciulli et al.,[151] and Galgani et al.[152] all point to a role for CHO availability in appetite regulation, the variance in eating behavior that this accounts for is often small (<10 percent). As such, a large proportion of the variance in these studies is often unexplained. This is again consistent with the idea that hedonic signals may act to override such biological regulation. This effect would be particularly evident in a real-world setting, where the artificial conditions of laboratory testing are not present. However, despite this there is evidence from long-term studies that CHO availability does play a role in regulating food intake and body weight.

Eckel et al.[153] reported that a negative CHO balance (which would promote glycogen depletion) was predictive of long-term weight gain. Energy balance was measured in thirty-nine participants following either a fifteen-day isocaloric high-fat (50 percent fat) or high-CHO diet (55 percent CHO). Body composition was subsequently tracked for four years. There was a significant inverse relationship between CHO balance following the high-CHO diet and the change in fat mass over the follow-up period ($r = -0.46$; $P = 0.007$). Individuals who had the highest positive twenty-four-hour CHO balance (indicative of CHO storage) following the high-CHO diet gained less body mass ($P = 0.024$), fat mass ($P = 0.001$), and percentage body fat ($P = 0.006$) over the follow-up period. It was suggested that these individuals were less susceptible to weight gain due to their ability to maintain glycogen levels within a set range. While it should be noted that CHO balances do not necessarily reflect stored glycogen availability, it could be argued that a habitual negative CHO balance would attenuate availability as a reduced amount of dietary CHO intake would be available for storage. Based on Flatt's model, the metabolic drive to increase EI to replenish glycogen stores would be attenuated due to higher habitual storage. However, given the longitudinal nature of the study, it is difficult to determine whether CHO balance represented a biological marker of divergent behavior or the mechanisms through which susceptibility was conferred.

These data are consistent with studies showing that impaired resting fat oxidation is a causal factor in the development of obesity, with a high resting RQ reported to predict weight gain.[89–91] For example, Zurlo et al.[89] reported in 111 Pima Indians that those with a high resting RQ (0.87; 90th percentile) were 2.5 times more likely to gain 5 kg or more over a 25- (± 11-) month follow-up period than those with a lower resting RQ (0.82; 10th percentile). Similarly, Seidell et al.[90] followed 775 men (aged 18–98 years) over ten years and found that resting RQ at the start of the study was positively correlated with weight gain. While the mean weight gain was only 0.07 ± 6.4 kg, 122 men (15 percent) gained more than 5 kg, while forty men (5 percent) gained more than 10 kg. The adjusted relative risk of gaining 5 kg or more in initially nonobese men with a fasting RQ of ≥ 0.85 was 2.42 times that of men with a RQ ≤ 0.76. One mechanism through which a higher resting RQ may contribute to elevated weight gain may be associated with a metabolically induced increase in EI. A higher twenty-four-hour RQ (indicative of higher CHO oxidation) may attenuate glycogen stores and increase the orexigenic drive.

Conclusions

Substrate metabolism has the capacity to influence weight gain through a number of different processes. While an impaired capacity to oxidize fat under fasting conditions has been linked to long-term weight gain,[89–91] the partitioning of dietary food, and in particular fat, during the postprandial period has also been implicated in the susceptibility to overconsumption and weight gain.[119] A stronger preference toward the storage of dietary fat has been observed in obese and formally obese individuals, which would encourage the accumulation of adipose tissue and intramuscular triglycerides.

Furthermore, rather than being simply a response to exercise or food intake, substrate metabolism may also play a role in the regulation of EI. The glycogenostatic theory, as proposed by Flatt,[116–118] suggests that glycogen availability may be central in eliciting negative feedback signals to restore energy homeostasis. As glycogen availability is tightly regulated, depletion of these stores may act as a biological cue to stimulate EI in an effort to restore glycogen levels to a predetermined set point.[116–118] As exercise can alter substrate metabolism and availability,[134] depletion of such energy stores may act a mechanism through which exercise-induced compensatory eating is driven. Evidence to suggest a direct link between substrate metabolism during acute exercise and compensatory eating is limited,[145,154] and primarily based on isolated evidence which retrospectively categorized of "high" and "low" fat oxidizers. Indeed, despite the attraction of the glycogenostatic model, short-term manipulations of glycogen stores via exercise or diet in humans have not produced consistent findings.[146–149]

Despite large intra- and interday fluctuations in glycogen stores, there is no consistent evidence to strongly support a direct role of glycogen level in the acute regulation of appetite and EI. The detection of this relationship will depend on the experimental protocol, the magnitude of glycogen depletion, and the sensitivity to measure adaptive changes in behavior (e.g., EI). However, recent evidence suggests that CHO balance plays a role in the short term, regulating EI and body weight.[151–153] A negative CHO balance, which indicates that oxidation exceeds dietary intake and would attenuate the amount stored, has been shown to be predictive of greater *ad libitum* feeding.[148,149,151,152] Furthermore, a negative carbohydrate balance has also been shown to be predictive of weight gain,[153] but again further research is needed to support these findings. As such, whatever the theoretical mechanism adopted to understand the operation of the psychobiological system, it appears that substrate metabolism mediates between diet and physical activity and their effects on appetite control and body weight.

REFERENCES

1. Macdiarmid, J., Blundell, J. Assessing dietary intake: who, what and why of under-reporting. *Nutrition Research Reviews*. 1998;11(02):231–53.
2. Blundell, J., Stubbs, R., Golding, C., Croden, F., Alam, R., Whybrow, S., et al. Resistance and susceptibility to weight gain: individual variability in response to a high-fat diet. *Physiology and Behavior*. 2005;86(5):614–22.
3. King, N., Hopkins, M., Caudwell, P., Stubbs, R., Blundell, J. Individual variability following 12 weeks of supervised exercise: identification and characterization of compensation for exercise-induced weight loss. *International Journal of Obesity*. 2008;32(1):177–84.
4. King, N., Caudwell, P., Hopkins, M., Stubbs, J., Naslund, E., Blundell, J. Dual-process action of exercise on appetite control: increase in orexigenic drive but improvement in meal-induced satiety. *American Journal of Clinical Nutrition*. 2009;90(4):921.
5. Hill, J., Wyatt, H., Reed, G., Peters, J. Obesity and the environment: where do we go from here? *Science*. 2003;299(5608):853.
6. Brown, W., Williams, L., Ford, J., Ball, K., Dobson, A. Identifying the energy gap: magnitude and determinants of 5-year weight gain in midage women. *Obesity*. 2005;13(8):1431–41.
7. Cutler, D., Glaeser, E., Shapiro, J. Why have Americans become more obese? *The Journal of Economic Perspectives*. 2003;17(3):93–118.
8. Prentice, A., Jebb, S. Obesity in Britain: gluttony or sloth? *British Medical Journal*. 1995;311(7002):437.
9. Philipson, T. The world-wide growth in obesity: an economic research agenda. *Health Econ*. 2001;10(1):1–7.
10. Westerterp, K., Speakman, J. Physical activity energy expenditure has not declined since the 1980s and matches energy expenditures of wild mammals. *International Journal of Obesity*. 2008;32(8):1256–63.
11. Swinburn, B., Sacks, G., Lo, S., Westerterp, K., Rush, E., Rosenbaum, M., et al. Estimating the changes in energy flux that characterize the rise in obesity prevalence. *American Journal of Clinical Nutrition*. 2009;89(6):1723.
12. Swinburn, B., Sacks, G., Ravussin, E. Increased food energy supply is more than sufficient to explain the U.S. epidemic of obesity. *American Journal of Clinical Nutrition*. 2009;90(6):1453.
13. Anderson, J., Luan, J., Hoie, L. Structured weight-loss programs: a meta-analysis of weight loss at 24 weeks and assessment of effects of intensity of intervention. *Advanced Therapies*. 2004;21:61–75.

14. Golay, A., Allaz, A., Morel, Y., De Tonnac, N., Tankova, S., Reaven, G. Similar weight loss with low- or high-carbohydrate diets. *American Journal of Clinical Nutrition.* 1996;63(2):174.
15. Hill, J. Understanding and addressing the epidemic of obesity: an energy balance perspective. *Endocrinlogy Review.* 2006;27(7):750.
16. NHLBI. Clinical guidelines on the identification, evaluation and treatment of overweight and obesity in adults: National Institutes of Health, National Heart, Lung and Blood Institute; 2002.
17. Miller, W., Koceja, D., Hamilton, E. A meta-analysis of the past 25 years of weight loss research using diet, exercise or diet plus exercise intervention. *International Journal of Obesity*. 1997;21(10):941–7.
18. Clifton, P. Dietary treatment for obesity. *Nature Clinical Practice Gastroenterology and Hepatology.* 2008;5(12):672–81.
19. Astrup, A., Grunwald, G., Melanson, E., Saris, W., Hill, J. The role of low-fat diets in body weight control: a meta-analysis of *ad libitum* dietary intervention studies. *International Journal of Obesity*. 2000;24(12):1545–52.
20. Samaha, F., Iqbal, N., Seshadri, P., Chicano, K., Daily, D., McGrory, J., et al. A low-carbohydrate as compared with a low-fat diet in severe obesity. *The New England Journal of Medicine.* 2003;348(21):2074.
21. Brehm, B., Seeley, R., Daniels, S., D'Alessio, D. A randomized trial comparing a very low carbohydrate diet and a calorie-restricted low fat diet on body weight and cardiovascular risk factors in healthy women. *Journal of Clinical Endocrinology and Metabolism.* 2003;88(4):1617.
22. Foster, G., Wyatt, H., Hill, J., McGuckin, B., Brill, C., Mohammed, B., et al. A randomized trial of a low-carbohydrate diet for obesity. *The New England Journal of Medicine.* 2003;348(21):2082.
23. Yancy, W., Olsen, M., Guyton, J., Bakst, R., Westman, E. A low-carbohydrate, ketogenic diet versus a low-fat diet to treat obesity and hyperlipidemia. *Annals of Internal Medicine*. 2004;140(10):769.
24. Stern, L., Iqbal, N., Seshadri, P., Chicano, K., Daily, D., McGrory, J., et al. The effects of low-carbohydrate versus conventional weight loss diets in severely obese adults: one-year follow-up of a randomized trial. *Annals of Internal Medicine*. 2004;140(10):778.
25. Hession, M., Rolland, C., Kulkarni, U., Wise, A., Broom, J. Systematic review of randomized controlled trials of low-carbohydrate vs. low-fat/low-calorie diets in the management of obesity and its comorbidities. *Obesity Reviews.* 2009;10(1):36–50.
26. Wu, T., Gao, X., Chen, M., Van Dam, R. Long-term effectiveness of diet-plus-exercise interventions versus diet-only interventions for weight loss: a meta-analysis. *Obesity Reviews.* 2009;10(3):313–23.
27. Rolls, B. The relationship between dietary energy density and energy intake. *Physiology and Behavior*. 2009;97(5):609–15.
28. Ello-Martin, J., Ledikwe, J., Rolls, B. The influence of food portion size and energy density on energy intake: implications for weight management. *American Journal of Clinical Nutrition*. 2005;82(1):236S.
29. Rolls, B., Roe, L., Meengs, J. Reductions in portion size and energy density of foods are additive and lead to sustained decreases in energy intake. *American Journal of Clinical Nutrition*. 2006;83(1):11.
30. Flood, J., Rolls, B. Soup preloads in a variety of forms reduce meal energy intake. *Appetite.* 2007;49(3):626–34.

31. Rolls, B., Bell, E., Castellanos, V., Chow, M., Pelkman, C., Thorwart, M. Energy density but not fat content of foods affected energy intake in lean and obese women. *American Journal of Clinical Nutrition*. 1999;69(5):863.
32. Anderson, J., Konz, E., Frederich, R., Wood, C. Long-term weight-loss maintenance: a meta-analysis of U.S. studies. *American Journal of Clinical Nutrition*. 2001;74(5):579.
33. Curioni, C., Lourenco, P. Long-term weight loss after diet and exercise: a systematic review. *International Journal of Obesity*. 2005;29(10):1168–74.
34. Ravussin, E., Lillioja, S., Anderson, T., Christin, L., Bogardus, C. Determinants of 24-hour energy expenditure in man. Methods and results using a respiratory chamber. *Journal of Clinical Investigation*. 1986;78(6):1568.
35. Donnelly, J., Blair, S., Jakicic, J., Manore, M., Rankin, J., Smith, B. American College of Sports Medicine Position Stand. Appropriate physical activity intervention strategies for weight loss and prevention of weight regain for adults. *Medicine and Science in Sports and Exercise*. 2009;41(2):459.
36. Jakicic, J., Clark, K., Coleman, E., Donnelly, J., Foreyt, J., Melanson, E., et al. American College of Sports Medicine position stand. Appropriate intervention strategies for weight loss and prevention of weight regain for adults. *Medicine and Science in Sports and Exercise*. 2001;33(12):2145.
37. Shaw, K., Gennat, H., O'Rourke, P., Del Mar, C. Exercise for overweight or obesity. *Cochrane Database of Systematic Reviews*. 2006;4:112–7.
38. Catenacci, V., Wyatt, H. The role of physical activity in producing and maintaining weight loss. *Nature Clinical Practice Endocrinology and Metabolism*. 2007;3(7):518.
39. Macfarlane, D., Thomas, G. Exercise and diet in weight management. *British Journal of Sports Medicine*. 2009.
40. Redman, L., Heilbronn, L., Martin, C., De Jonge, L., Williamson, D., Delany, J., et al. Metabolic and behavioral compensations in response to caloric restriction: implications for the maintenance of weight loss. *PLoS ONE*. 2009;4(2):4377.
41. Cox, K., Burke, V., Morton, A., Beilin, L., Puddey, I. Independent and additive effects of energy restriction and exercise on glucose and insulin concentrations in sedentary overweight men. *American Journal of Clinical Nutrition*. 2004;80(2):308.
42. Nieman, D., Brock, D., Butterworth, D., Utter, A., Nieman, C. Reducing diet and/or exercise training decreases the lipid and lipoprotein risk factors of moderately obese women. *Journal of the American College of Nutrition*. 2002;21(4):344.
43. Larson-Meyer, D., Redman, E., Heilbronn, L., Martin, C., Ravussin, E. Caloric restriction with or without exercise: the fitness vs. fatness debate. *Medicine and Science in Sports and Exercise*. 2010;42(1):152–9.
44. Schoeller, D. The energy balance equation: looking back and looking forward are two very different views. *Nutrition Reviews*. 2009;67(5):249–54.
45. Boutcher, S., Dunn, S. Factors that may impede the weight loss response to exercise-based interventions. *Obesity Reviews*. 2009;10(6):671–80.
46. King, N., Caudwell, P., Hopkins, M., Byrne, N., Colley, R., Hills, A., et al. Metabolic and behavioral compensatory responses to exercise interventions: barriers to weight loss. *Obesity*. 2007;15(6):1373–83.
47. Schwartz, A., Doucet, É. Relative changes in resting energy expenditure during weight loss: a systematic review. *Obesity Reviews*.11(7):531–47.
48. Westerterp, K. Physical activity, food intake, and body weight regulation: insights from doubly labeled water studies. *Nutrition Reviews*. 2010;68(3):148–54.
49. Mayer, J. Glucostatic mechanism of regulation of food intake. *The New England Journal of Medicine*. 1953;249(1):13.
50. Ballor, D., Keesey, R. A meta-analysis of the factors affecting exercise-induced changes in body mass, fat mass and fat-free mass in males and females. *International Journal of Obesity*. 1991;15(11):717.

51. Fogelholm, M., Kukkonen-Harjula, K. Does physical activity prevent weight gain? A systematic review. *Obesity Reviews*. 2000;1(2):95–111.
52. Elder, S., Roberts, S. The effects of exercise on food intake and body fatness: a summary of published studies. *Nutrition Reviews*. 2007;65(1):1–19.
53. Borer, K. How effective is exercise in producing fat loss? *Kinesiology*. 008;40(2):126–37.
54. Jakicic, J., Otto, A. Treatment and prevention of obesity: what is the role of exercise? *Nutrition Reviews*. 2006;64(s1):S57–S61.
55. Goele, K., Bosy-Westphal, A., Rümcker, B., Lagerpusch, M., Müller, M. Influence of changes in body composition and adaptive thermogenesis on the difference between measured and predicted weight loss in obese women. *Obesity Facts*. 2009;2(2):105–9.
56. Jakicic, J., Winters, C., Lang, W., Wing, R. Effects of intermittent exercise and use of home exercise equipment on adherence, weight loss, and fitness in overweight women: a randomized trial. *JAMA*. 1999;282(16):1554.
57. Jakicic, J., Wing, R., Butler, B., Robertson, R. Prescribing exercise in multiple short bouts versus one continuous bout: effects on adherence, cardiorespiratory fitness, and weight loss in overweight women. *International Journal of Obesity*. 1995;19(12):893–901.
58. Colley, R., Hills, A., King, N., Byrne, N. Exercise-induced energy expenditure: Implications for exercise prescription and obesity. *Patient Education and Counseling*. 2010;79(3):327–32.
59. Colley, R., Hills, A., O'Moore-Sullivan, T., Hickman, I., Prins, J., Byrne, N. Variability in adherence to an unsupervised exercise prescription in obese women. *International Journal of Obesity*. 2008;32(5):837–44.
60. Manthou, E., Gill, J., Wright, A., Malkova, D. Behavioural compensatory adjustments to exercise training in overweight women. *Medicine and Science in Sports and Exercise*. 2010;42(6):1121–28..
61. Barwell, N., Malkova, D., Leggate, M., Gill, J. Individual responsiveness to exercise-induced fat loss is associated with change in resting substrate utilization. *Metabolism*. 2009;58(9):1320–8.
62. Martins, C., Kulseng, B., King, N., Holst, J., Blundell, J. The effects of exercise-induced weight loss on appetite-related peptides and motivation to eat. *Journal of Clinical Endocrinology and Metabolism*. 2010;doi:10.1210/jc.2009-2082.
63. Church, T., Martin, C., Thompson, A., Earnest, C., Mikus, C., Blair, S. Changes in weight, waist circumference and compensatory responses with different doses of exercise among sedentary, overweight postmenopausal women. *PLoS ONE*. 2009;4(2):4515.
64. Bouchard, C., Rankinen, T. Individual differences in response to regular physical activity. *Medicine and Science in Sports and Exercise*. 2001;33(6):S446.
65. Rankinen, T., Bray, M., Hagberg, J., PéRusse, L., Roth, S., Wolfarth, B., et al. The human gene map for performance and health-related fitness phenotypes: the 2005 update. *Medicine and Science in Sports and Exercise*. 2006;38(11):1863.
66. Sisson, S., Katzmarzyk, P., Earnest, C., Bouchard, C., Blair, S., Church, T. Volume of exercise and fitness non-response in sedentary, post-menopausal women. *Medicine and Science in Sports and Exercise*. 2009;41(3):539.
67. Bouchard, C. Individual differences in the response to regular exercise. *International Journal of Obesity and Related Metabolic Disorders*. 1995;19(4):S5–S8.
68. Boulé, N., Weisnagel, S., Lakka, T., Tremblay, A., Bergman, R., Rankinen, T., et al. Effects of exercise training on glucose homeostasis. *Diabetes Care*. 2005;28(1):108.
69. Dulloo, A. Suppressed thermogenesis as a cause for resistance to slimming and obesity rebound: adaptation or illusion? *International Journal of Obesity*. 2007;31(2):201–3.
70. Leibel, R., Rosenbaum, M., Hirsch, J. Changes in energy expenditure resulting from altered body weight. *The New England Journal of Medicine*. 1995;332(10):621.
71. Stiegler, P., Cunliffe, A. The role of diet and exercise for the maintenance of fat-free mass and resting metabolic rate during weight loss. *Sports Medicine*. 2006;36(3):239–62.

72. Jonge, L., Bray, G. The thermic effect of food is reduced in obesity. *Nutrition Reviews*. 2002;60(9):295–7.
73. Doucet, E., Imbeault, P., St-Pierre, S., Alméras, N., Mauriège, P., Després, J., et al. Greater than predicted decrease in energy expenditure during exercise after body weight loss in obese men. Clinical *Science*. 2003;105(1):89–96.
74. Schneider, K., Spring, B., Pagoto, S. Exercise and energy intake in overweight, sedentary individuals. *Eating Behaviors*. 2009;10(1):29–35.
75. Heilbronn, L., de Jonge, L., Frisard, M., DeLany, J., Larson-Meyer, D., Rood, J., et al. Effect of 6-month calorie restriction on biomarkers of longevity, metabolic adaptation, and oxidative stress in overweight individuals: a randomized controlled trial. *JAMA*. 2006;295(13):1539.
76. Tremblay, A., Major, G., Doucet, É., Trayhurn, P., Astrup, A. Role of adaptive thermogenesis in unsuccessful weight-loss intervention. *Future Lipidology*. 2007;2(6):651–8.
77. Martin, C., Heilbronn, L., De Jonge, L., DeLany, J., Volaufova, J., Anton, S., et al. Effect of calorie restriction on resting metabolic rate and spontaneous physical activity. *Obesity*. 2007;15(12):2964–73.
78. Tremblay, A., Chaput, J. Adaptive reduction in thermogenesis and resistance to lose fat in obese men. *British Journal of Nutrition*. 2009;102(04):488–92.
79. Doucet, E., St-Pierre, S., Alméras, N., Després, J., Bouchard, C., Tremblay, A. Evidence for the existence of adaptive thermogenesis during weight loss. *British Journal of Nutrition*. 2001;85(06):715–23.
80. Major, G., Doucet, E., Trayhurn, P., Astrup, A., Tremblay, A. Clinical significance of adaptive thermogenesis. *International Journal of Obesity*. 2007;31(2):204–12.
81. Rosenbaum, M., Hirsch, J., Gallagher, D., Leibel, R. Long-term persistence of adaptive thermogenesis in subjects who have maintained a reduced body weight. *American Journal of Clinical Nutrition*. 2008;88(4):906.
82. Flatt, J. Exaggerated claim about adaptive thermogenesis. *International Journal of Obesity*. 2007;31(10):1626.
83. Weyer, C., Pratley, R., Salbe, A., Bogardus, C., Ravussin, E., Tataranni, P. Energy expenditure, fat oxidation, and body weight regulation: a study of metabolic adaptation to long-term weight change. *Journal of Clinical Endocrinology and Metabolism*. 2000;85(3):1087.
84. MacLean, P., Jackman, M., Bessesen, D. Compensatory response to reducing body weight. *Drug Discovery Today: Disease Mechanisms*. 2005;2(3):313–9.
85. Amati, F., Dube, J., Shay, C., Goodpaster, B. Separate and combined effects of exercise training and weight loss on exercise efficiency and substrate oxidation. *Journal of Applied Physiology*. 2008;105(3):825.
86. Talanian, J., Galloway, S., Heigenhauser, G., Bonen, A., Spriet, L. Two weeks of high-intensity aerobic interval training increases the capacity for fat oxidation during exercise in women. *Journal of Applied Physiology*. 2007;102(4):1439.
87. Berggren, J., Boyle, K., Chapman, W., Houmard, J. Skeletal muscle lipid oxidation and obesity: influence of weight loss and exercise. *American Journal of Physiology—Endocrinology and Metabolism*. 2008;294(4):E726.
88. Speakman, J., Selman, C. Physical activity and resting metabolic rate. *Proceedings of the Nutrition Society*. 2003;62(03):621–34.
89. Zurlo, F., Lillioja, S., Esposito-Del Puente, A., Nyomba, B., Raz, I., Saad, M., et al. Low ratio of fat to carbohydrate oxidation as predictor of weight gain: study of 24-h RQ. *American Journal of Physiology—Endocrinology and Metabolism*. 1990;259(5):E650.
90. Seidell, J., Muller, D., Sorkin, J., Andres, R. Fasting respiratory exchange ratio and resting metabolic rate as predictors of weight gain: the Baltimore Longitudinal Study on Aging. *International Journal of Obesity and Related Metabolic Disorders*. 1992;16(9):667.

91. Marra, M., Scalfi, L., Covino, A., Esposito-Del Puente, A., Contaldo, F. Fasting respiratory quotient as a predictor of weight changes in non-obese women. *International Journal of Obesity and Related Metabolic Disorders*. 1998.
92. Froidevaux, F., Schutz, Y., Christin, L., Jequier, E. Energy expenditure in obese women before and during weight loss, after refeeding, and in the weight-relapse period. *American Journal of Clinical Nutrition*. 1993;57(1):35.
93. Valtuena, S., Salas-Salvado, J., Lorda, P. The respiratory quotient as a prognostic factor in weight-loss rebound. *International Journal of Obesity*. 1997;21(9):811–7.
94. Ranneries, C., Bulow, J., Buemann, B., Christensen, N., Madsen, J., Astrup, A. Fat metabolism in formerly obese women. *American Journal of Physiology—Endocrinology and Metabolism*. 1998;274(1):E155.
95. King, N., Blundell, J. High-fat foods overcome the energy expenditure induced by high-intensity cycling or running. *European Journal of Clinical Nutrition*. 1995;49(2):114.
96. King, N., Snell, L., Smith, R., Blundell, J. Effects of short-term exercise on appetite responses in unrestrained females. *European Journal of Clinical Nutrition*. 1996; 50(10):663.
97. King, N., Lluch, A., Stubbs, R., Blundell, J. High dose exercise does not increase hunger or energy intake in free living males. *European Journal of Clinical Nutrition*. 1997;51(7):478–83.
98. Blundell, J., King, N. Effects of exercise on appetite control: loose coupling between energy expenditure and energy intake. *International Journal of Obesity and Related Metabolic Disorders*. 1998;22:S22–S29.
99. Levitsky, D. The non-regulation of food intake in humans: hope for reversing the epidemic of obesity. *Physiology and Behavior*. 2005;86(5):623–32.
100. Stubbs, R., Sepp, A., Hughes, D., Johnstone, A., King, N., Horgan, G., et al. The effect of graded levels of exercise on energy intake and balance in free-living women. *International Journal of Obesity and Related Metabolic Disorders*. 2002;26(6):866–9.
101. Whybrow, S., Hughes, D., Ritz, P., Johnstone, A., Horgan, G., King, N., et al. The effect of an incremental increase in exercise on appetite, eating behaviour and energy balance in lean men and women feeding *ad libitum*. *British Journal of Nutrition*. 2008;100(05):1109–15.
102. Stubbs, R., Sepp, A., Hughes, D., Johnstone, A., Horgan, G., King, N., et al. The effect of graded levels of exercise on energy intake and balance in free-living men, consuming their normal diet. *European Journal of Clinical Nutrition*. 2002;56(2):129–40.
103. Stubbs, R., Hughes, D., Johnstone, A., Whybrow, S., Horgan, G., King, N., et al. Rate and extent of compensatory changes in energy intake and expenditure in response to altered exercise and diet composition in humans. *American Journal of Physiology—Regulatory, Integrative and Comparative Physiology*. 2004;286(2):350.
104. Blundell, J. E. A psychobiological approach to appetite and weight control. In: Brownell, K., Fairburn, C. *Eating disorders and obesity: a comprehensive handbook*. 1995.
105. Green, S., Delargy, H., Joanes, D., Blundell, J. A satiety quotient: a formulation to assess the satiating effect of food. *Appetite*. 1997;29(3):291–304.
106. Levine, J. Nonexercise activity thermogenesis (NEAT): environment and biology. *American Journal of Physiology—Endocrinology and Metabolism*. 2004;286(5):E675.
107. Levine, J., Kotz, C. NEAT—non-exercise activity thermogenesis—egocentric and geocentric environmental factors vs. biological regulation. *Acta Physiologica Scandinavica*. 2005;184(4):309–18.
108. Levine, J. Non-exercise activity thermogenesis (NEAT). *Nutrition Reviews*. 2004; 62(s2):S82–S97.
109. Levine, J., Lanningham-Foster, L., McCrady, S., Krizan, A., Olson, L., Kane, P., et al. Interindividual variation in posture allocation: possible role in human obesity. *Science*. 2005;307(5709):584.

110. Levine, J., Eberhardt, N., Jensen, M. Role of nonexercise activity thermogenesis in resistance to fat gain in humans. *Science.* 1999;283(5399):212.
111. Taylor, H., Keys, A., Hylander, C., Meyerhoff, H., Taylor, R. Adapting to caloric restriction. *Science.* 1950;112(2904).
112. Heyman, M., Young, V., Fuss, P., Tsay, R., Joseph, L., Roberts, S. Underfeeding and body weight regulation in normal-weight young men. *American Journal of Physiology—Regulatory, Integrative and Comparative Physiology.* 1992;263(2):250.
113. Velthuis-te Wierik, E., Meijer, P., Kluft, C., van den Berg, H. Beneficial effect of a moderately energy-restricted diet on fibrinolytic factors in non-obese men. *Metabolism.* 1995;44(12):1548–52.
114. Sénéchal, M., Arguin, H., Bouchard, D., Carpentier, A., Ardilouze, J., Dionne, I., et al. Interindividual variations in resting metabolic rate during weight loss in obese postmenopausal women: a pilot study. *Metabolism.* 2010;59(4):478–85.
115. Blundell, J. E., Rogers, P. J., Hill, A. J. (1987) Evaluating the satiating power of foods: implications for acceptance and consumption. In: Solms, J., Booth, D., Pangborn, R., Raunhardt, O., editors. *Food acceptance and nutrition.* London: Academic Press; 1987.
116. Flatt, J. Dietary fat, carbohydrate balance, and weight maintenance: effects of exercise. *American Journal of Clinical Nutrition.* 1987;45(1):296.
117. Flatt, J. The difference in the storage capacities for carbohydrate and for fat, and its implications in the regulation of body weight. *Annals of the New York Academy of Science.* 1987;499:104–23.
118. Flatt, J. Glycogen levels and obesity. *International Journal of Obesity.* 1996;20:1–11.
119. Bessesen, D., Bull, S., Cornier, M. Trafficking of dietary fat and resistance to obesity. *Physiology and Behavior.* 2008;94(5):681–8.
120. Westerterp, K. Dietary fat oxidation as a function of body fat. *Current Opinion in Lipidology.* 2009;20:45–9.
121. Astrup, A., Buemann, B., Christensen, N., Tourbro, S., Raben, A. Failure to increase lipid oxidation in response to increasing dietary fat content in formally obese women. *American Journal of Physiology.* 1994;29:592–9.
122. Raben, A., Kiens, B., Richter, E. Differences in glycaemia, hormonal response and energy expenditure after a meal rich in mono- and disaccharides compared to a meal rich in polysaccharides in physically fit and sedentary subjects. *Clinical Physiology and Functional Imaging.* 1994;14(3):267–80.
123. Faraj, M., Jones, P., Sniderman, A., Cianflone, K. Enhanced dietary fat clearance in postobese women. *The Journal of Lipid Research.* 2001;42(4):571.
124. Blaak, E., Hul, G., Verdich, C., Stich, V., Martinez, A., Petersen, M., et al. Fat oxidation before and after a high fat load in the obese insulin-resistant state. *Journal of Clinical Endocrinology and Metabolism.* 2006;91(4):1462.
125. Marrades, M., Martinez, J., Moreno-Aliaga, M. ZAG, a lipid mobilizing adipokine, is downregulated in human obesity. *Journal of Physiology and Biochemistry.* 2008; 64(1):61–6.
126. Westerterp, K., Smeets, A., Lejeune, M., Wouters-Adriaens, M., Westerterp-Plantenga, M. Dietary fat oxidation as a function of body fat. *American Journal of Clinical Nutrition.* 2008;87(1):132.
127. Giacco, R., Clemente, G., Busiello, L., Lasorella, G., Rivieccio, A., Rivellese, A., et al. Insulin sensitivity is increased and fat oxidation after a high-fat meal is reduced in normal-weight healthy men with strong familial predisposition to overweight. *International Journal of Obesity.* 2004;28(2):342–8.
128. Goodpaster, B., Katsiaras, A., Kelley, D. Enhanced fat oxidation through physical activity is associated with improvements in insulin sensitivity in obesity. *Diabetes.* 2003;52(9):2191.

129. Bergouignan, A., Trudel, G., Simon, C., Chopard, A., Schoeller, D., Momken, I., et al. Physical inactivity differentially alters dietary oleate and palmitate trafficking. *Diabetes*. 2009;58(2):367.
130. Toubro, S., Sorensen, T., Hindsberger, C., Christensen, N., Astrup, A. Twenty-four-hour respiratory quotient: the role of diet and familial resemblance. *Journal of Clinical Endocrinology and Metabolism*. 1998;83(8):2758.
131. Helge, J., Fraser, A., Kriketos, A., Jenkins, A., Calvert, G., Ayre, K., et al. Interrelationships between muscle fibre type, substrate oxidation and body fat. *Medicine and Science in Sports and Exercise*. 1999;31(5):S372.
132. Goedecke, J., Gibson, A., Grobler, L., Collins, M., Noakes, T., Lambert, E. Determinants of the variability in respiratory exchange ratio at rest and during exercise in trained athletes. *American Journal of Physiology—Endocrinology and Metabolism*. 2000;279(6):E1325.
133. Venables, M., Achten, J., Jeukendrup, A. Determinants of fat oxidation during exercise in healthy men and women: a cross-sectional study. *Journal of Applied Physiology*. 2005;98(1):160.
134. Achten, J., Jeukendrup, A. Optimizing fat oxidation through exercise and diet. *Nutrition*. 2004;20(7–8):716–27.
135. Leonhardt, M., Langhans, W. Fatty acid oxidation and control of food intake. *Physiology and Behavior*. 2004;83(4):645–51.
136. Friedman, M. Control of energy intake by energy metabolism. *American Journal of Clinical Nutrition*. 1995;62(5):1096S.
137. Langhans, W. Fatty acid oxidation in the energostatic control of eating—a new idea. *Appetite*. 2008;51(3):446–51.
138. Langhans, W. Control of eating by hepatic oxidation of fatty acids: a note of caution. Author's reply. *Appetite*. 2009;53(2):274–6.
139. Allen, M., Bradford, B. Control of eating by hepatic oxidation of fatty acids: a note of caution. *Appetite*. 2009;53(2):272–3.
140. Scharrer, E. Control of food intake by fatty acid oxidation and ketogenesis. *Nutrition*. 1999;15(9):704–14.
141. Gatta, B., Zuberbuehler, C., Arnold, M., Aubert, R., Langhans, W., Chapelot, D. Acute effects of pharmacological modifications of fatty acid metabolism on human satiety. *British Journal of Nutrition*. 2008;101(12):1867–77.
142. Woods, S., Ramsay, D. Homeostasis: beyond Curt Richter. *Appetite*. 2007;49(2):388–98.
143. Björntorp, P., Sjöström, L. Carbohydrate storage in man: speculations and some quantitative considerations. *Metabolism*. 1978;27(12):1853–65.
144. Casey, A., Mann, R., Banister, K., Fox, J., Morris, P., Macdonald, I., et al. Effect of carbohydrate ingestion on glycogen resynthesis in human liver and skeletal muscle, measured by 13C MRS. *American Journal of Physiology—Endocrinology and Metabolism*. 2000;278(1):E65.
145. Almeras, N., Lavallee, N., Despres, J., Bouchard, C., Tremblay, A. Exercise and energy intake: effect of substrate oxidation. *Physiology and Behavior*. 1995;57(5):995.
146. Stubbs, R., Murgatroyd, P., Goldberg, G., Prentice, A. Carbohydrate balance and the regulation of day-to-day food intake in humans. *American Journal of Clinical Nutrition*. 1993;57(6):897.
147. Shetty, P., Prentice, A., Goldberg, G., Murgatroyd, P., McKenna, A., Stubbs, R., et al. Alterations in fuel selection and voluntary food intake in response to isoenergetic manipulation of glycogen stores in humans. *American Journal of Clinical Nutrition*. 1994;60(4):534.
148. Stubbs, R., Harbron, C., Murgatroyd, P., Prentice, A. Covert manipulation of dietary fat and energy density: effect on substrate flux and food intake in men eating *ad libitum*. *American Journal of Clinical Nutrition*. 1995;62(2):316.

149. Snitker, S., Larson, D., Tataranni, P., Ravussin, E. *Ad libitum* food intake in humans after manipulation of glycogen stores. *American Journal of Clinical Nutrition*. 1997; 65(4):941.
150. Sparti, A., Windhauser, M., Champagne, C., Bray, G. Effect of an acute reduction in carbohydrate intake on subsequent food intake in healthy men. *American Journal of Clinical Nutrition*. 1997;66(5):1144.
151. Pannacciulli, N., Salbe, A., Ortega, E., Venti, C., Bogardus, C., Krakoff, J. The 24-h carbohydrate oxidation rate in a human respiratory chamber predicts *ad libitum* food intake. *American Journal of Clinical Nutrition*. 2007;86(3):625.
152. Galgani, J., de Jonge, L., Most, M., Bray, G., Smith, S. Effect of a 3-day high-fat feeding period on carbohydrate balance and *ad libitum* energy intake in humans. *International Journal of Obesity*. 2010.
153. Eckel, R., Hernandez, T., Bell, M., Weil, K., Shepard, T., Grunwald, G., et al. Carbohydrate balance predicts weight and fat gain in adults. *American Journal of Clinical Nutrition*. 2006;83(4):803.

6 The Reward Deficiency Hypothesis

Implications for Obesity and Other Eating Disorders

Brenda M. Geiger, Erin N. Umberg, and Emmanuel N. Pothos

CONTENTS

Introduction ... 103
Epidemiology of Obesity ... 103
Food Reward ... 105
Food as a Dependence-Inducing Substance ... 106
The Reward Deficiency Hypothesis ... 107
D-amphetamine and Reward Deficiency ... 107
Neurobiological Evidence for Reward Deficiency in Obesity ... 107
Genetic Evidence of Reward Deficiency in Obesity ... 112
Transgenerational Obesity and Reward Deficiency ... 112
Central Modulators of Dopamine Signaling That Are Altered in Obesity ... 113
Peripheral Signals Influencing Hedonic Aspects of Food Intake ... 114
Evidence Disputing the Reward Deficiency Hypothesis in Obesity ... 115
The Reward Deficiency Hypothesis in Eating Disorders ... 116
Binge-Eating Disorder ... 116
Bulimia Nervosa (BN) ... 118
Anorexia Nervosa ... 120
Conclusions ... 122
Acknowledgments ... 123
References ... 123

INTRODUCTION

Epidemiology of Obesity

Obesity has grown to epidemic proportions in the United States and around the world. Current estimates are that more than 30 percent of the US adult population is obese and as many as two-thirds of the population are overweight (Ogden, Carroll, et al. 2006; Ford, Li, et al. 2010). In addition, the obesity problem is not limited to adults. The prevalence of overweight children has increased to more than 15 percent over the last several decades with more than 30 percent of children at risk for being overweight with a Body Mass Index (BMI) z-score greater than the 85th percentile (Ogden, Carroll et al. 2008). Because obesity not only is associated with a number of cardiovascular diseases and diabetes, but also is related to cancers, mental disorders, fertility, and orthopedic issues, the annual direct medical cost burden of obesity in the United States is estimated to be close to $150 billion in adults and another $14 billion in children (Hammond and Levine 2010).

Obesity develops as the result of a positive energy balance over time. To date, a great deal of research has focused on the molecular mechanisms responsible for the homeostatic control of food intake and energy expenditure. However, the extraordinarily large and constantly increasing number of patients with obesity in recent times indicates that other systems, either peripheral or neural, must override the homeostatic system of energy balance to induce the hyperphagia associated with obesity. One potential overriding system is the central reward system that codes for the hedonic value of palatable food. The common pathway for central reward for both drugs of abuse and natural rewards, like that for food, is the central dopamine pathway. In fact, central dopamine has been known to play a role in the regulation of food intake since the early 1970s (Smith and Schneider 1988). In this review, we focus on how changes in the neurobiology of this pathway, particularly alterations that lead to deficiencies in reward-associated central dopaminergic neurotransmission, might influence the decision to continue eating beyond one's energy needs. Table 6.1 summarizes the findings discussed in this chapter. In addition, we address how central dopamine deficits may contribute to the pathology of other eating disorders including binge eating disorder, bulimia nervosa, and anorexia nervosa.

TABLE 6.1
Evidence for Reward Deficiency Syndrome in Obesity

Neurobiological Evidence	Effect on Dopamine	References
Evoked and extracellular central dopamine release are low in rats fed a cafeteria-style diet, ob/ob mice, and obesity-prone rats.	↓ dopamine ↓ TH ↓ DAT ↓ VMAT2	Pothos et al. (1998) Geiger et al. (2009) Fulton et al. (2006) Geiger et al. (2008)
Cafeteria diet-fed rats have deficits in their reward threshold.	↓ D2R activity	Johnson and Kenny (2010)
High-fat diet-fed rats have decreased response to amphetamine.	↓ dopamine	Davis J.F. et al. (2008)
Rats selected for obesity proneness exhibit decreased reward responses.	↓ dopamine	Rada et al. (2010)
Taq1a1 allele is associated with obesity.	↓ D2R activity ↓ dopamine	Blum et al. (1995b) Stice et al. (2008a)
Decreased activity by fMRI and PET in obese patients.	↓ D2R activity	Wang et al. (2001) Stice et al. (2008b)

Food Reward

Even though hyperphagia associated with dietary obesity is not listed in the DSM-IV as an addictive disorder and it is unlikely to be categorized as such in DSM-V, the chronic consumption of excess calories has repeatedly been shown to have multiple overlapping neurobiological mechanisms as those seen with addiction to drugs of abuse, beginning with the activation of central dopamine signaling. In contrast to the role of hypothalamic dopamine in regulating homeostatic aspects of food intake, mesolimbic dopamine signaling has been implicated in hedonic eating ("hyperphagia"). Like addictive drugs, palatable foods have been shown to persistently increase nucleus accumbens (NAc) dopamine (e.g., Rada et al. 2005). Because obesity is associated with lower basal dopamine function, it is plausible that increased food intake in these individuals is a means to compensate for the hyporesponsive dopamine system.

Food reward is at least partly mediated by opioid signaling pathways (Yeomans and Gray 2002; Berthoud and Morrison 2008). Similar to its effect on dopamine, palatable food increases levels of endogenous opiates. While opiate enhancement via agonist injections and exogenous opioid administration induces hyperphagia (Holtzman 1974; Erlanson-Albertsson 2005), opiate blockage through antagonists such as naloxone decreases food intake (Levine et al. 1995, 2003). Interestingly, this effect is specific to palatable (sweet and fatty) food consumption, suggesting that opioids modulate the hedonic aspects of food. Furthermore, opioids increase dopamine release in the nucleus accumbens, and conversely opioid-induced hyperphagia is at least partially dependent on the activation of dopamine receptor signaling in the accumbens (Elman et al. 2006). Therefore, dopamine-dependent modulation of opioid signaling is a pathway through which dopamine may affect food intake (Vucetic and Reyes 2010).

While palatable foods are associated with increases in release of accumbens dopamine, it is important to note that such increases have also been shown to occur in conjunction with motor activity (Salamone et al. 1994; Cousins and Salamone 1996), reward learning (Balleine and Dickinson 1998; Di Chiara et al. 1999), and salience attribution (Berridge and Robinson 1998; Gallagher et al. 1999). Furthermore, a number of studies using functional magnetic resonance imaging (fMRI) with event-related designs have indicated that the accumbens is activated preferentially during reward anticipation (Knutson et al. 2001; O'Doherty et al. 2002), while reward outcome is marked by medial prefrontal cortex activity. Because temporal precision distinguishing anticipation from outcome was not taken into account in the animal models, an anticipatory (versus reward outcome) role for the sucrose-induced dopamine release cannot be excluded. Hence, studies that pair sucrose presentation with a predictable cue (e.g., tone or light) and measure accumbens dopamine before the ingestion and tasting of sweet solution in "sugar-dependent" rats would better parse out anticipation- versus outcome-responsive dopamine release. Although several researchers (Phillips et al. 1993; Richardson and Gratton 1996) have found that dopamine release was triggered by conditioned stimuli before actual food intake, their models did not induce sugar dependency in the rats prior to experimental manipulations. It would be interesting to see if sugar-dependent rats display similar dopamine activity in response to conditioned stimuli and whether these "anticipatory" increases are modulated by reward outcome.

Food as a Dependence-Inducing Substance

While it is evident that food, in particular palatable food, can activate the central reward systems, one question that is still debated is whether a person can become dependent on food itself in the same way that classical drugs of abuse like heroin, cocaine, alcohol, or amphetamines induce dependence in some people. According to the DSM-IV, the definition for dependence to a substance includes a pattern of use of the substance that leads to clinical impairment and can include any three of seven criteria over a twelve-month period. These criteria include the development of tolerance, the exhibition of withdrawal syndrome, excess intake over long periods, lack of success in quitting, excessive time or effort to obtain the substance or recover from its use, changes in social activities due to the substance use, and continued use despite knowledge of the risks associated with use of the substance (American Psychiatric Association Task Force on DSM-IV 1994).

Some scientists argue that because food is required for survival, it is unlikely that one could actually become dependent on it and self-reports of "addiction" to various food substances misuse the term "addiction" (Rogers and Smit 2000). However, a growing body of evidence indicates that the specific macronutrient compositions of certain palatable foods may in fact cause neurobiological changes that could cause psychological dependence. For example, the salted food addiction hypothesis suggests that people consume salted food because it is an opiate receptor agonist and they use it as a form of "self-medication" (Cocores and Gold 2009). Others argue that refined food can cause dependence based on its high concentration of components with significant effects on central dopamine like refined carbohydrates, fat, salt, and caffeine (Ifland et al. 2009).

Specific macronutrients appear to have properties that can lead to dependence. In animals given intermittent access to sucrose, it was found that sucrose consumption can lead to opioid withdrawal-like symptoms (Avena 2007; Avena et al. 2008). In humans, evaluation of the behavior of self-designated carbohydrate cravers indicates that these people also have trouble controlling their intake of high-carb beverages (Corsica and Spring 2008; Spring et al. 2008). An evaluation of the properties of fructose also indicates that its effects are very similar to those of ethanol, a highly addictive substance (Lustig 2010). In some cases, not only sugar but also fat may lead to addiction-like behaviors. In a study using high-fat/high-sucrose food, animals showed withdrawal-like symptoms (Pickering, et al. 2009). Together, the evidence supports that at least some of the most palatable and high-energy food components elicit hyperphagic responses due to a physical or emotional dependence on their rewarding properties. This dependence may be manifested through modifications of central reward processing that override homeostatic energy balance and the interaction of central systems with peripheral signals of satiety such as leptin, cholecystokinin (CCK), and others.

THE REWARD DEFICIENCY HYPOTHESIS

D-Amphetamine and Reward Deficiency

A current hypothesis for how dopamine dysregulation induces addictive behavior is through the "reward deficiency syndrome" (Blum et al. 1995). In applying this theory to obesity research, this hypothesis can be reinstated to emphasize that humans and animals with low dopamine levels in the nucleus accumbens and other targeted projections of the midbrain dopamine neurons are prone to excessive consumption of palatable food in an effort to increase central dopamine signaling as a compensatory response, particularly since their dopamine systems do not respond to powerful stimulants such as d-amphetamine at the same level as in subjects with higher basal dopamine levels (Pothos et al. 1998; Geiger et al. 2009). In fact, amphetamine can be considered the historical foundation of the reward deficiency hypothesis in obesity as it was marketed as benzedrine in the 1930s and became one of the first effective weight loss drugs (Harris, et al. 1947). However, it was known back then as an ephedrine analog, and its effects on central dopamine systems were unknown. Since then, numerous other agonists of the dopamine system have been shown to reduce food intake and/or weight loss (Cooper et al. 1990, 2006; HA and Cooper 1994; van der Hoek and Cooper 1994). In addition, deficiencies in the dopamine system have been shown in a number of studies looking at the effects of drugs of abuse on central reward pathways. In imaging studies, substance-dependent humans consistently show decreased D2 receptor activity (Volkow, Fowler et al. 2004; Volkow, Fowler et al. 2004), which matches the decreased D2 activity seen in obese humans (Wang et al. 2001; Stice et al. 2008). The Taql A1 allele of the DRD2 gene, which codes for the D2 receptor, has also been associated with an increased susceptibility to alcoholism, drug use, and addictive behaviors like excess gambling or sex (Blum et al. 1990, 1995, 2000). This allele results in decreased D2 activity and hyporesponsiveness of the central dopamine system (Blum et al. 2000) as well as obesity (Stice et al. 2008). These results implicate depleted dopamine as a significant contributing factor to substance dependence.

Neurobiological Evidence for Reward Deficiency in Obesity

Recently several studies in both humans and animals have linked obesity to low levels of central dopamine. In our studies of dietary obese animals, we have shown that rats fed a cafeteria-style diet consisting of a number of high-sugar and high-fat components have depleted central dopamine signal (Figure 6.1) (Geiger et al. 2009). Furthermore, the response of the central dopamine system in these animals to amphetamine was reduced in terms of absolute levels of dopamine released when compared to the response in chow-fed controls. Furthermore, the animals that had been consuming the cafeteria diet had a dopamine response in their nucleus accumbens to a palatable food challenge and not to normal lab chow. This result suggested that prolonged exposure to highly palatable food induces the central dopamine system in these animals to preferentially respond to these types of food (Figure 6.2) (Geiger et al. 2009).

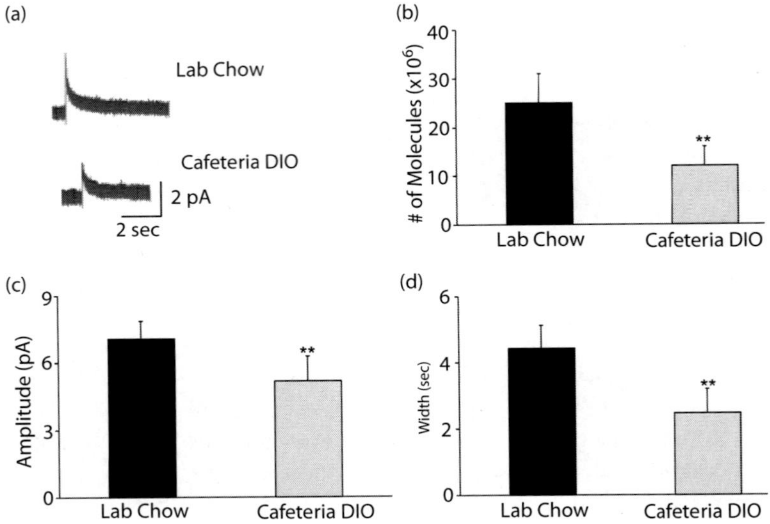

FIGURE 6.1 Evoked dopamine release from the dorsal striatum in brain slices. (a) Representative traces from acute coronal dorsal striatum slices of chow-fed animals (top; $n = 31$ stimulations in seven slices) and cafeteria DIO animals (bottom; $n = 15$ stimulations in four slices). The average number of molecules released (b), event amplitude (c), and width (d) from the dietary obese animals were significantly lower than in chow-fed animals (*$p < 0.01$ by one-way ANOVA). Data from B. M. Geiger et al. 2009. Reprinted with permission.

Other studies investigating the role of central dopamine in dietary obese or high-fat diet feeding models have also shown that obese animals have deficits in dopamine signaling. A model of brain stimulation reward, where rats are taught to self-administer an electrical stimulus that activates their central reward pathways, has shown that cafeteria diet-fed rats have deficits in their responses to reward compared to chow-fed controls. The deficits in reward are linked to deficits in striatal D2 receptors in the obese animals (Johnson and Kenny 2010). In studies of rats fed a high-fat diet, dopamine turnover, dopamine release, and response to dopamine-associated cues, like amphetamine, were reduced in the nucleus accumbens even before a weight difference developed (Davis et al. 2008; York et al. 2010). Taken together, all of these studies indicate that either the diet or the weight gain itself leads to deficits in central dopamine signaling, which compels obese animals to consume more food as a compensatory response. However, they do not address the relative contributions of the genetic background, the diet, and the excess body weight on these effects.

In the selectively inbred rat model of obesity predisposition and its obesity-resistant counterpart, macronutrient composition can be controlled for as the obesity-prone rats gain more weight than the obesity-resistant rats without increasing the fat or carbohydrate levels of their normal laboratory chow (Levin et al. 1997). Using this model, we have shown depletions in dopamine signal in these animals very early in

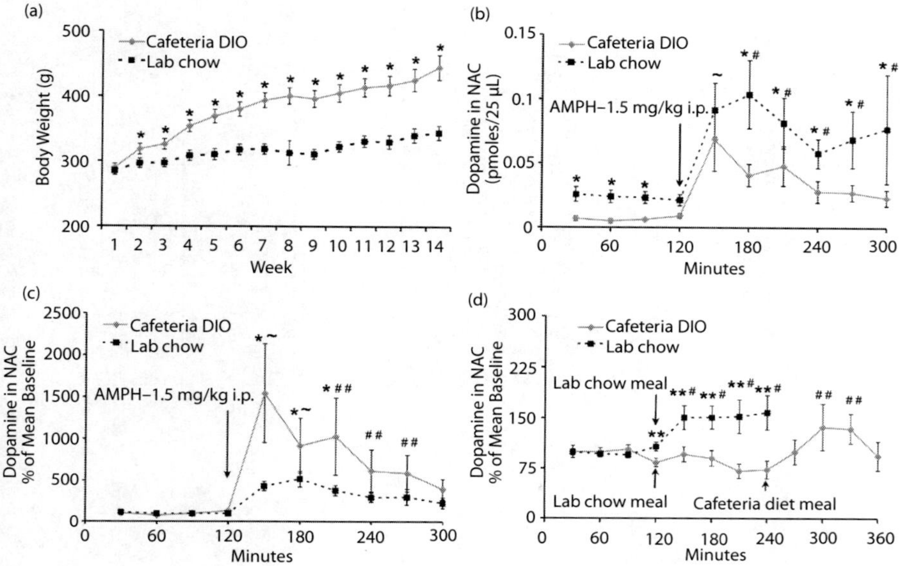

FIGURE 6.2 Basal, amphetamine- and laboratory chow meal–challenged nucleus accumbens dopamine levels are decreased in dietary obese rats. (a) Body weight of cafeteria DIO rats during a fourteen-week period was significantly more than the laboratory chow-fed group beginning at week 2 of the dietary regimen ($^*p < 0.01$ by one-way ANOVA). (b) Basal and amphetamine-challenged extracellular dopamine levels during week 14 in the nucleus accumbens of cafeteria DIO rats ($n = 9$) was significantly lower than in chow-fed rats ($n = 13$). (c) The percent increase from baseline after amphetamine was higher in the cafeteria DIO rats than in the chow-fed rats ($^*p < 0.01$ between groups, $\sim p < 0.05$ within both groups, $^{\#}p < 0.05$ within laboratory chow fed group only, $^{\#\#}p < 0.05$ within DIO group only relative to baseline prior to the amphetamine injection). (d) A plain chow meal was presented to cafeteria DIO ($n = 18$) and chow-fed groups ($n = 22$) after four baseline samples. Only the chow-fed group showed significant increases in dopamine after the meal. The cafeteria meal that was presented to a subset ($n = 8$) of the cafeteria DIO group 2.5 hours after the regular chow meal resulted in a significant increase in dopamine release ($^{**}p < 0.05$ between groups, $^{\#}p < 0.05$ within the laboratory chow fed group only, $^{\#\#}p < 0.05$ within the cafeteria DIO group only relative to baseline prior to the laboratory chow or cafeteria meal). Data from B. M. Geiger et al. 2009. Reprinted with permission.

life, before a difference in body weight develops, that persist into adulthood. These deficits were associated with reduced activity levels and decreased hyperactivity response to amphetamine injection (Figure 6.3). Furthermore, we were able to link these dopamine deficits to deficits in mRNA and protein expression of key regulators of central dopamine signaling including tyrosine hydroxylase (TH), the rate-limiting enzyme for dopamine synthesis, and the vesicular monoamine transporter (VMAT2), which packages cytoplasmic dopamine into synaptic vesicles (Figure 6.4). In a similar model, outbred rats were selected for their potential to develop obesity based on their body weight gain following five days of exposure to a high-fat diet. Prior to the

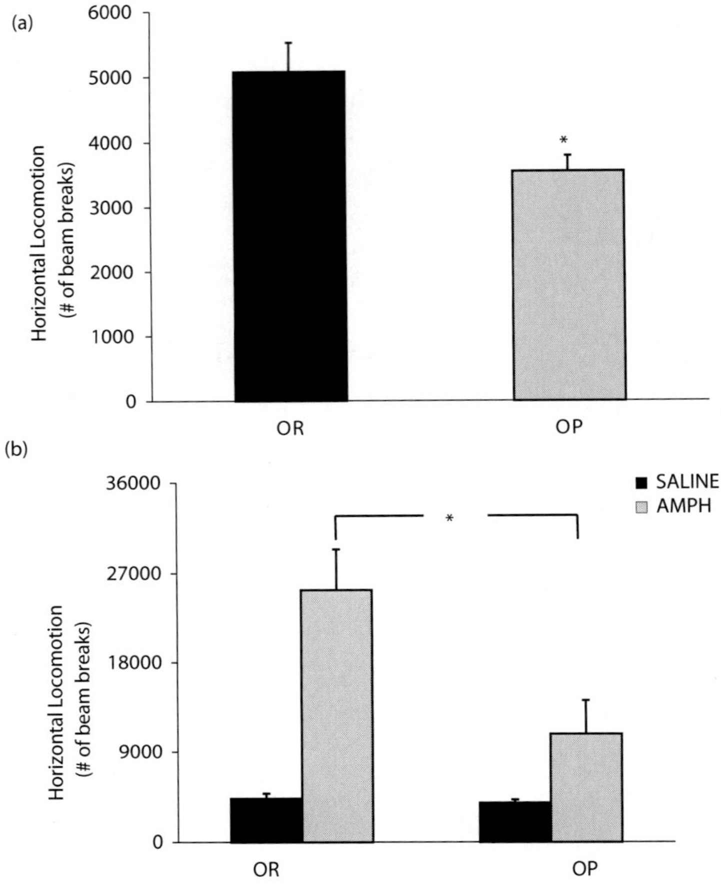

FIGURE 6.3 Reduced locomotion and response to amphetamine injection in juvenile obesity-prone (OP) rats. (a) Basal levels of horizontal locomotion were decreased in OP rats compared to obesity-resistant (OR) rats. (b) Response to amphetamine injection was also decreased in the OP rats compared to the OR rats. $n = 4$ animals per treatment group. $*p < 0.01$ by one-way ANOVA.

development of obesity in these rats, they exhibited deficits in basal dopamine levels in the nucleus accumbens and decreased dopaminergic response to a food challenge indicating that increased quantities of palatable food are necessary for them to have the same reward response as their obesity-resistant counterparts (Rada et al. 2010).

Imaging studies in obese humans support the findings so far in animal models. A fMRI study of adolescent girls following consumption of a palatable milkshake compared to a tasteless solution showed a negative correlation between BMI and activation of the caudate nucleus possibly due to depletions in dopamine receptor availability in this area (Stice et al. 2008). Positron emission tomography (PET) studies looking at D2 receptor activity indicate that activity at this receptor is decreased

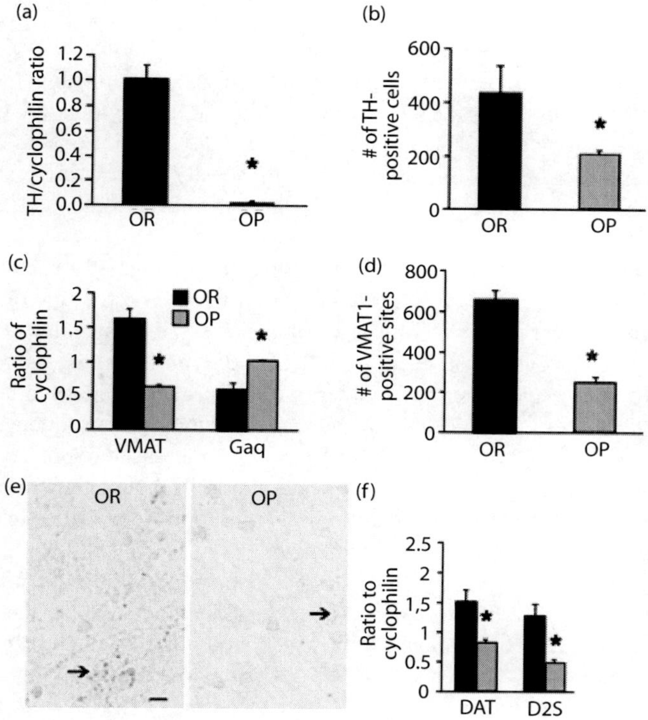

FIGURE 6.4 Lower mRNA and protein expression regulators of dopamine synthesis and exocytosis in VTA cell cultures from obesity-prone (OP) neonatal animals. (a) mRNA expression of TH was significantly lower in dopamine neuronal cultures derived from OP rats (gray bar) than in those from obesity-resistant (OR) rats (black bar, $n = 5$ pooled cultures/group in triplicate). (b) TH-immunopositive cells in cultures from OP rats ($n = 11$ cultures) were significantly less than those in cultures from OR rats ($n = 6$ cultures). (c) mRNA expression of the neuronal vesicular monoamine transporter VMAT2 was significantly lower while mRNA expression of endogenous VMAT2 downregulator $G\alpha_q$ was significantly higher in cultures from OP rats than in those from OR rats ($n = 5$ cultures/group). (d) VMAT2-immunopositive sites in cultures from OP rats ($n = 12$ cultures) were significantly less than those in sites from OR rats ($n = 16$ cultures). (e) Representative VMAT2 immunostaining in VTA cultures from OR and OP pups, seven days postplating. Arrows point to VMAT2-positive sites. Scale bar = 100 μm; ×20 view. (f) Lower mRNA expression of the DAT transporter and the presynaptic dopamine autoreceptor D2S in cultures from OP (gray bars) than OR neonates (black bars, $n = 5$ cultures/group). $*p < 0.01$. Data from B. M. Geiger et al. 2008. Reprinted with permission.

in obese subjects, similar to the decreased activity seen in people who are drug addicts (Wang et al. 2001, 2004). However, the ability to test the function of central dopamine in humans is still rudimentary, and more sophisticated techniques are required to fully understand whether the deficiencies in dopamine signaling seen in rodents are also present in obese humans.

Genetic Evidence of Reward Deficiency in Obesity

Studies investigating the contributing factors for susceptibility to obesity suggest that it is a result of a complex set of gene–gene and gene–environment interactions, including interactions with the diet (Loos 2009; Hetherington and Cecil 2010). Through various epidemiological and genome-wide association studies, several genes and gene loci have been associated with obesity. Interestingly, a number of the genes identified in these studies have also been linked to the central dopamine system or addiction.

Genome-wide association studies have shown the strongest link between the fat mass and obesity-associated (FTO) locus and obesity. While the role of this gene in central reward processing is still unknown, it is expressed in the nucleus accumbens shell at a low level (Olszewski et al. 2009). In the hypothalamus, its general function seems to be orexigenic since food deprivation has been shown to upregulate its expression (Olszewski et al. 2009). Because the functional consequence of the risk allele for increased BMI associated with the FTO is increased expression of FTO, it is possible that in obese models, FTO is upregulated in the nucleus accumbens as a result of a decreased dopamine signal coding for a food-deprived state and induces increased food intake through this mechanism.

Perhaps the most commonly linked gene to both addiction and obesity is the dopamine receptor D2 gene. In particular, the Taql A1 allele that has been associated with addiction has also been shown to be associated with a reduced striatal activity in obese individuals (Blum, et al. 1996; Stice et al. 2008). This finding is consistent with the growing body of evidence that dysregulation of the D2 receptor leads to attenuations in dopamine signaling in numerous models of dietary obesity (Wang et al. 2004; Geiger et al. 2008; Johnson and Kenny 2010). In fact, the D2 receptor knockout mouse exhibits increased food intake and other behavioral phenotypes that are typically attributed to dopamine deficiencies like reduced locomotor response to cocaine and amphetamine, even though dopamine levels in the midbrain seem to be normal in these animals (Holmes et al. 2004; Garcia-Tornadu et al. 2010). However, this could be due to compensatory mechanisms like increased expression of the plasma membrane dopamine active transporter (DAT) as a compensatory response, which is sufficient to replace dopamine but not the functional rewarding properties of dopamine mediated by the D2 receptor (Parish et al. 2002; Holmes et al. 2004).

Transgenerational Obesity and Reward Deficiency

Evidence of transmission of the obese phenotype from mother to child through the maternal diet indicates that epigenetic modifications of genes associated with central dopamine signaling may be a contributing factor to the development of decreased dopamine release in obesity (Levin et al. 1997; Boney et al. 2005; Buckley et al. 2005; Gorski et al. 2006; Chen et al. 2008; Geiger et al. 2008). One modification that is directly linked to diet is DNA methylation. Promoter regions of genes often include CpG islands where DNA methylation may occur. Diet composition may modify the amount of free methyl donors available and by changing the level of methylation in a gene's promoter region, the level of transcription of a specific gene may be altered.

A recent study in mice of the offspring of mothers fed a high-fat diet indicates that these offspring also have deficits in central dopamine signaling (Vucetic et al. 2010). An analysis of the DNA methylation status of key dopamine- and opioid-related genes in these animals indicate that in specific genes, like the dopamine reuptake transporter, there is decreased methylation of the promoter region which leads to increased expression of this gene. The overall effect of increased DAT levels is decreased synaptic dopamine, supporting the idea that deficiencies in this system can lead to hyperphagia, particularly in an environment of highly palatable food (Vucetic et al. 2010).

Paternal influences may also be important in the transmission of obesity from one generation to the next and therefore in central dopamine signaling. It has been recently shown that diet-induced obesity and excessive food intake can be transmitted through the paternal lineage (Yazbek et al. 2010). Because caloric restriction in these offspring seems to be the most effective mechanism by which diet-induced obesity is avoided, it is feasible that the affected allele results in changes in central dopamine signaling in the protected animals. Other evidence of paternal influences on the food reward system includes the relationship between obese children and alcoholic fathers (Fortuna 2010). While it is difficult to separate environmental influences in this case, a meta-analysis of the studies reporting an association of the Taq1a1 allele and alcoholism indicates that this is indeed a significant correlation (Munafo et al. 2007). Therefore, the link between alcoholism and obesity could be through this mechanism.

Central Modulators of Dopamine Signaling That Are Altered in Obesity

Recent studies have begun to evaluate interactions between homeostatic signaling pathways for energy homeostasis and hedonic pathways of food reward. In particular, receptors for both orexin and melanocortin, which are considered homeostatic energy balance signals, have been identified in the ventral tegmental area (VTA). Orexin has been shown to influence food intake and stimulates dopamine signaling. Orexin neurons project from the lateral hypothalamus to the VTA of the midbrain (Fadel and Deutch 2002). Within the VTA, orexin receptors have been identified on TH-positive neurons, and activation of orexin neurons has been shown to increase locomotor activity and dopamine release in mice (Korotkova et al. 2003; Narita et al. 2006). Furthermore, elevated levels of orexin have been linked to increased locomotion in obesity-resistant rats and daily injections of orexin into the paraventricular nucleus result in weight loss due to increased activity with little impact on food intake (Teske et al. 2006; Novak and Levine 2009). Obesity-prone animals are less sensitive to the locomotor effects of orexin resulting in decreased energy expenditure (Novak et al. 2006). Because dopamine also contributes to the amount of spontaneous locomotion in an animal, these differences could indicate that altered orexin signaling in obesity leads to lower dopaminergic tone. Further studies are needed to confirm this link.

Melanocortin-concentrating hormone (MCH) is another hypothalamic signal modifying food intake. The importance of melanocortin signaling on dietary obesity is evident due to the link between obesity and mutations in the MC4 receptor gene. Mutations in this gene account for almost 5 percent of early onset obesity in children, making it by far the most common single gene linked to obesity (Loos 2009; Vimaleswaran and Loos 2010). Furthermore, links between MCH signaling and dopamine have been described. The MCH –/– mouse is a lean mouse that exhibits elevated dopamine signal and sensitivity to psychostimulant activation of central dopamine pathways (Pissios et al. 2008). These results indicate that in addition to its known hypothalamic roles in energy balance, MCH also interacts with the food reward pathways in the midbrain.

Brain-derived neurotrophic factor (BDNF) deficiency is related to obesity in both humans and animals through mechanisms that trigger hyperphagia (Unger et al. 2007; Cordeira et al. 2010; Vimaleswaran and Loos 2010). Furthermore, studies in mice lacking BDNF in the ventral tegmental area indicate that these mice eat higher levels of high-fat food and their evoked dopamine release is lower than in wild-type controls. Peripheral administration of a D1 receptor agonist to these animals showed reduced consumption of the high-fat diet (Cordeira et al. 2010). These results indicate that deficiencies in BDNF are linked to deficiencies in central dopamine signaling, hyperphagia, and weight gain.

Peripheral Signals Influencing Hedonic Aspects of Food Intake

Peripheral signals may be contributing to the changes in central dopamine levels seen in obesity. Recent studies of humans and rats that have had bariatric surgery indicate that interventions in the gut structure change the perceived rewarding properties of food. Surgery patients have indicated that their sensitivity to sweet taste is altered following surgery (Burge et al. 1995; Miras and le Roux 2010). Along these same lines, two recent studies in rats have shown that following gastric bypass surgery and subsequent weight loss, rats have altered responses to sweet taste that likely make it less pleasant at higher concentrations (Hajnal et al. 2010; Shin et al. 2010). These results may indicate that the obese rats did not sense the rewarding properties of the sucrose solution at lower concentrations. However, how the central dopamine signaling pathways might change in the surgery group and what peripheral signals might interact with central dopamine in this model still need investigation.

One peripheral signal that has been shown to influence central dopamine neurotransmission is leptin. Leptin-deficient rodents and humans are known to exhibit excessive hyperphagia. Recently, the leptin receptor has been identified in the ventral tegmental area and substantia nigra including tyrosine-hydroxylase positive neurons (Fulton et al. 2006; Hommel et al. 2006; Leinninger et al. 2009; Opland, et al. 2010). The specific role of leptin in dopamine signaling is not entirely clear. In acute studies, it has been shown to increase activity of the central dopamine system (Perry et al. 2010). Furthermore, we have shown that evoked dopamine release and TH activity are reduced in the leptin deficient ob/ob mouse (Fulton et al. 2006). These deficits can be reversed with leptin treatment. Leptin treatment has also been shown to decrease food intake in models of leptin deficiency; however, leptin resistance remains a significant hurdle to treating common dietary obesity with leptin replacement therapy. Regardless, chronic depletions of leptin signaling appear to result in overall decreased dopaminergic tone and excessive hyperphagia supporting the reward deficiency hypothesis in these animals.

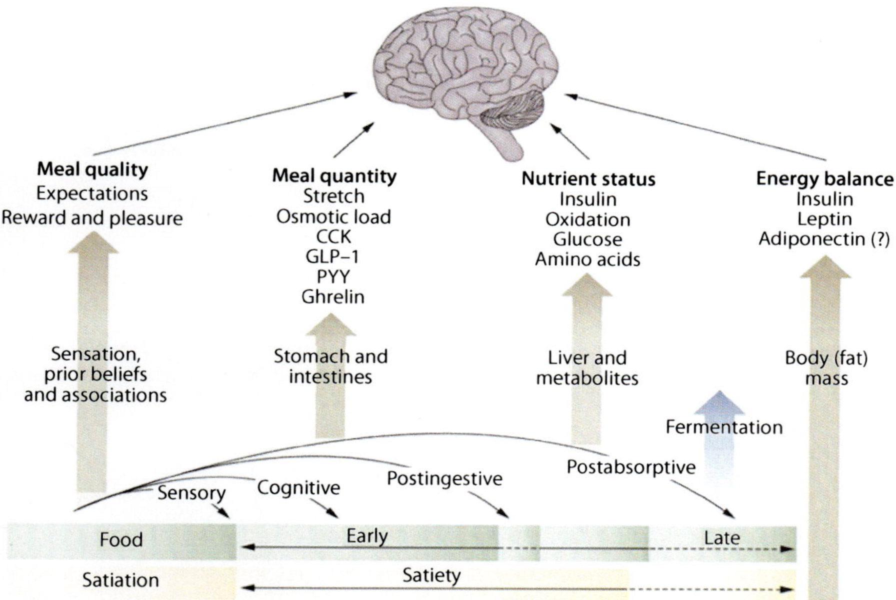

COLOR FIGURE 5.4 Behavioral and physiological events that occur following food intake. Satiety cascade from J. Blundell modified by D. Mela. Blundell et al.[115]

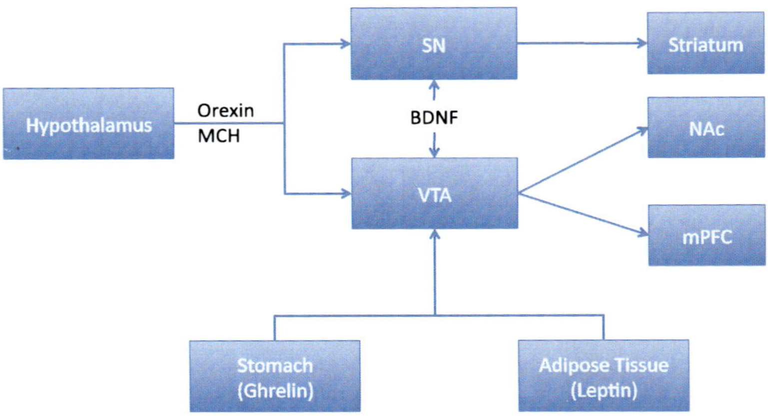

COLOR FIGURE 6.5 Schematic of the most significant central and peripheral signals influencing central dopamine neurotransmission linked to food intake. NAc = nucleus accumbens, mPFC = medial prefrontal cortex

COLOR FIGURE 9.6 Model depicting the influence of negative diet perceptions on the relationship between sweet food cravings and dietary adherence in GDM.

A second peripheral signal that may contribute to reward deficiency is ghrelin. Ghrelin is a 28 amino acid peptide that is released in the stomach prior to a meal and is involved in the control of food intake, growth hormone secretion and the development of adiposity (Kojima et al. 1999; Tschop et al. 2000; van der Lely et al. 2004; Abizaid et al. 2006). Studies looking at the effects of ghrelin in the midbrain have shown that acute treatment with ghrelin increases central dopamine signaling, likely in anticipation of food reward (Abizaid et al. 2006; Malik et al. 2008). Interestingly, circulating levels of ghrelin in obesity are decreased and evidence for ghrelin resistance in the hypothalamic NPY/AgRP neurons has been shown (Briggs et al. 2010). Because of ghrelin's influence on central dopamine signaling, the potential effect of decreased plasma ghrelin and/or ghrelin resistance in the ventral tegmental area in obese individuals would be an overall decreased dopaminergic tone.

Evidence Disputing the Reward Deficiency Hypothesis in Obesity

In some cases, it appears that reward deficiency does not explain hyperphagia and the development of obesity. It has been well established that several D2 receptor antagonists, like olanzapine and clozapine, used for the treatment of various mood disorders result in excessive amounts of weight gain. The accepted mechanism of action for these agents is that they block the D2 autoreceptor, which is a negative regulator of TH activity. This action allows for increased dopamine synthesis and, as a result, increased levels of dopamine in the synapse and indicates that elevated levels of dopamine can result in hyperphagia and excessive weight gain. However, a recent study in rodents shows that the weight gain associated with these drugs is related to a combined effect between the dopamine receptor antagonism and alterations in serotonin neurotransmission (Kirk et al. 2009; Reynolds and Kirk 2010). These results indicate that, in this case, the hyperphagia associated with these drugs is related to off-target effects of the drug on the serotonin system, likely in the hypothalamus.

A second hypothesis that is often thought to compete with the reward deficiency hypothesis is that those people who are susceptible to overconsumption of rewarding substances have an enhanced sensitivity to the rewarding properties of these substances. People who are overweight or obese would be more able to sense the rewarding properties of palatable food and therefore consume more than someone who maintains a normal weight (Davis and Fox 2008). One mutant mouse model that supports this hypothesis is the CLOCK mutant mouse. This mouse does not have a functional CLOCK gene and exhibits an increased response to the rewarding properties of cocaine. In addition, these animals exhibit increased food intake and increased TH levels. Increased sensitivity to the reward seems to result in the obese phenotype of these animals (McClung et al. 2005; Turek et al. 2005; Manev and Uz 2006). A recent study by Davis and Fox in overweight and obese humans may help reconcile the differences between the reward sensitization and reward deficiency hypothesis. Using a self-report questionnaire evaluating participants' sensitivity to reward, they found that BMI was positively correlated to reward sensitivity in overweight individuals, but as BMI reached obese levels, reward sensitivity decreased, indicating reward deficiencies in participants with the development of obesity (Davis and Fox 2008).

THE REWARD DEFICIENCY HYPOTHESIS IN EATING DISORDERS

Binge-Eating Disorder

Binge-eating disorder (BED) is characterized by recurrent episodes of distressing, uncontrollable eating of excessively large amounts of food (binge eating) without the inappropriate compensatory weight loss behaviors of bulimia nervosa or anorexia nervosa (Association 2000; Dingemans et al. 2002). The prevalence of BED in the U.S. general population is estimated to be 1–2 percent (Striegel-Moore and Franko 2003; Hudson et al. 2007), with a lifetime prevalence of 3 percent (Hudson et al. 2007). Up to 30 percent of obese individuals seeking weight loss treatment suffer from BED, and extreme weight gain is a common consequence of BED (Spitzer et al. 1993).

The type of food seems to be of importance for bingeing in BED. The afflicted patients prefer high-caloric food with high fat and sugar content (de Zwaan et al. 1992; Yanovski et al. 1992). This preference might be mediated by the heightened reward value of these types of foods. A number of researchers have demonstrated that reward functioning is altered in animal models of binge eating as well as in clinical studies of humans with BED. While animal models of binge eating match the criteria for consumption of a large quantity of food in a defined period of time (Corwin and Buda-Levin 2004), it is difficult to capture the subjective feelings of distress or loss of control seen in BED. Nevertheless, researchers try to mimic environmental precursors that lead to binge eating in humans, such as stress, dietary restriction, and intermittent exposure to palatable foods.

A number of researchers posit that some form of dietary restriction is a critical precipitant of binge eating (Howard and Porzelius 1999). These models employ restricting daily intake of caloric ration (Hagan et al. 2003), shortening the duration of daily food access (Giraudo et al. 1993; Inoue et al. 2004), combining food restriction and environmental stress (Hagan et al. 2002), or exposing animals to intermittent sugar and chow (Avena 2007). Restriction models of binge eating are designed to simulate dieting and cessation or "breaking" of a diet with palatable food that is typical of binge eating.

The intermittent exposure paradigm has the most robust effects on neurochemistry, leading to long-lasting changes in dopamine and opioid transmission (Colantuoni et al. 2001, 2002; Avena 2007). Colantuoni and colleagues have shown that binge-eating animals on an intermittent feeding schedule with twelve-hour access to sucrose solution and chow, followed by twelve-hour deprivation, show neurobiological changes similar to those evoked by drugs of abuse (Colantuoni et al. 2002). Specifically, these binge-eating rats have increased mu-opioid and D1 receptor binding (Colantuoni et al. 2001), as well as increased D3 mRNA in the NAc (Spangler et al. 2004). Moreover, rats maintained on a food deprivation and refeeding schedule are more responsive to the hyperphagic effects of the kappa opioid agonist, butorphanol, demonstrating that this pattern of eating leads to alterations within the endogenous opioid system (Hagan and Moss 1991; Boggiano et al. 2005). Kelley and colleagues have demonstrated a decrease in enkephalin, an endogenous opioid, gene expression in the NAc of binge-eating rats given restricted access to highly palatable food for two weeks (Kelley et al. 2003). After thirty-six hours of food deprivation, rats trained to binge eat sugar display significantly decreased dopamine release within the nucleus accumbens compared to rats that did not binge eat the sugar (Avena et al. 2008).

Taken together, these alterations in the dopamine and opioid systems of binge-eating animals may reflect changes in reward sensitivity. In particular, the decreased dopamine release and enkephalin gene expression align with the central reward deficits identified in obesity (Pothos et al. 1998; Geiger et al. 2008; Geiger et al. 2009) and other forms of addiction (Blum et al. 2000). The increased D1 receptor binding and D3 expression seen in these binge-eating animals are a plausibly compensatory upregulation in response to decreased neurotransmitter release.

In humans, alterations within the dopamine and endogenous opioid systems have been identified in individuals with BED. A significantly higher frequency of binge eaters present with short alleles (seven or nine repeats) of the variable number tandem repeats (VNTR) polymorphism in the 3′ untranslated region of the dopamine transporter (DAT1) gene (Shinohara et al. 2004). Subjects who carry a genotype containing a short allele demonstrate significantly lower dopamine transporter binding in the striatum than carriers of a long allele (Heinz et al. 2000). With less transporter binding, extracellular dopamine is left to bathe the synapse longer possibly resulting in desensitization and subsequent decrease in dopamine release. This would render predisposed individuals *hypo*-responsive to reward.

Alternatively, increases in extracellular dopamine could result in hypersensitivity to reward which some researchers posit as a contributing factor to binge eating (Davis et al. 2004, 2007; Franken and Muris 2005). For example, individuals with BED have been shown to have increased reward sensitivity when measured on the Sensitivity to Reward scale of the Sensitivity to Punishment and Sensitivity to Reward Questionnaire (Torrubia et al. 2002). Other researchers have demonstrated positive correlations between the degree of bingeing, BMI, and self-reports on reward sensitivity (Loxton and Dawe 2001; Franken and Muris 2005). A recent study by Schienle and colleagues used fMRI to explore whether BED patients would be characterized by an elevated reward sensitivity and increased activation in reward-processing brain areas while viewing pictures depicting high-caloric food (Schienle et al. 2009). They found heightened medial OFC reactivity in response to food cues in BED individuals as compared to both lean and obese nonbingeing controls. These authors suggest that reward drive may be translated into compulsive overeating in BED patients. Interestingly, an important role of the OFC has been implicated in drug-addictive states, especially in drug craving (Wang et al. 2004; Volkow and Wise 2005).

Polymorphisms in the dopamine D4 receptor gene (DRD4) have been associated with BED. Specifically, Levitan and colleagues found that carriers of the 7R allele had a significantly greater risk of binge eating over their lifetime than did 7R allele noncarriers (Levitan et al. 2004). The D4 receptor is considered to be D2-like in which the activated receptor inhibits the enzyme adenylate cyclase, thereby reducing the intracellular concentration of the second messenger cyclic AMP (Neve et al. 2004). The 7R allele of this receptor (also termed the "long" variant) results in decreased intracellular signaling (Asghari et al. 1995). Therefore, these authors hypothesized that individuals with low intrinsic dopamine activity in brain reward pathways attempt to compensate by using various reinforcing behaviors, including increased food consumption.

Other genetic studies have investigated the role of the dopamine D2 receptor in binge eating. There is reasonable evidence that individuals with the *Taq1*A+ allele (i.e., A1/A1 and A1/A2 genotypes) of the D2 receptor gene have reduced brain dopamine function compared to those with the A1– allele (i.e., the A2/A2 genotype) due to a 30–40 percent reduction in D2 receptor density in the striatal region (Jonsson et al. 1999; Noble 2003). While several studies have demonstrated a link between obesity and the *Taq1*A+ polymorphism, a recent study that broke down obese individuals by presence of BED found that only the group *without* BED showed this association (Davis et al. 2009). Dopamine receptor binding studies using PET could shed further light on whether individuals with BED demonstrate altered dopamine transmission.

Researchers have also looked at polymorphisms in the mu-opioid receptor (*OPRM1*) gene in obese individuals with and without BED (Davis et al. 2009). The rarer G118 allele of this gene has shown greater affinity for β-endorphin and morphine, but reduced mRNA and protein expression in vitro (Kroslak et al. 2007). In vivo evidence suggests a "gain-of-function" for those possessing the G allele, as seen, for instance, by increased reward from maternal attachment in rhesus macaque infants (Barr et al. 2008), increased alcohol stimulation in human subjects (Filbey et al. 2008), and a greater tendency for drug use and abuse in general (van den Wildenberg et al. 2007). The BED group had a greater than expected frequency of the "gain of function" G allele of the *OPRM1* (mu receptor). The authors suggest that the exaggerated response to opioid activation seen in this genotype may translate to heightened reward sensitivity to palatable foods which have also been shown to activate mu opioid receptors (Dum et al. 1983; Kirkham and Cooper 1988; Colantuoni et al. 2001, 2002; Kelley et al. 2002; Nathan and Bullmore 2009). Moreover, this begs the question of distinct mechanisms driving weight gain in obese populations with or without BED. The opposing evidence for both reward deficiency versus reward sensitivity in obese individuals may be reconciled by recognizing that perhaps both ends of the reward sensitivity spectrum predispose an individual to overeating and subsequent weight gain. Along these lines, obesity risk as a function of dopamine activity may be represented by an inverted "U" function with BED subjects polarizing on the upward sloping arm of the curve. Future studies may want to look at dopaminergic function in children at risk for obesity (demographically) and follow them longitudinally to see if initial dopamine or opioid receptor binding differences can predict those individuals who will later develop BED.

Bulimia Nervosa (BN)

Individuals with BN share a similar predisposition for compulsive overeating as those with BED. However, the binge in BN is followed by a compensatory behavior such as self-induced vomiting or excessive exercising. Animal models for this eating disorder are the same as those seen for BED and aim to mimic the binge-eating behavior. There have been several studies that have attempted to include the effects of purging in their models of binge eating. For example, Avena and colleagues measured dopaminergic function in "sham-fed" rats that were trained to binge eat using the previously described intermittent feeding schedule (Avena et al. 2006). The sham-feeding protocol involved gastric cannulas being placed in the stomachs of the animals and opened during food intake to empty ingested material. Even without macronutrient absorption, there were significant rises of dopamine in the NAc of these rats. Moreover, previous studies have indicated that acetylcholine (ACh) in the NAc is involved in the satiation process when dopamine rises during a meal (Mark et al. 1992). This rise in ACh was significantly reduced in sham-feeding rats (Avena et al. 2006), perhaps perpetuating binge-eating behavior.

Other protocols that have controlled for lack of postingestive effects have looked at changes in neurochemistry using orosensory stimuli alone. Rats that are fed saccharine, a noncaloric sugar substitute, still exhibit significant rises in accumbens dopamine (Mark et al. 1991). Hajnal et al. (2004) demonstrated the same effect of sweet taste on NAc dopamine when rats obtain sucrose by merely licking. Therefore, even with attenuated macronutrient absorption, binge eating in BN may alter dopamine transmission. Importantly, binge foods for individuals with BN are often sweet tasting and are high in sugar and fat (Drewnowski et al. 1992; Gendall et al. 1997), mimicking the palatable diet seen in binge-eating animal models. Furthermore, the intermittent feeding schedule is of particular relevance to individuals with BN where individuals often restrict intake in between bouts of bingeing.

In humans, several studies looking at dopaminergic function in BN suggest a hyporesponsive system similar to that seen in obesity. Jimerson and colleagues (1992) demonstrated that more frequently bingeing patients have reduced levels of the dopamine metabolite homovanillic acid (HVA) in cerebrospinal fluid and decreased plasma HVA concentrations. Although specific dopamine receptor studies in the CNS have not been performed with BN individuals, there is an association between BN and the Taq I A1 allele of the D2 receptor gene (Nisoli et al. 2007), and this polymorphism has been shown to correlate with lower D2 receptor density (Pohjalainen et al. 1998). Notably, both the Taq I A1 allele and the short *s* allele polymorphism discussed below are also significantly associated with substance abuse disorders (Blum et al. 1990; Comings et al. 1994; Noble 1994; Sander et al. 1997; Brody et al. 2006). Peripherally, individuals with BN have been shown to have reduced D2 receptor gene subscript which may be indicative of central receptor levels (Frieling et al. 2010). Individuals with lower levels of D2 receptors may be hyporesponsive to reward and in turn seek out rewarding stimuli with more intensity to compensate for this hypoactive reward circuitry (Wang et al. 2004). Moreover, the high frequency of the 7R allele of the DRD4 gene in BED is associated with greater BMI in women with BN (Kaplan et al. 2008). Greater weight gain may reflect more severe bingeing behavior in these individuals with this polymorphism that confers lower dopaminergic activity.

A number of researchers have also indicated aberrations in the dopamine transporter DAT in individuals with BN. This eating disorder is associated with the short *s* allele of the DAT gene (Shinohara et al. 2004) which confers lower transporter binding (Heinz et al. 2000). Concurrent with this genotype, Tauscher and colleagues used single photon emission computed tomography (SPECT) to demonstrate that subjects with BN have reduced striatal DAT availability compared to healthy controls and this difference was more pronounced with longer duration of illness (Tauscher et al. 2001). However, Frieling and colleagues more recently demonstrated that peripheral levels of DAT are significantly higher in individuals with BN compared to healthy controls (Frieling et al. 2010). This could reflect a difference in central versus peripheral dopamine signaling. Alternatively, levels of DAT may upregulate as the illness progresses to compensate for repeated dopamine surges during years of bingeing. This would effectively counteract extra levels of dopamine bathing the synapse. Indeed, BN subjects in Frieling's study reported nearly twice as long an illness duration (mean = 9 years) as subjects in Tauscher's study (mean = 5 years). Along these lines, BN may begin as a disease of hypersensitivity to reward, but over time result in a hyporesponsive state where binge eating continues as a means to compensate for lower basal levels of dopamine.

Alterations in reward function may also reflect manipulations in the opioid system. Individuals with BN have been found to have lower levels of β-endorphin, an endogenous opioid (Brewerton 1992). Furthermore, μ-opioid receptor binding in the insular cortex is decreased in BN subjects (Bencheriff 2005). The insular cortex has been repeatedly implicated in other reward-driven behaviors, including drug abuse (Garavan et al. 2000) and gambling (Elliott et al. 2000). This downregulation could be a compensatory response to repeated bingeing which has been shown to activate the opioid system (Dum et al. 1983; Kirkham and Cooper 1988; Colantuoni et al. 2002).

Unfortunately, most studies looking at bulimic individuals collapse purging and nonpurging groups into one cohort. This may mask two distinct disease states and may account for polarizing differences in reward sensitivity in these individuals. It would be interesting to see if the presence of purging predicts differences in dopaminergic activity in BN.

Anorexia Nervosa

Although the majority of this section has focused on eating disorders of overconsumption, it should also be noted that reward deficits may play a role in anorexia nervosa (AN). AN is marked by extreme weight loss, with individuals maintaining a body weight with a BMI of less than 18 kg/m^2. Individuals with AN have anhedonic, overcontrolled, perfectionist personalities, and demonstrate reduced novelty seeking (Sohlberg and Strober 1994; Anderluh et al. 2003). They seem to find little that is rewarding in life, aside from losing weight, suggesting that these individuals may suffer from the so-called reward deficiency syndrome (Blum et al. 1996). Furthermore, there is some indication that individuals with AN respond to typical and atypical neuroleptics (Cassano et al. 2003; Brewerton 2004), supporting a role for dopamine in the etiology of this disorder.

Several polymorphisms of the aforementioned DRD2 gene, including the −141 Indel variant (Bergen et al. 2005) and the Taq1 a1 allele (Nisoli et al. 2007), are associated with AN. Associations with the DRD4 gene have been met with mixed results. While some researchers found no significant link (Hinney et al. 1999; Karwautz et al. 2001), others report that a significant association exists between the C-521T SNP and AN (Bachner-Melman et al. 2007). This C allele has been shown to be associated with less transcriptional activity (Okuyama et al. 2000) and may reflect a hyporesponsive state in AN. During acute illness, Frieling and colleagues found significantly increased levels of DAT mRNA and significantly decreased levels of D2DR mRNA in blood of individuals with AN (Frieling et al. 2010). Increased DAT expression indicates greater dopamine uptake and, hence, shorter duration of action for released dopamine in the synapse. Furthermore, plasma levels of norepinephrine, a downstream molecule in the dopamine pathway; 24-h urinary excretion of HVA, a dopamine metabolite; and 3-methoxy-4-hydroxyphenylglycol (MHPG), a metabolite of norepinephrine, have been found to be lower in AN individuals during illness (Halmi et al. 1978; Gross et al. 1979). However, these researchers found that abnormalities normalized after weight restoration. Although it is possible that peripheral measures may not be in accordance with brain levels, these findings indicate decreased dopamine activity at least in acute AN.

Although starvation in itself induces changes in neurochemical and metabolic molecules, some aberrations may persist after recovery and weight restoration, indicating that they are not simply due to malnutrition, but rather a predisposing trait disturbance. For example, women recovered from restricting-type AN (AN-R) had a reduction of the dopamine metabolite homovanillic acid (HVA) in cerebrospinal fluid (CSF) compared with control women (CW) (Kaye et al. 1999). More recently, Frank and colleagues demonstrated that individuals recovered from AN have increased dopamine D2/D3 receptor binding in the antero-ventral striatum as measured by PET using [^{11}C]raclopride (Frank et al. 2005). Because PET measures of [^{11}C]raclopride binding are sensitive to endogenous dopamine concentrations (Drevets et al. 2001), this difference could be accounted for either by a reduction in the intrasynaptic dopamine concentration or by an elevation of the density and/or affinity of the D2/D3 receptors in this region. Increased receptor binding may reflect an upregulation to compensate for lower basal levels of catecholamine these individuals may have during illness (Halmi et al. 1978). However, it should be noted that this finding contrasts to the fact that obese individuals have been shown to have decreased D2 receptor binding (Wang, Volkow et al. 2001). Therefore, although patients in Frank's study were in recovery, it cannot be ruled out that D2/D3 receptor binding might be inversely related to weight and eating.

Wagner and colleagues have demonstrated that individuals recovered from AN also have altered reward processing of salient stimuli (Wagner et al. 2007). They measured striatal responses to a simple monetary reward task using event-related fMRI. In contrast to healthy controls, recovered AN subjects had similar activations to both positive and negative feedback suggesting that these individuals may have difficulty in distinguishing rewarding versus punishing stimuli. These alterations in reward processing may therefore be reflective of dopaminergic deficits seen in AN. Furthermore, because these individuals were weight restored and in recovery, it is plausible that aberrations in reward pathways may be a predisposing risk factor for this disease.

Excessive exercise is another common feature of AN. Individuals often report being "addicted" to activity and report that it is difficult to refrain from exercise despite many adverse physical consequences (e.g., an unhealthy decrease in body weight, decreased bone density, and stress fractures). Additionally, symptoms reminiscent of drug withdrawal, including anxiety, depression, and irritability, often develop when these individuals are unable to exercise (Adams 2002; Aidman and Woolard 2003; Allegre et al. 2006). Importantly, physical activity activates the mesolimbic dopamine system, increases dopamine release within reward pathways (Chaouloff 1989; Foley and Fleshner 2008; Smith et al. 2008), and causes release of endogenous opioid peptides in the brain (Hoffmann et al. 1990). In addition to the endogenous opioids released during starvation, the rewarding effects of exercise may further reinforce these maladaptive behaviors in a manner reminiscent of addiction.

As with BED and BN, limitations clearly exist using cross-sectional studies. All of these disorders wreak havoc on the body, and thus it cannot be ruled out that neurobiological differences seen in these individuals are the result of unhealthy dieting styles rather than predisposing risk factors. Future research should focus on collecting data in longitudinal cohorts to determine which neurochemical aberrations are present before disease onset.

CONCLUSIONS

The reward deficiency hypothesis appears to be a direction worth pursuing in current obesity and eating disorder research. An ever-growing body of literature indicates that decreased central dopamine signaling plays a significant role in the development of obesity as the result of chronic hyperphagia. Furthermore, alterations in this system are also present in other eating disorders and influenced by a number of central and peripheral signals that have been associated with metabolism and control of food intake (Figure 6.5). Efforts to combat these disorders

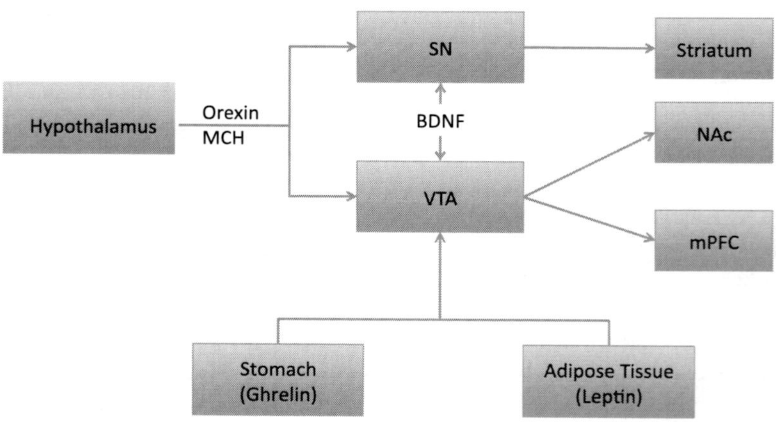

FIGURE 6.5 (*See color insert.*) Schematic of the most significant central and peripheral signals influencing central dopamine neurotransmission linked to food intake. NAc = nucleus accumbens, mPFC = medial prefrontal cortex.

through pharmacological and lifestyle interventions targeted at elevation of central dopamine may be critical as most other treatments have not been successful. For example, using drugs that are currently approved in other diseases where dopamine deficits occur, like Parkinson's disease, might provide viable strategies for treating obesity and eating disorders. Finally, by identifying early biomarkers of obesity predisposition that are related to dopamine signaling, we may be able to target dopaminergic interventions early in life to prevent or inhibit the onset of obesity in children who are otherwise susceptible to it.

ACKNOWLEDGMENTS

Studies from our laboratory and results presented in figures were made possible through the generous support of the National Institutes of Health (DK065872, ARRA 3R01DK065872, and F31DA023760) and a Smith Family Foundation Award of Excellence in Biomedical Research (ENP).

REFERENCES

Abizaid, A., Z. W. Liu, et al. 2006. Ghrelin modulates the activity and synaptic input organization of midbrain dopamine neurons while promoting appetite. *J Clin Invest* 116 (12): 3229–39.

Adams, B. B. 2002. Exercise-induced anaphylaxis in a marathon runner. *Int J Dermatol* 41 (7): 394–6.

Aidman, E. V., and S. Woolard. 2003. The influence of self-reported exercise addiction on acute emotional an physiological responses to brief exercise deprivation. *Psychol Sports Exerc* 4: 224–36.

Allegre, B., M. Souville, et al. 2006. Definitions and measures of exercise dependence. *Addict Res Theory* 14: 631–46.

American Psychiatric Association, ed. 2000. *Diagnostic and Statistical Manual of Mental Disorders* Washington, DC: Author.

American Psychiatric Association and American Psychiatric Association, Task Force on DSM-IV. 2000. *Diagnostic and Statistical Manual of Mental Disorders: DSM-IV*. Washington, DC: American Psychiatric Association.

Anderluh, M. B., K. Tchanturia, et al. 2003. Childhood obsessive-compulsive personality traits in adult women with eating disorders: defining a broader eating disorder phenotype. *Am J Psychiatry* 160 (2): 242–7.

Asghari, V., S. Sanyal, et al. 1995. Modulation of intracellular cyclic AMP levels by different human dopamine D4 receptor variants. *J Neurochem* 65 (3): 1157–65.

Avena, N. M. 2007. Examining the addictive-like properties of binge eating using an animal model of sugar dependence. *Exp Clin Psychopharmacol* 15 (5): 481–91.

Avena, N. M., P. Rada, et al. 2008. Evidence for sugar addiction: behavioral and neurochemical effects of intermittent, excessive sugar intake. *Neurosci Biobehav Rev* 32 (1): 20–39.

Avena, N. M., P. Rada, et al. 2006. Sucrose sham feeding on a binge schedule releases accumbens dopamine repeatedly and eliminates the acetylcholine satiety response. *Neuroscience* 139 (3): 813–20.

Bachner-Melman, R., E. Lerer, et al. 2007. Anorexia nervosa, perfectionism, and dopamine D4 receptor (DRD4). *Am J Med Genet B Neuropsychiatr Genet* 144B (6): 748–56.

Balleine, B. W., and A. Dickinson. 1998. Goal-directed instrumental action: contingency and incentive learning and their cortical substrates. *Neuropharmacology* 37 (4–5): 407–19.

Barr, C. S., M. L. Schwandt, et al. 2008. Variation at the mu-opioid receptor gene (OPRM1) influences attachment behavior in infant primates. *Proc Natl Acad Sci U S A* 105 (13): 5277–81.

Bergen, A. W., M. Yeager, et al. 2005. Association of multiple DRD2 polymorphisms with anorexia nervosa. *Neuropsychopharmacology* 30 (9): 1703–10.

Berridge, K. C., and T. E. Robinson. 1998. What is the role of dopamine in reward: hedonic impact, reward learning, or incentive salience? *Brain Res Brain Res Rev* 28 (3): 309–69.

Berthoud, H. R., and C. Morrison. 2008. The brain, appetite, and obesity. *Annu Rev Psychol* 59: 55–92.

Blum, K., E. R. Braverman, et al. 2000. Reward deficiency syndrome: a biogenetic model for the diagnosis and treatment of impulsive, addictive, and compulsive behaviors. *J Psychoactive Drugs* 32 (suppl. i–iv): 1–112.

Blum, K., E. R. Braverman, et al. 1996. Increased prevalence of the Taq I A1 allele of the dopamine receptor gene (DRD2) in obesity with comorbid substance use disorder: a preliminary report. *Pharmacogenetics* 6 (4): 297–305.

Blum, K., E. P. Noble, et al. 1990. Allelic association of human dopamine D2 receptor gene in alcoholism. *Jama* 263 (15): 2055–60.

Blum, K., P. J. Sheridan, et al. 1995. Dopamine D2 receptor gene polymorphisms in Scandinavian chronic alcoholics: a reappraisal. *Eur Arch Psychiatry Clin Neurosci* 245 (1): 50–2.

Blum, K., P. J. Sheridan, et al. 1995. Dopamine D2 receptor gene variants: association and linkage studies in impulsive-addictive-compulsive behaviour. *Pharmacogenetics* 5 (3): 121–41.

Blum, K., P. J. Sheridan, et al. 1996. The D2 dopamine receptor gene as a determinant of reward deficiency syndrome. *J R Soc Med* 89 (7): 396–400.

Boggiano, M. M., P. C. Chandler, et al. 2005. Combined dieting and stress evoke exaggerated responses to opioids in binge-eating rats. *Behav Neurosci* 119 (5): 1207–14.

Boney, C. M., A. Verma, et al. 2005. Metabolic syndrome in childhood: association with birth weight, maternal obesity, and gestational diabetes mellitus. *Pediatrics* 115 (3): e290–6.

Brewerton, T. D. 2004. 9th annual meeting of the Eating Disorders Research Society. *Expert Opin Investig Drugs* 13 (1): 73–8.

Briggs, D. I., P. J. Enriori, et al. 2010. Diet-induced obesity causes ghrelin resistance in arcuate NPY/AgRP neurons. *Endocrinology* 151 (10): 4745–55.

Brody, A. L., M. A. Mandelkern, et al. 2006. Gene variants of brain dopamine pathways and smoking-induced dopamine release in the ventral caudate/nucleus accumbens. *Arch Gen Psychiatry* 63 (7): 808–16.

Buckley, A. J., B. Keseru, et al. 2005. Altered body composition and metabolism in the male offspring of high fat-fed rats. *Metabolism* 54 (4): 500–7.

Burge, J. C., J. Z. Schaumburg, et al. 1995. Changes in patients' taste acuity after Roux-en-Y gastric bypass for clinically severe obesity. *J Am Diet Assoc* 95 (6): 666–70.

Cassano, G. B., M. Miniati, et al. 2003. Six-month open trial of haloperidol as an adjunctive treatment for anorexia nervosa: a preliminary report. *Int J Eat Disord* 33 (2): 172–7.

Chaouloff, F. 1989. Physical exercise and brain monoamines: a review. *Acta Physiol Scand* 137 (1): 1–13.

Chen, H., D. Simar, et al. 2008. Maternal and postnatal overnutrition differentially impact appetite regulators and fuel metabolism. *Endocrinology* 149 (11): 5348–56.

Cocores, J. A., and M. S. Gold. 2009. The Salted Food Addiction Hypothesis may explain overeating and the obesity epidemic. *Med Hypotheses* 73 (6): 892–9.

Colantuoni, C., P. Rada, et al. 2002. Evidence that intermittent, excessive sugar intake causes endogenous opioid dependence. *Obes Res* 10 (6): 478–88.

Colantuoni, C., J. Schwenker, et al. 2001. Excessive sugar intake alters binding to dopamine and mu-opioid receptors in the brain. *Neuroreport* 12 (16): 3549–52.

Comings, D. E., D. Muhleman, et al. 1994. The dopamine D2 receptor gene: a genetic risk factor in substance abuse. *Drug Alcohol Depend* 34 (3): 175–80.

Cooper, S. J., H. A. Al-Naser, et al. 2006. The anorectic effect of the selective dopamine D1-receptor agonist A-77636 determined by meal pattern analysis in free-feeding rats. *Eur J Pharmacol* 532 (3): 253–7.

Cooper, S. J., J. Francis, et al. 1990. The anorectic effect of SK&F 38393, a selective dopamine D1 receptor agonist: a microstructural analysis of feeding and related behaviour. *Psychopharmacology (Berl)* 100 (2): 182–7.

Cordeira, J. W., L. Frank, et al. 2010. Brain-derived neurotrophic factor regulates hedonic feeding by acting on the mesolimbic dopamine system. *J Neurosci* 30 (7): 2533–41.

Corsica, J. A., and B. J. Spring. 2008. Carbohydrate craving: a double-blind, placebo-controlled test of the self-medication hypothesis. *Eat Behav* 9 (4): 447–54.

Corwin, R. L., and A. Buda-Levin. 2004. Behavioral models of binge-type eating. *Physiol Behav* 82 (1): 123–30.

Cousins, M. S., and J. D. Salamone. 1996. Involvement of ventrolateral striatal dopamine in movement initiation and execution: a microdialysis and behavioral investigation. *Neuroscience* 70 (4): 849–59.

Davis, C., and J. Fox. 2008. Sensitivity to reward and body mass index (BMI): evidence for a non-linear relationship. *Appetite* 50 (1): 43–9.

Davis, C., K. Patte, et al. 2007. From motivation to behaviour: a model of reward sensitivity, overeating, and food preferences in the risk profile for obesity. *Appetite* 48 (1): 12–19.

Davis, C., S. Strachan, et al. 2004. Sensitivity to reward: implications for overeating and overweight. *Appetite* 42 (2): 131–8.

Davis, C. A., R. D. Levitan, et al. 2009. Dopamine for wanting and opioids for liking: a comparison of obese adults with and without binge eating. *Obesity (Silver Spring)* 17 (6): 1220–5.

Davis, J. F., A. L. Tracy, et al. 2008. Exposure to elevated levels of dietary fat attenuates psychostimulant reward and mesolimbic dopamine turnover in the rat. *Behav Neurosci* 122 (6): 1257–63.

de Zwaan, M., D. O. Nutzinger, et al. 1992. Binge eating in overweight women. *Compr Psychiatry* 33 (4): 256–61.

Di Chiara, G., G. Tanda, et al. 1999. Drug addiction as a disorder of associative learning. Role of nucleus accumbens shell/extended amygdala dopamine. *Ann N Y Acad Sci* 877: 461–85.

Dingemans, A. E., M. J. Bruna, et al. 2002. Binge eating disorder: a review. *Int J Obes Relat Metab Disord* 26 (3): 299–307.

Drevets, W. C., C. Gautier, et al. 2001. Amphetamine-induced dopamine release in human ventral striatum correlates with euphoria. *Biol Psychiatry* 49 (2): 81–96.

Drewnowski, A., D. D. Krahn, et al. 1992. Taste responses and preferences for sweet high-fat foods: evidence for opioid involvement. *Physiol Behav* 51 (2): 371–9.

Dum, J., C. Gramsch, et al. 1983. Activation of hypothalamic beta-endorphin pools by reward induced by highly palatable food. *Pharmacol Biochem Behav* 18 (3): 443–7.

Elliott, R., K. J. Friston, et al. 2000. Dissociable neural responses in human reward systems. *J Neurosci* 20 (16): 6159–65.

Elman, I., D. Borsook, et al. 2006. Food intake and reward mechanisms in patients with schizophrenia: implications for metabolic disturbances and treatment with second-generation antipsychotic agents. *Neuropsychopharmacology* 31 (10): 2091–2120.

Erlanson-Albertsson, C. 2005. How palatable food disrupts appetite regulation. *Basic Clin Pharmacol Toxicol* 97 (2): 61–73.

Fadel, J., and A. Y. Deutch. 2002. Anatomical substrates of orexin-dopamine interactions: lateral hypothalamic projections to the ventral tegmental area. *Neuroscience* 111 (2): 379–87.

Filbey, F. M., L. Ray, et al. 2008. Differential neural response to alcohol priming and alcohol taste cues is associated with DRD4 VNTR and OPRM1 genotypes. *Alcohol Clin Exp Res* 32 (7): 1113–23.

Foley, T. E., and M. Fleshner. 2008. Neuroplasticity of dopamine circuits after exercise: implications for central fatigue. *Neuromolecular Med* 10 (2): 67–80.

Ford, E. S., C. Li, et al. 2010. Trends in obesity and abdominal obesity among adults in the United States from 1999–2008. *Int J Obes (Lond)* 35 (5):736–43.

Fortuna, J. L. 2010. Sweet preference, sugar addiction and the familial history of alcohol dependence: shared neural pathways and genes. *J Psychoactive Drugs* 42 (2): 147–51.

Frank, G. K., U. F. Bailer, et al. 2005. Increased dopamine D2/D3 receptor binding after recovery from anorexia nervosa measured by positron emission tomography and [11c] raclopride. *Biol Psychiatry* 58 (11): 908–12.

Franken, I. H., and P. Muris. 2005. Individual differences in reward sensitivity are related to food craving and relative body weight in healthy women. *Appetite* 45 (2): 198–201.

Frieling, H., K. D. Romer, et al. 2010. Epigenetic dysregulation of dopaminergic genes in eating disorders. *Int J Eat Disord* 43 (7): 577–83.

Fulton, S., P. Pissios, et al. 2006. Leptin regulation of the mesoaccumbens dopamine pathway. *Neuron* 51 (6): 811–22.

Gallagher, M., R. W. McMahan, et al. 1999. Orbitofrontal cortex and representation of incentive value in associative learning. *J Neurosci* 19 (15): 6610–4.

Garavan, H., J. Pankiewicz, et al. 2000. Cue-induced cocaine craving: neuroanatomical specificity for drug users and drug stimuli. *Am J Psychiatry* 157 (11): 1789–98.

Garcia-Tornadu, I., M. I. Perez-Millan, et al. 2010. New insights into the endocrine and metabolic roles of dopamine D2 receptors gained from the Drd2 mouse. *Neuroendocrinology* 92 (4): 207–14.

Geiger, B. M., G. G. Behr, et al. 2008. Evidence for defective mesolimbic dopamine exocytosis in obesity-prone rats. *FASEB J* 22 (8): 2740–6.

Geiger, B. M., M. Haburcak, et al. 2009. Deficits of mesolimbic dopamine neurotransmission in rat dietary obesity. *Neuroscience* 159 (4): 1193–9.

Gendall, K. A., P. E. Sullivan, et al. 1997. The nutrient intake of women with bulimia nervosa. *Int J Eat Disord* 21 (2): 115–27.

Giraudo, S. Q., M. K. Grace, et al. 1993. Naloxone's anorectic effect is dependent upon the relative palatability of food. *Pharmacol Biochem Behav* 46 (4): 917–21.

Gorski, J. N., A. A. Dunn-Meynell, et al. 2006. Postnatal environment overrides genetic and prenatal factors influencing offspring obesity and insulin resistance. *Am J Physiol Regul Integr Comp Physiol* 291 (3): R768–78.

Gross, H. A., C. R. Lake, et al. 1979. Catecholamine metabolism in primary anorexia nervosa. *J Clin Endocrinol Metab* 49 (6): 805–9.

Ha, A. I.-N., and S. J. Cooper. 1994. A-68930, a novel, potent dopamine D1 receptor agonist: a microstructural analysis of its effects on feeding and other behaviour in the rat. *Behav Pharmacol* 5 (2): 210–18.

Hagan, M. M., P. C. Chandler, et al. 2003. The role of palatable food and hunger as trigger factors in an animal model of stress induced binge eating. *Int J Eat Disord* 34 (2): 183–97.

Hagan, M. M., and D. E. Moss. 1991. An animal model of bulimia nervosa: opioid sensitivity to fasting episodes. *Pharmacol Biochem Behav* 39 (2): 421–2.

Hagan, M. M., P. K. Wauford, et al. 2002. A new animal model of binge eating: key synergistic role of past caloric restriction and stress. *Physiol Behav* 77 (1): 45–54.

Hajnal, A., P. Kovacs, et al. 2010. Gastric bypass surgery alters behavioral and neural taste functions for sweet taste in obese rats. *Am J Physiol Gastrointest Liver Physiol* 299 (4): G967–79.

Hajnal, A., G. P. Smith, et al. 2004. Oral sucrose stimulation increases accumbens dopamine in the rat. *Am J Physiol Regul Integr Comp Physiol* 286 (1): R31–7.

Halmi, K. A., H. Dekirmenjian, et al. 1978. Catecholamine metabolism in anorexia nervosa. *Arch Gen Psychiatry* 35 (4): 458–60.

Hammond, R. A., and R. Levine. 2010. The economic impact of obesity in the United States. *Diabetes, Metabolic Syndrome and Obesity: Targets and Therapy* 3: 285–95.

Harris, S. C., A. C. Ivy, et al. 1947. The mechanism of amphetamine-induced loss of weight; a consideration of the theory of hunger and appetite. *J Am Med Assoc* 134 (17): 1468–75.

Heinz, A., D. Goldman, et al. 2000. Genotype influences in vivo dopamine transporter availability in human striatum. *Neuropsychopharmacology* 22 (2): 133–9.

Hetherington, M. M., and J. E. Cecil. 2010. Gene-environment interactions in obesity. *Forum Nutr* 63: 195–203.

Hinney, A., J. Schneider, et al. 1999. No evidence for involvement of polymorphisms of the dopamine D4 receptor gene in anorexia nervosa, underweight, and obesity. *Am J Med Genet* 88 (6): 594–7.

Hoffmann, P., L. Terenius, et al. 1990. Cerebrospinal fluid immunoreactive beta-endorphin concentration is increased by voluntary exercise in the spontaneously hypertensive rat. *Regul Pept* 28 (2): 233–9.

Holmes, A., J. E. Lachowicz, et al. 2004. Phenotypic analysis of dopamine receptor knockout mice; recent insights into the functional specificity of dopamine receptor subtypes. *Neuropharmacology* 47 (8): 1117–34.

Holtzman, S. G. 1974. Behavioral effects of separate and combined administration of naloxone and d-amphetamine. *J Pharmacol Exp Ther* 189 (1): 51–60.

Hommel, J. D., R. Trinko, et al. 2006. Leptin receptor signaling in midbrain dopamine neurons regulates feeding. *Neuron* 51 (6): 801–10.

Howard, C. E., and L. K. Porzelius. 1999. The role of dieting in binge eating disorder: etiology and treatment implications. *Clin Psychol Rev* 19(1): 25–44.

Hudson, J. I., E. Hiripi, et al. 2007. The prevalence and correlates of eating disorders in the National Comorbidity Survey Replication. *Biol Psychiatry* 61 (3): 348–58.

Ifland, J. R., H. G. Preuss, et al. 2009. Refined food addiction: a classic substance use disorder. *Med Hypotheses* 72 (5): 518–26.

Inoue, K., E. P. Zorrilla, et al. 2004. Reduction of anxiety after restricted feeding in the rat: implication for eating disorders. *Biol Psychiatry* 55 (11): 1075–81.

Jimerson, D. C., M. D. Lesem, et al. 1992. Low serotonin and dopamine metabolite concentrations in cerebrospinal fluid from bulimic patients with frequent binge episodes. *Arch Gen Psychiatry* 49 (2): 132–8.

Johnson, P. M., and P. J. Kenny. 2010. Dopamine D2 receptors in addiction-like reward dysfunction and compulsive eating in obese rats. *Nat Neurosci* 13 (5): 635–41.

Jonsson, E. G., M. M. Nothen, et al. 1999. Polymorphisms in the dopamine D2 receptor gene and their relationships to striatal dopamine receptor density of healthy volunteers. *Mol Psychiatry* 4 (3): 290–6.

Kaplan, A. S., R. D. Levitan, et al. 2008. A DRD4/BDNF gene-gene interaction associated with maximum BMI in women with bulimia nervosa. *Int J Eat Disord* 41 (1): 22–8.

Karwautz, A., S. Rabe-Hesketh, et al. 2001. Individual-specific risk factors for anorexia nervosa: a pilot study using a discordant sister-pair design. *Psychol Med* 31 (2): 317–29.

Kaye, W. H., G. K. Frank, et al. 1999. Altered dopamine activity after recovery from restricting-type anorexia nervosa. *Neuropsychopharmacology* 21 (4): 503–6.

Kelley, A. E., V. P. Bakshi, et al. 2002. Opioid modulation of taste hedonics within the ventral striatum. *Physiol Behav* 76 (3): 365–77.

Kelley, A. E., M. J. Will, et al. 2003. Restricted daily consumption of a highly palatable food (chocolate Ensure®) alters striatal enkephalin gene expression. *Eur J Neurosci* 18 (9): 2592–8.

Kirk, S. L., J. Glazebrook, et al. 2009. Olanzapine-induced weight gain in the rat: role of 5-HT2C and histamine H1 receptors. *Psychopharmacology (Berl)* 207 (1): 119–25.

Kirkham, T. C., and S. J. Cooper. 1988. Attenuation of sham feeding by naloxone is stereospecific: evidence for opioid mediation of orosensory reward. *Physiol Behav* 43 (6): 845–7.

Knutson, B., C. M. Adams, et al. 2001. Anticipation of increasing monetary reward selectively recruits nucleus accumbens. *J Neurosci* 21 (16): RC159.

Kojima, M., H. Hosoda, et al. 1999. Ghrelin is a growth-hormone-releasing acylated peptide from stomach. *Nature* 402 (6762): 656–60.

Korotkova, T. M., O. A. Sergeeva, et al. 2003. Excitation of ventral tegmental area dopaminergic and nondopaminergic neurons by orexins/hypocretins. *J Neurosci* 23 (1): 7–11.

Kroslak, T., K. S. Laforge, et al. 2007. The single nucleotide polymorphism A118G alters functional properties of the human mu opioid receptor. *J Neurochem* 103 (1): 77–87.

Leinninger, G. M., Y. H. Jo, et al. 2009. Leptin acts via leptin receptor-expressing lateral hypothalamic neurons to modulate the mesolimbic dopamine system and suppress feeding. *Cell Metab* 10 (2): 89–98.

Levin, B. E., A. A. Dunn-Meynell, et al. 1997. Selective breeding for diet-induced obesity and resistance in Sprague-Dawley rats. *Am J Physiol* 273 (2 pt. 2): R725–30.

Levine, A. S., C. M. Kotz, et al. 2003. Sugars and fats: the neurobiology of preference. *J Nutr* 133 (3): 831S–834S.

Levine, A. S., D. T. Weldon, et al. 1995. Naloxone blocks that portion of feeding driven by sweet taste in food-restricted rats. *Am J Physiol* 268 (1 pt. 2): R248–52.

Levitan, R. D., M. Masellis, et al. 2004. The dopamine-4 receptor gene associated with binge eating and weight gain in women with seasonal affective disorder: an evolutionary perspective. *Biol Psychiatry* 56 (9): 665–9.

Loos, R. J. 2009. Recent progress in the genetics of common obesity. *Br J Clin Pharmacol* 68 (6): 811–29.

Loxton, N. J., and S. Dawe. 2001. Alcohol abuse and dysfunctional eating in adolescent girls: the influence of individual differences in sensitivity to reward and punishment. *Int J Eat Disord* 29 (4): 455–62.

Lustig, R. H. 2010. Fructose: metabolic, hedonic, and societal parallels with ethanol. *J Am Diet Assoc* 110 (9): 1307–21.

Malik, S., F. McGlone, et al. 2008. Ghrelin modulates brain activity in areas that control appetitive behavior. *Cell Metab* 7 (5): 400–9.

Manev, H., and T. Uz. 2006. Clock genes as a link between addiction and obesity. *Eur J Hum Genet* 14 (1): 5.

Mark, G. P., D. S. Blander, et al. 1991. A conditioned stimulus decreases extracellular dopamine in the nucleus accumbens after the development of a learned taste aversion. *Brain Res* 551 (1–2): 308–10.

Mark, G. P., P. Rada, et al. 1992. Effects of feeding and drinking on acetylcholine release in the nucleus accumbens, striatum, and hippocampus of freely behaving rats. *J Neurochem* 58 (6): 2269–74.

McClung, C. A., K. Sidiropoulou, et al. 2005. Regulation of dopaminergic transmission and cocaine reward by the Clock gene. *Proc Natl Acad Sci USA* 102 (26): 9377–81.

Miras, A. D., and C. W. le Roux. 2010. Bariatric surgery and taste: novel mechanisms of weight loss. *Curr Opin Gastroenterol* 26 (2): 140–5.

Munafo, M. R., I. J. Matheson, et al. 2007. Association of the DRD2 gene Taq1A polymorphism and alcoholism: a meta-analysis of case-control studies and evidence of publication bias. *Mol Psychiatry* 12 (5): 454–61.

Narita, M., Y. Nagumo, et al. 2006. Direct involvement of orexinergic systems in the activation of the mesolimbic dopamine pathway and related behaviors induced by morphine. *J Neurosci* 26 (2): 398–405.

Nathan, P. J., and E. T. Bullmore. 2009. From taste hedonics to motivational drive: central mu-opioid receptors and binge-eating behaviour. *Int J Neuropsychopharmacol* 1–14.

Neve, K. A., J. K. Seamans, et al. 2004. Dopamine receptor signaling. *J Recept Signal Transduct Res* 24 (3): 165–205.

Nisoli, E., A. Brunani, et al. 2007. D2 dopamine receptor (DRD2) gene Taq1A polymorphism and the eating-related psychological traits in eating disorders (anorexia nervosa and bulimia) and obesity. *Eat Weight Disord* 12 (2): 91–6.

Noble, E. P. 1994. Polymorphisms of the D2 dopamine receptor gene and alcoholism and other substance use disorders. *Alcohol Alcohol Suppl* 2: 35–43.

Noble, E. P. 2003. D2 dopamine receptor gene in psychiatric and neurologic disorders and its phenotypes. *Am J Med Genet B Neuropsychiatr Genet* 116B (1): 103–25.

Novak, C. M., C. M. Kotz, et al. 2006. Central orexin sensitivity, physical activity, and obesity in diet-induced obese and diet-resistant rats. *Am J Physiol Endocrinol Metab* 290 (2): E396–403.

Novak, C. M., and J. A. Levine. 2009. Daily intraparaventricular orexin-A treatment induces weight loss in rats. *Obesity (Silver Spring)* 17 (8): 1493–8.

O'Doherty, J. P., R. Deichmann, et al. 2002. Neural responses during anticipation of a primary taste reward. *Neuron* 33 (5): 815–26.

Ogden, C. L., M. D. Carroll, et al. 2006. Prevalence of overweight and obesity in the United States, 1999–2004. *JAMA* 295 (13): 1549–55.

Ogden, C. L., M. D. Carroll, et al. 2008. High body mass index for age among US children and adolescents, 2003–2006. *JAMA* 299 (20): 2401–5.

Okuyama, Y., H. Ishiguro, et al. 2000. Identification of a polymorphism in the promoter region of DRD4 associated with the human novelty seeking personality trait. *Mol Psychiatry* 5 (1): 64–9.

Olszewski, P. K., R. Fredriksson, et al. 2009. Hypothalamic FTO is associated with the regulation of energy intake not feeding reward. *BMC Neurosci* 10: 129.

Opland, D. M., G. M. Leinninger, et al. 2010. Modulation of the mesolimbic dopamine system by leptin. *Brain Res* 1350: 65–70.

Parish, C. L., D. Stanic, et al. 2002. Effects of long-term treatment with dopamine receptor agonists and antagonists on terminal arbor size. *Eur J Neurosci* 16 (5): 787–94.

Perry, M. L., G. M. Leinninger, et al. 2010. Leptin promotes dopamine transporter and tyrosine hydroxylase activity in the nucleus accumbens of Sprague-Dawley rats. *J Neurochem* 114 (3): 666–74.

Phillips, A. G., L. J. Atkinson, et al. 1993. Increased extracellular dopamine in the nucleus accumbens of the rat elicited by a conditional stimulus for food: an electrochemical study. *Can J Physiol Pharmacol* 71 (5–6): 387–93.

Pickering, C., J. Alsio, et al. 2009. Withdrawal from free-choice high-fat high-sugar diet induces craving only in obesity-prone animals. *Psychopharmacology (Berl)* 204 (3): 431–43.

Pissios, P., L. Frank, et al. 2008. Dysregulation of the mesolimbic dopamine system and reward in MCH-/- mice. *Biol Psychiatry* 64 (3): 184–91.

Pohjalainen, T., J. O. Rinne, et al. 1998. The A1 allele of the human D2 dopamine receptor gene predicts low D2 receptor availability in healthy volunteers. *Mol Psychiatry* 3 (3): 256–60.

Pothos, E. N., D. Sulzer, et al. 1998. Plasticity of quantal size in ventral midbrain dopamine neurons: possible implications for the neurochemistry of feeding and reward. *Appetite* 31: 405.

Rada, P., N. M. Avena, et al. 2005. Daily bingeing on sugar repeatedly releases dopamine in the accumbens shell. *Neuroscience* 134 (3): 737–44.

Rada, P., M. E. Bocarsly, et al. 2010. Reduced accumbens dopamine in Sprague-Dawley rats prone to overeating a fat-rich diet. *Physiol Behav* 101 (3): 394–400.

Reynolds, G. P., and S. L. Kirk. 2010. Metabolic side effects of antipsychotic drug treatment: pharmacological mechanisms. *Pharmacol Ther* 125 (1): 169–79.

Richardson, N. R., and A. Gratton. 1996. Behavior-relevant changes in nucleus accumbens dopamine transmission elicited by food reinforcement: an electrochemical study in rat. *J Neurosci* 16 (24): 8160–9.

Rogers, P. J., and H. J. Smit. 2000. Food craving and food addiction: a critical review of the evidence from a biopsychosocial perspective. *Pharmacol Biochem Behav* 66 (1): 3–14.

Salamone, J. D., M. S. Cousins, et al. 1994. Nucleus accumbens dopamine release increases during instrumental lever pressing for food but not free food consumption. *Pharmacol Biochem Behav* 49 (1): 25–31.

Sander, T., H. Harms, et al. 1997. Allelic association of a dopamine transporter gene polymorphism in alcohol dependence with withdrawal seizures or delirium. *Biol Psychiatry* 41 (3): 299–304.

Schienle, A., A. Schafer, et al. 2009. Binge-eating disorder: reward sensitivity and brain activation to images of food. *Biol Psychiatry* 65 (8): 654–61.

Shin, A. C., H. Zheng, et al. 2010. Roux-en-Y gastric bypass surgery changes food reward in rats. *Int J Obes (Lond).*

Shinohara, M., H. Mizushima, et al. 2004. Eating disorders with binge-eating behaviour are associated with the s allele of the 3′-UTR VNTR polymorphism of the dopamine transporter gene. *J Psychiatry Neurosci* 29 (2): 134–7.

Smith, G. P., and L. H. Schneider. 1988. Relationships between mesolimbic dopamine function and eating behavior. *Ann N Y Acad Sci* 537: 254–61.

Smith, M. A., S. R. Gergans, et al. 2008. Chronic exercise increases sensitivity to the conditioned rewarding effects of cocaine. *Pharmacol Rep* 60 (4): 561–5.

Sohlberg, S., and M. Strober. 1994. Personality in anorexia nervosa: an update and a theoretical integration. *Acta Psychiatr Scand Suppl* 378: 1–15.

Spangler, R., K. M. Wittkowski, et al. 2004. Opiate-like effects of sugar on gene expression in reward areas of the rat brain. *Brain Res Mol Brain Res* 124 (2): 134–42.

Spitzer, R. L., S. Yanovski, et al. 1993. Binge eating disorder: its further validation in a multisite study. *Int J Eat Disord* 13 (2): 137–53.

Spring, B., K. Schneider, et al. 2008. Abuse potential of carbohydrates for overweight carbohydrate cravers. *Psychopharmacology (Berl)* 197 (4): 637–47.

Stice, E., S. Spoor, et al. 2008. Relation between obesity and blunted striatal response to food is moderated by TaqIA A1 allele. *Science* 322 (5900): 449–52.

Stice, E., S. Spoor, et al. 2008. Relation of reward from food intake and anticipated food intake to obesity: a functional magnetic resonance imaging study. *J Abnorm Psychol* 117 (4): 924–35.

Striegel-Moore, R. H., and D. L. Franko. 2003. Epidemiology of binge eating disorder. *Int J Eat Disord* 34 (suppl.): S19–29.

Tauscher, J., W. Pirker, et al. 2001. [123I] beta-CIT and single photon emission computed tomography reveal reduced brain serotonin transporter availability in bulimia nervosa. *Biol Psychiatry* 49 (4): 326–32.

Teske, J. A., A. S. Levine, et al. 2006. Elevated hypothalamic orexin signaling, sensitivity to orexin A, and spontaneous physical activity in obesity-resistant rats. *Am J Physiol Regul Integr Comp Physiol* 291 (4): R889–99.

Torrubia, R., C. Avila, et al. 2002. The Sensitivity to Punishment and Sensitivity to Reward Questionnaire (SPSRQ) as a measure of Gray's anxiety and impulsivity dimensions. *Personality and Individual Differences* 31: 837–62.

Tschop, M., D. L. Smiley, et al. 2000. Ghrelin induces adiposity in rodents. *Nature* 407 (6806): 908–13.

Turek, F. W., C. Joshu, et al. 2005. Obesity and metabolic syndrome in circadian *Clock* mutant mice. *Science* 308 (5724): 1043–5.

Unger, T. J., G. A. Calderon, et al. 2007. Selective deletion of Bdnf in the ventromedial and dorsomedial hypothalamus of adult mice results in hyperphagic behavior and obesity. *J Neurosci* 27 (52): 14265–74.

van den Wildenberg, E., R. W. Wiers, et al. 2007. A functional polymorphism of the mu-opioid receptor gene (OPRM1) influences cue-induced craving for alcohol in male heavy drinkers. *Alcohol Clin Exp Res* 31 (1): 1–10.

van der Hoek, G. A. and S. J. Cooper. 1994. The selective dopamine uptake inhibitor GBR 12909: its effects on the microstructure of feeding in rats. *Pharmacol Biochem Behav* 48 (1): 135–40.

van der Lely, A. J., M. Tschop, et al. 2004. Biological, physiological, pathophysiological, and pharmacological aspects of ghrelin. *Endocr Rev* 25 (3): 426–57.

Vimaleswaran, K. S. and R. J. Loos. 2010. Progress in the genetics of common obesity and type 2 diabetes. *Expert Rev Mol Med* 12:e7.

Volkow, N. D., J. S. Fowler, et al. 2004. The addicted human brain viewed in the light of imaging studies: brain circuits and treatment strategies. *Neuropharmacology* 47 (suppl. 1): 3–13.

Volkow, N. D., J. S. Fowler, et al. 2004. Dopamine in drug abuse and addiction: results from imaging studies and treatment implications. *Mol Psychiatry* 9 (6): 557–69.

Volkow, N. D. and R. A. Wise. 2005. How can drug addiction help us understand obesity? *Nat Neurosci* 8 (5): 555–60.

Vucetic, Z., J. Kimmel, et al. 2010. Maternal high-fat diet alters methylation and gene expression of dopamine and opioid-related genes. *Endocrinology* 151 (10): 4756–64.

Vucetic, Z. and T. M. Reyes. 2010. Central dopaminergic circuitry controlling food intake and reward: implications for the regulation of obesity. *Wiley Interdiscip Rev Syst Biol Med* 2 (5): 577–93.

Wagner, A., H. Aizenstein, et al. 2007. Altered reward processing in women recovered from anorexia nervosa. *Am J Psychiatry* 164 (12): 1842–9.

Wang, G. J., N. D. Volkow, et al. 2001. Brain dopamine and obesity. *Lancet* 357 (9253): 354–7.

Wang, G. J., N. D. Volkow, et al. 2004. Similarity between obesity and drug addiction as assessed by neurofunctional imaging: a concept review. *J Addict Dis* 23 (3): 39–53.

Yanovski, S. Z., M. Leet, et al. 1992. Food selection and intake of obese women with binge-eating disorder. *Am J Clin Nutr* 56 (6): 975–80.

Yazbek, S. N., S. H. Spiezio, et al. 2010. Ancestral paternal genotype controls body weight and food intake for multiple generations. *Hum Mol Genet* 19 (21): 4134–44.

Yeomans, M. R., and R. W. Gray. 2002. Opioid peptides and the control of human ingestive behaviour. *Neurosci Biobehav Rev* 26(6): 713–28.

York, D. A., L. Teng, et al. 2010. Effects of dietary fat and enterostatin on dopamine and 5-hydroxytrytamine release from rat striatal slices. *Brain Res* 1349: 48–55.

7 Potential Consequences of Obesity on Cognitive Behavior

Nicole A. Jurdak and Robin B. Kanarek

CONTENTS

Introduction 133
Associations between Obesity and Brain Function in Human Populations 134
Adults 134
Distribution of Body Fat and Cognitive Behavior 135
Sex Differences 136
Obesity and Dementia 136
Children and Adolescents 137
Obesity and Brain Function in Experimental Animals 139
Cognitive Assessments 139
Diet Composition 140
Duration of Feeding 141
Sex Differences 141
Potential Mechanisms Underlying the Interactions among Obesity, Brain Function, and Behavior 141
Obesity, Brain Structure, and Function 142
Hippocampus 142
Obesity, Glucose Regulation, and Brain Functioning 143
Obesity, Oxidative Stress, and Inflammation 144
Brain-Derived Neurotrophic Factor (BDNF) 144
Lipids and Cognitive Functioning 145
Future Directions 145
References 146

INTRODUCTION

Obesity is one of the most important public health challenges facing our nation. Almost 70 percent of adult Americans are overweight; and of those individuals, approximately one-half can be classified as obese (Flegal et al. 2010). Younger individuals are not immune to the development of obesity. Over the past two decades, the prevalence of obesity in this country has more than doubled for children ages 2–5 (5.0 percent to 12.4 percent) and 6–11 years (6.5 percent to 17.0 percent); and more than tripled for adolescents ages 12–19 years (5.0 percent to 17.6 percent) (Centers for Disease Control and Prevention 2009). Moreover, while once primarily limited to the United States and other Western societies, obesity is rapidly becoming a global problem. It is estimated that worldwide 300 million adults are obese and 1 billion are overweight (World Health Organization 2009). As in the United States, obesity is not limited to adults. In 2010, 42 million children under the age of five were reported to be overweight with close to 35 million of these children living in developing countries (World Health Organization 2011).

Obesity is a predisposing factor for a number of chronic illnesses, including cardiovascular disease, hypertension, type 2 diabetes, asthma, and cancer (Duncan 2006; Saland 2007; Finer 2011). Moreover, there is growing evidence that obesity is linked to impairments in brain functioning, and consequently cognitive behavior. These alterations in mental functioning may be mediated, at least in part, by the comorbid conditions associated with obesity. For example, it is well documented that both type 2 diabetes and cardiovascular disease increase the risk of cognitive dysfunctions. There is also emerging support for the proposal that obesity can directly impair mental activity. These impairments may have lifelong consequences. As described in this chapter, obesity has been associated with learning deficits in children, problems with executive functioning in adults, and accelerated cognitive decline in older individuals.

This chapter will evaluate the evidence for a relationship between obesity, brain function, and behavior. The first portion of the chapter will review data from studies examining this relationship in human populations. Second, to provide information on the effects of obesity on neural functioning and behavior, without the confounds of the social and environmental variables which can influence the results of human research, studies using animal models of obesity will be discussed. Finally, proposed mechanisms for the relationship between excess adiposity and decrements in cognitive behavior will be presented.

ASSOCIATIONS BETWEEN OBESITY AND BRAIN FUNCTION IN HUMAN POPULATIONS

A variety of strategies have been used to assess the potential effects of obesity on cognitive behavior in our own species. Some researchers have used epidemiological data to determine the relationship between body weight and cognitive functioning, while others have compared performance of lean and obese individuals on a variety of cognitive tasks measuring attention, decision making, learning, and memory. Additionally, with increasing frequency, researchers are making use of neuroimaging technology (e.g., functional magnetic resonance imaging [fMRI] and positron emission tomography [PET]) to more directly examine brain functioning in obese individuals.

Adults

During the first decade of the twenty-first century, awareness that obesity may negatively impact brain functioning and cognitive behavior in adults increased. Initial evidence of the detrimental effects of obesity on mental functioning came from epidemiological studies comparing the relationship between obesity and cognitive behavior in healthy men and women (Elias et al. 2003; Cournot et al. 2006; Dore et al. 2008). For example, Elias and colleagues (2003) reported that obesity and hypertension in late middle-aged and elderly men, but not women, were associated with poorer cognitive functioning as measured by standard neuropsychological tests. Additionally, the data revealed that the effects of obesity and hypertension were additive, that is, the presence of both risk factors led to greater cognitive deficits than the presence of either or none (Elias et al. 2003). Similarly, Cournot and collaborators (2006) reported that performance on tests of selective attention and short-term memory in healthy middle-aged men and women were negatively associated with Body Mass Index (BMI).

Studies directly comparing the performance of obese and lean individuals on tests of cognitive behavior support the results of epidemiological research. Obese individuals are reported to do more poorly than lean controls on tasks measuring what has been termed "executive functioning" which includes decision making, problem solving, planning, and mental flexibility (Gunstad et al. 2007; Boeka and Lokken, 2008; Brogna et al. 2011). As these studies controlled for potentially confounding variables (e.g., hypertension, type 2 diabetes, and sleep apnea), the preceding findings have led to the proposal that obesity itself is related to impairments in executive functioning.

Distribution of Body Fat and Cognitive Behavior

Most studies examining the role of obesity in cognitive functioning have used body weight or BMI as the measure of obesity. However, the accumulation of fat in the abdominal region, or what is clinically termed "central obesity," is a better predictor of the adverse physiological consequences of obesity than total body fat. Central obesity may also be more detrimental for brain functioning and behavior than global obesity (Jagust et al. 2005; Dore et al. 2008; Whitmer et al. 2008; Gustafson et al. 2009; Debette et al. 2010). Independent of diabetes and cardiovascular disease, measures of central obesity are better predictors of cognitive functioning and dementia later in life than BMI (Dore et al. 2008; Whitmer et al. 2008). Moreover, central obesity is more likely to be associated with reductions in brain volume than global obesity (Jagust et al. 2005; Anan et al. 2010; Debette et al. 2010).

Recent studies have identified a subset of individuals who are obese without experiencing the metabolic abnormalities typically ascribed to the obese status. This condition has been termed "metabolically healthy (or benign) obesity." Despite excess adipose tissue, individuals with metabolically benign obesity have favorable metabolic profiles characterized by normal insulin sensitivity and blood pressure and normal lipid, hormonal, and inflammatory profiles. Additionally, in comparison to obese individuals without this favorable profile, individuals with metabolically benign obesity do not have an increased risk of developing type 2 diabetes or cardiovascular disease. It is estimated that approximately one-third of obese individuals in this country are metabolically healthy (Karelis 2008; Stefan et al. 2008; Wildman et al. 2008; Wildna et al. 2008; Velho et al. 2010).

The specific variables which mediate the metabolic consequences of obesity remain to be elucidated. However, it has been suggested that individuals with metabolically benign obesity have lower levels of visceral, liver, and muscle fat than obese individuals with more metabolic abnormalities (Stefan et al. 2008). Another factor which may contribute to metabolically benign obesity is physical activity. Obese individuals who engage in regular physical activity are more likely to display metabolically benign obesity than those who are less active (Velho et al. 2010)

To date, no studies have specifically investigated cognitive behavior in individuals with metabolically benign obesity. However, based on previous data, it can be hypothesized that individuals with metabolically benign obesity will be less likely to display deficits in brain functioning and behavior than obese individuals with less favorable metabolic profiles.

Sex Differences

With few exceptions, studies assessing the relationship between obesity and cognitive behavior have used either men or women, or, if using both sexes, have failed to evaluate the data separately for men and women. However, there are indications in the literature of sex differences in the effects of increased adiposity on mental functioning. Both Elias and colleagues (2003) and Kanaya and colleagues (2009) reported that obesity was associated with impairments in cognitive behavior in men, but not in women. In contrast, other investigators have reported that obese women have a greater risk of displaying cognitive deficits than obese men (Gustafson et al. 2003; Gorospe and Dave 2007), while others have failed to find sex-related differences in the effects of obesity on cognitive performance (Cournot et al. 2006). These disparate findings indicate that future investigations should evaluate sex differences in the effects of obesity on cognitive functioning.

Obesity and Dementia

As individuals are living longer, the risk of developing neurodegenerative diseases is increasing. In 2000, there were 4.5 million people in the United States with Alzheimer's disease, the most common form of age-related dementia. It is estimated that by 2050 this number will increase almost threefold to 13.2 million (Hebert et al. 2003). Alzheimer's disease is characterized by brain atrophy, the loss of neurons and synapses, and the accumulation of amyloid plaques and neurofibrillary tangles. These changes in brain morphology are concentrated in areas of the brain associated with learning and memory, including the hippocampus and the association cortices of the frontal, temporal, and parietal lobes (Braak and Braak 1991). Cognitive symptoms of Alzheimer's disease include primary amnesia, as well as deficits in language, semantic knowledge, abstract reasoning, executive functioning, attention, and visospatial abilities (Perry and Hodges 1999). In addition to Alzheimer's disease, many elderly individuals suffer from vascular dementia, which results from impaired blood flow to the brain, and is accompanied by behavioral symptoms which are often difficult to differentiate from those of Alzheimer's disease.

Results of longitudinal epidemiological studies have led to the conclusion that obesity during middle age is a risk factor for both Alzheimer's and vascular dementia (Beydoun et al. 2008; Profenno et al. 2010). It has been hypothesized that the negative metabolic consequences of obesity, including hyperglycemia, insulin resistance, dyslipidemia, and hypertension, may contribute to the cognitive deficits observed in

Alzheimer's disease and vascular dementia (e.g., Beydoun et al. 2008; Craft 2009; Naderali et al. 2009; Abbatecola et al. 2010; Profenno et al. 2010). However, research has shown that independent of its metabolic consequences, obesity in middle and late life increases the risk for age-related dementias (Whitmer et al. 2005; Gorospe and Dave 2007; Beydoun et al. 2008; Profenno et al. 2010). For example, several investigators have reported that after adjustment for age, smoking, and comorbid conditions such as type 2 diabetes and cardiovascular disease, the risk of developing dementia was directly related to BMI in middle age (Whitmer et al. 2005; Gorospe and Dave 2007; Hassing et al. 2009). However, as for many of the metabolic consequences of obesity, results of recent studies suggest that measures of central obesity, such as waist circumference, may be even better predictors of dementia than BMI (Beydoun et al. 2008; Luchsinger 2010; Luchsinger and Gustafson 2009).

The association between obesity and age-related declines in cognitive functioning could have a major impact on global health. The growing prevalence in obesity coupled with a longer life span could ultimately contribute to a significant increase in the incidence of dementia.

Children and Adolescents

The relationship between obesity and cognitive function in children and adolescents is gaining attention due to the rise in the prevalence of juvenile obesity, and the increased number of children and adolescents with obesity-related metabolic disturbances (Botero and Wolfsdorf 2005; Voss and Welsh 2010). As in adults, increased adiposity has been associated with cognitive impairments in children and adolescents (Cserjesi et al. 2007; Li et al. 2008; Lokken et al. 2009; Roberts et al. 2010; Verdejo-Garcia et al. 2010: Cosgrove et al. 2009; Ramaswany et al. 2010; Yau et al. 2010). In particular, obese children and adolescents are reported to do more poorly on tests measuring attention, mental flexibility, and decision making than their lean counterparts (Cserjesi et al. 2007; Lokken et al. 2009; Verdejo-Garcia et al. 2010). Moreover, increased body weight has been linked to decreased visuospatial organization and a reduction in general mental ability in children (Li et al. 2008).

In children and adolescents, as well as adults, obesity is a risk factor for a number of metabolic issues including glucose intolerance, insulin resistance, type 2 diabetes, and sleep apnea, which could affect cognitive behavior. Thus, it is difficult to distinguish the effects of obesity per se on mental performance from those of its associated comorbidities. For example, results of a study demonstrating that obese adolescents with type 2 diabetes did not perform as well on tests measuring verbal memory, psychomotor efficiency, executive functioning, and academic achievement as well-matched nondiabetic obese adolescents suggest that the metabolic consequences of obesity, rather than weight, may be important in mediating cognitive behavior in younger individuals (Yau et al. 2010). Additionally, obese children and adolescents have an increased risk of developing obstructive sleep apnea relative to their normal-weight peers (Arens and Muzumdar 2010). As it has been established that sleep apnea itself is associated with deficits in cognitive behavior, it is possible that sleep disturbances, rather than just weight, are leading to poor performance on tests in mental abilities in young obese individuals.

One question which arises from the preceding findings is, do the deficits in cognitive functioning observed in obese children and adolescents translate into poorer academic performance? Studies investigating this question have produced inconsistent results. While some investigators have reported no differences in academic performance between obese and lean students (Canning and Mayer 1967; Ramaswamy et al. 2010; Verdejo-Garcia et al. 2010), others have noted that academic performance is lower in obese than lean students (Datar et al. 2004). In a recent study, Ramaswamy and colleagues (2010) found that BMI was not significantly related to academic performance. However, BMI was negatively correlated with teachers' evaluations of effort in the classroom. It was proposed that the reductions in effort associated with obesity may ultimately result in suboptimal performance in academic settings.

Changes in brain functioning in obese children and adolescents remain to be elucidated. However, recent work using magnetic resonance imaging (MRI) revealed that in comparison to obese adolescents without type 2 diabetes, obese adolescents with type 2 diabetes displayed enlargement of the cerebral ventricles and reductions in white matter volume, particularly in the frontal lobes. Further, assessments employing diffusion tensor imaging brought to light a reduction in the integrity of the microstructure of the white and gray matter in obese adolescents with type 2 diabetes (Yau et al. 2010). Although neuroimaging studies of brain function in younger obese individuals are limited, as noted below, studies using older participants indicate that obesity may impair the functions of the prefrontal cortex, an area of the brain known to mediate executive functioning (Volkow et al. 2009). Thus, convergent evidence from behavioral and imaging studies points to impairments in the functions of the prefrontal executive system in overweight children and adolescents.

Results of the studies cited above (e.g., Cserjesi et al. 2007; Li et al. 2008; Yau et al. 2010) clearly demonstrate a negative relationship between body weight and cognitive performance. However, the direction of this relationship is not clear from these findings. It has been proposed that increased body weight results in poorer performance of cognitive tasks. However, it is also possible that poorer cognitive abilities contribute to the development of obesity in children and adolescences. Children and adolescents with poorer cognitive abilities may be more likely to engage in behaviors which promote weight gain, such as high television viewing, poor dietary intake, and low physical activity (Li et al. 2008). It is also possible that psychosocial variables such as family income, parental education, school environment, and access to extracurricular activities (e.g., sports teams and summer camp) contribute to both the development of obesity and impairments in cognitive behavior. Moreover, it is well established that overweight children and adolescents suffer from the negative societal view of excess weight. Obese individuals are the most discriminated-against subgroup in our society. Obese children are frequently described as lazy, evil, unattractive, stupid, cruel, and gluttonous, whereas lean children are viewed in a more positive light through trait portrayals of attractiveness, happiness, intelligence, and kindness (Puhl and Latner 2007). These negative stereotypes can lead to isolation, depression, and low self-esteem and can create a burden for individuals from the stereotyped groups which can undermine their performance on cognitive tasks (Schmadar et al. 2008).

Independent of the direction of the relationship between obesity and cognitive functioning, the question of the interaction of obesity and mental abilities in children and adolescents is an important one. As yet, no studies have assessed the long-term effects of early obesity on cognitive abilities or whether the detrimental effects of early obesity can be reversed by weight loss or other strategies such as physical activity.

Clinical studies have provided important insights into the role of obesity on cognition and brain functioning. However, in studies using humans, it is difficult to separate the effects of obesity from those of other variables which may also influence cognitive behavior (e.g., health status, environmental conditions, dietary habits, physical activity, and experience with cognitive tasks). Additionally, human studies are as yet unable to detect the cellular and molecular changes that may mediate obesity-related alterations in mental abilities. To address these issues, investigators have turned to research employing experimental animals.

OBESITY AND BRAIN FUNCTION IN EXPERIMENTAL ANIMALS

The use of animal models has contributed significantly to our understanding of obesity. Research employing animals with spontaneous mutations (e.g., the ob/ob mouse, db/db mouse, and Zucker fatty rat) and, more recently, genetically engineered transgenic animals has helped to identify some of the genes which may be linked to human obesity, while animals suffering from obesity as a result of experimentally induced brain damage have provided information about neural structures important in the regulation of food intake and body weight. However, with respect to relevance to the human condition, animal models of diet-induced obesity may be the most informative. Both rats and mice fed palatable diets containing large proportions of fats and/or sugars consume more than their counterparts given diets containing lesser amounts of these nutrients. Although on a daily basis, overconsumption of these palatable diets is small, amounting to only a 5–10 percent increase in total daily energy intake, over time, this pattern of intake leads to obesity and a number of the physiological aberrations seen in human obesity, including glucose intolerance, insulin resistance, and hypertriglyceridemia (Kanarek and Hirsch 1977; Molteni et al. 2002; Jurdak and Kanarek 2008; Jurdak et al. 2009; Kanoski and Davidson 2011).

Evidence for the negative consequences of obesity on cognitive behavior initially came from studies demonstrating that animals with diet-induced obesity did more poorly on tasks measuring learning and memory than nonobese animals. For example, Winocur and Greenwood (Greenwood and Winocur 1990; Winocur and Greenwood 1999, 2005) reported that relative to rats fed a low-fat diet, rats who had consumed a diet high in saturated fat for several months gained more weight and displayed deficits in both short- and long-term memory. Subsequent studies have confirmed the deleterious effects of increased energy intake on body weight and cognitive behavior, and have characterized a number of factors which can influence the effects of these diets on cognitive behavior (Molteni et al. 2002; Goldbart et al. 2006; Jurdak and Kanarek 2008; Valladoid-Acebes et al. 2011).

COGNITIVE ASSESSMENTS

Most studies assessing the effects of diet-induced obesity on cognitive behavior have used tests of spatial memory such as the Morris water maze. The Morris water maze is the most commonly used test of learning and memory in animals (Morris 1984). In the Morris water maze, an animal is placed into a circular pool of water from which it learns to escape by swimming to a platform hidden beneath the surface of the water. Performance on the Morris water maze differs as a function of a wide range of independent variables including exposure to stress, drug treatment, brain damage, nutritional deficiencies, and age. The Morris water maze also has been used to assess performance in a number of animal models of neurocognitive disorders, including cerebrovascular disease, neurotrauma, and Alzheimer's disease. Performance on the Morris water maze is particularly sensitive to alterations in hippocampal functioning (D'Hooge and De Deyn 2001; Vorhees and Williams 2006).

Diet-induced obesity has been associated with impairments in performance on the Morris water maze and other tests of spatial memory. These findings have led to the proposal that obesity may particularly compromise hippocampal-dependent learning and memory (Kanoski and Davidson 2011). And indeed, as described below, changes in the structure and activity of the hippocampus are frequently observed in animals with diet-induced obesity. However, it should be noted that intake of energy-dense diets can also detrimentally affect tasks believed to be mediated by areas of the brain other than the hippocampus (Jurdak and Kanarek 2009; Kanoski and Davidson 2010). More specifically, studies suggest that tasks dependent on the prefrontal cortex may also be adversely affected by diet-induced obesity (Kanoski et al. 2007; Jurdak and Kanarek 2009; Kanoski and Davidson 2010). Recent work demonstrating that intake of a high-fat diet impaired spatial learning after just three days, but only impaired nonspatial learning after sixty days, indicates that longer term exposure to energy-dense diets may be required for negative effects on hippocampal-independent tasks to be observed (Kanoski and Davidson 2010).

Diet Composition

Studies reporting deleterious effects of diet-induced obesity on cognitive behavior have used diets varying widely in nutrient composition. Some researchers have used diets high in fat (e.g., Greenwood and Winocur 1990; Pistell et al. 2010; Valladolid et al. 2011), while others have used diets high in sugar (e.g., Jurdak and Kanarek 2008; Jurdak et al. 2009; Ross et al. 2009) or high in both fat and sugar (e.g., Moltini et al. 2002; Wu et al. 2003; Goldbart et al. 2006). Additionally, even within a diet category (e.g., high-fat diets), there has been little consistency in the type or exact quantity of nutrient used. These findings suggest that excess energy intake may be more important in mediating decrements in cognitive behavior than the exact composition of the diet. However, there is evidence that diets high in saturated fats (e.g., lard or butterfat) are more likely to result in impairments in cognitive performance, as well as obesity and symptoms resembling human metabolic syndrome, than diets high in polyunsaturated fats (e.g., flaxseed oil and safflower oil) (Greenwood and Winocur 1990; Buettner et al. 2007; Pistell et al. 2010; Yu et al. 2010).

While it appears that excess caloric intake and body weight gain may be most important in determining decrements in cognitive behavior in animals, a need for systematic evaluation of the effects of dietary composition on metabolic responses and cognitive outcomes remains. Such studies would help to elucidate which nutrients may be most likely to contribute to obesity-related cognitive dysfunctions.

Duration of Feeding

In general, the longer the time that an animal has consumed a high-energy diet, the greater the decrements observed in brain functioning and cognitive behavior (Molteni et al. 2002; Kanoski and Davidson 2010). For example, Molteni and colleagues (2002) reported that rats that ate a high-fat/high-sucrose diet for more than six months displayed poorer performance on a test of spatial memory and experienced greater reductions in levels of brain-derived neurotrophic factor than animals that ate the diet for less than six months.

While chronic intake may magnify the detrimental effects of an energy-dense diet on cognitive behavior, short-term consumption of a high-fat diet can also impair cognitive performance. Rats fed a high-fat diet for just nine days performed more poorly on a test of working memory than animals maintained on a low-fat diet (Murray et al. 2009). Moreover, Kanoski and Davidson (2010) reported that rats fed a diet high in fat and glucose for just three days made more errors on a spatial memory task than those fed a standard laboratory diet. Interestingly, in the same study, deficits in nonspatial learning were not observed until the animals had consumed the high-energy diet for sixty days. These findings suggest that areas of the brain associated with spatial learning and memory, such as the hippocampus, may be more sensitive to the detrimental effects of high-fat diets than areas associated with other types of learning.

Sex Differences

Conclusions about the effects of diet-induced obesity on cognitive performance have been generated almost exclusively from studies using male animals. However, the results of the few studies that have included females indicate that there may be sex differences in the effects of diet-induced obesity on cognitive behavior (Hwang et al. 2010; Jurdak and Kanarek 2010). For example, Hwang and coworkers (2010) reported that while both male and female mice fed a high-fat diet became obese, males gained more weight and displayed more severe hyperglycemia and hypercholesterolemia than females. Additionally, the obese male mice displayed poorer avoidance learning and greater abnormalities in hippocampal functioning than the obese female mice. Recent studies in rats also point toward the conclusion that males may be more vulnerable than females to the adverse effects of diet-induced obesity on cognitive behavior (Jurdak and Kanarek 2010). These results emphasize the importance of including males and females in studies assessing the role of obesity in mediating brain functioning and behavior.

POTENTIAL MECHANISMS UNDERLYING THE INTERACTIONS AMONG OBESITY, BRAIN FUNCTION, AND BEHAVIOR

The mechanisms underlying the association between obesity and impairments in cognitive functioning remain speculative. It has been assumed that the comorbid disorders which accompany obesity, including type 2 diabetes and cardiovascular disease, contribute to these impairments (Reijmer et al. 2010; Uranga et al. 2010). However, alterations in brain function and cognitive behavior have been observed in obesity prior to the overt development of these disorders (Valladolid-Acebes et al. 2011). These findings indicate that subtle physiological changes which accompany obesity also may be important for mediating changes in neural activity and behavior.

Obesity, Brain Structure, and Function

Brain-imaging techniques (e.g., computer tomography [CT], MRI, and fMRI) have provided researchers with the ability to assess structural and functional changes in the brain associated with increased body weight. In an initial study using CT scans, Gustafson and colleagues (2004) reported increased temporal lobe atrophy in women as a function of BMI. More recent studies have confirmed a relationship between increased adiposity and a reduction in brain size in young, middle-aged, and elderly individuals (e.g., Ward et al. 2005; Pannacciulli et al. 2006; Gazdzinski et al. 2008; Taki et al. 2008; Debette et al. 2010; Raji et al. 2010; Walther et al. 2010). Additionally, these studies have identified specific areas of the brain that may be altered in obesity. For example, Pannacciulli et al. (2006) reported that relative to lean individuals, obese individuals display reductions in gray matter density in the frontal lobe and postcentral gyrus, areas of the brain implicated in taste, the regulation of food intake, and reward and cognitive behavior. Moreover, reductions in the volume of the inferior and frontal cortex, the posterior cortex including the parahippocampal gyrus, and the cerebellum have also been observed in obese individuals relative to their lean counterparts (Taki et al. 2008; Walther et al. 2010). In contrast to the observed reductions in gray matter in obesity, larger volumes of both posterior and anterior white matter have been reported in obese individuals (Walther et al. 2010).

Body weight may moderate neuronal activity as well as brain structure (Volkow et al. 2009; Gazdzinski et al. 2010; Gonzales et al. 2010). Volkow and colleagues (2009) observed a negative association between BMI and baseline brain glucose metabolism in the anterior cingulate gyrus and prefrontal region of the brains of healthy middle-aged adults. Additionally, others have found that obesity in middle age is related to decreased brain activation during a cognitive challenge (Gonzales et al. 2010) and that a higher BMI is associated with a reduction in neuronal integrity in the anterior cingulate cortex of healthy elderly individuals (Gazdzinski et al. 2008, 2010). Taken together, the results of the previously described studies seem to indicate that obesity is most often associated with reductions in the size and activity of the frontal lobe and parahippocampal region, areas of the brain that are closely tied to processes related to executive functioning and that are most commonly impaired in age-related dementias.

Hippocampus

The hippocampus is one of the key brain structures involved in cognitive behavior. It plays an essential role in a host of memory processes, including the consolidation of short-term memories into long-term memories, and is particularly important for spatial memory (Shimamura 2010). Damage to the hippocampus resulting, for example, from trauma, stress, or aging leads to deficits in learning and memory. Impairments in hippocampal function may also mediate obesity-related deficits in cognitive behavior (Kanoski and Davidson 2011). For example, consumption of energy-dense diets is associated not only with obesity and deficits in spatial memory but also with reductions in dendritic spine density and impairments in synaptic plasticity within the hippocampus (Molteni et al. 2002; Stranahan et al. 2008; Hwang et al. 2010). Additionally, intake of energy-dense diets has been associated with reduced hippocampal neurogenesis (Lindqvist et al. 2006; Park et al, 2010).

Obesity, Glucose Regulation, and Brain Functioning

Abnormalities in glucose homeostasis, including peripheral hyperglycemia and insulin resistance, are frequently observed concomitants of obesity. As glucose is the primary fuel for the brain, it has been proposed that these abnormalities contribute to obesity-related deficits in cognitive behavior. Consistent with this proposal, among nondiabetic healthy middle-aged and elderly individuals, those with poorer glucose regulation perform more poorly on tasks of memory and executive functioning and have smaller head size-adjusted hippocampal volumes than those with better glucose regulation (Convit et al. 2003; Messier et al. 2010). Moreover, improvement in glucose tolerance has been associated with improvements in performance on cognitive tasks (Convit et al. 2003; Convit 2005).

Insulin resistance is one of the most common metabolic problems associated with obesity. Insulin receptors are widely distributed in the central nervous system, particularly in areas involved in cognitive behavior including the hippocampus and cerebral cortex. Given that insulin resistance decreases glucose uptake and cellular functioning, decreased sensitivity to insulin in hippocampal neurons has been proposed as one possible mechanism for obesity-related deficits in cognitive behavior. According to this proposal, in obesity, neurons in the hippocampus become less able to use glucose, resulting in disruptions in neurotransmitter transport and release. This cellular disturbance is thought to hinder neural transmission leading to impairments in learning and memory (Hoyer 2003; Bissells et al. 2004; Convit 2005). In support of this proposal, manipulations which improve insulin utilization, such as central administration of insulin or administration of a drug which increases the effectiveness of insulin, facilitate performance on cognitive tasks in both humans and animals (Park et al. 2000; Ryan et al. 2007; Reagan 2007; Pathan et al. 2008; McNay et al. 2010). In contrast, blockade of endogenous intrahippocampal signaling can impede memory performance, while chronic intake of a high-fat diet impairs cognitive functioning and attenuates both the cognitive and metabolic responses to hippocampal insulin administration (McNay et al. 2010). Further evidence of a deficit in hippocampal functioning in obesity comes from work demonstrating that rats chronically fed energy-dense diets displayed reduced insulin signaling in the hippocampus (Meilke et al. 2006).

In addition to moderating glucose metabolism, insulin can affect brain microvasculature. Insulin resistance is associated with vasoconstriction, which can ultimately reduce blood flow to the brain. This reduction in blood flow could contribute to obesity-related impairments in cognition. Taken together, the results of these studies are robust indicators that one way in which obesity could alter cognitive behavior is by interfering with insulin actions in the central nervous system.

Obesity, Oxidative Stress, and Inflammation

The detrimental effects of energy-dense diets on brain functioning and cognitive behavior may be mediated, at least in part, by oxidative stress (Zhang et al. 2009; Morrison et al. 2010; Park et al. 2010). Oxidative stress is one of the earliest negative consequences of obesity and may trigger the development of insulin resistance. Oxidative stress resulting from intake of energy-dense diets could result in damage to nucleic acids, proteins, and lipids important for normal brain functioning (Furukawa et al., 2004; Zhang et al. 2009). In support of this possibility, Morrison and colleagues (2010) reported that in aged mice, consumption of a high-fat diet impaired long-term spatial memory and increased oxidative damage to proteins integral to the hippocampus. Moreover, they found that both cognitive dysfunction and oxidative stress were highly correlated with body weight gain across the experiment. Other types of oxidative stress could also influence brain functioning. For example, obesity could also stimulate the production of a variety of lipid peroxidation products which could be toxic to the nervous system (Zhang et al. 2009; Park et al. 2010). In support of this possibility, Park and colleagues (2010) reported that mice consuming a high-fat diet gained more weight, had larger fat depots, and had higher hippocampal levels of malondialdehyde, a measure of lipid peroxidation, than animals given a standard diet.

Increases in oxidative stress can ultimately lead to inflammation. Epidemiological studies have revealed that obesity is often accompanied by chronic, low-level inflammation (Schoelson et al. 2007). These inflammatory responses are correlated with the degree of obesity and insulin resistance, and may result from increased levels of serum lipids (Pistell et al. 2010). Based on studies demonstrating that inflammation can disrupt brain pathways involved in cognitive behavior, it has been proposed that the cognitive impairments observed in animals fed high-fat diets are associated with brain inflammation. In support of this proposal, mice fed a diet high in saturated fat displayed impaired cognitive performance on a test of spatial memory and significant increases in the expression of a number of cytokines (e.g., TNFα, and IL-6), as well as increased reactive astrocytosis and microgliosis.

Brain-Derived Neurotrophic Factor (BDNF)

Brain-derived neurotrophic factor (BDNF) is one of a family of proteins that is important for a host of essential neural processes, such as neural differentiation, proliferation, survival, and plasticity. BDNF is synthesized predominantly by neurons in the hippocampus, and is hypothesized to be a key part of the neuronal processes that mediate hippocampal-dependent learning and memory (Kanoski and Davidson 2011). In support of this hypothesis, the expression of BDNF is elevated in the hippocampus of animals that have learned a spatial task (Mizuno et al. 2000). In contrast, impairments in learning and memory have been observed in transgenic mice animals with reduced expression of BDNF (Linnarsson et al. 1997).

Reductions in the expression of BDNF within the hippocampus may contribute to the detrimental effects of diet-induced obesity on learning and memory. Molteni and colleagues (2002) reported that, relative to rats chronically consuming a low-fat/complex-carbohydrate diet, rats consuming a high-fat/high-sugar diet did more poorly on the Morris water maze, a test of spatial learning, and displayed a reduction in hippocampal levels of BDNF. Moreover, the longer that animals were maintained on the high-fat/high-sucrose diet, the poorer their performance on the Morris water maze and the greater the reduction in hippocampal BDNF. Subsequent studies have confirmed that animals with diet-induced obesity display impairments in spatial and nonspatial learning and reductions in hippocampal levels of BDNF (Wu et al. 2003; Moltini et al. 2004; Kanoski et al. 2007; Stranahan et al. 2008).

Lipids and Cognitive Functioning

Obesity-related alterations in serum lipids could also contribute to impairments in cognitive behavior. Chronic intake of an energy-dense diet can lead to elevations in cholesterol and triglyceride levels. In turn, these elevations could result in decreased blood flow to the brain, alterations in the brain's fatty acid content, and decrements in cognitive functioning. In support of a role of lipids in determining obesity-related cognitive deficits, recent studies have shown that administration of triglycerides into the hippocampus can impair long-term potentiation and acquisition of cognitive tasks. In contrast, administration of a drug which reduces hypertriglyceridemia improved cerebral blood flow and cognitive capacity (Farr et al. 2008).

It is unlikely that there is one critical factor mediating the effect of obesity on brain functioning and behavior. Rather, it is more likely that multiple factors contribute to the reported decrements in cognitive capacity observed in obesity.

FUTURE DIRECTIONS

One important question which has not been extensively addressed is whether the association between obesity and decrements in cognitive performance is permanent. More specifically, can weight loss or other strategies, such as physical activity, reverse the deficits in cognitive behavior observed in obese individuals? Although the answer to this question is important for individuals of all ages, it is particularly important for children and adolescents. Early life is a critical time for learning, and early deficits in learning could translate into lifelong problems that could ultimately impact career advancement, social interactions, and quality of life.

With respect to the preceding question, there is substantial evidence that physical activity is beneficial not only for maintaining physical health but also for preserving mental performance. Results of studies in both experimental animals and humans indicate that physical activity can protect against age-related declines in learning and memory. Moreover, there is now evidence that physical activity can increase hippocampal volume and improve memory (Erickson et al. 2011). Emerging work also indicates that physical activity could reverse the harmful effects of obesity on brain functioning and behavior. In experimental animals, voluntary wheel running reversed the detrimental effects of a high-fat diet on spatial learning and hippocampal levels on brain-derived neurotrophic factor (Molteni et al. 2004). In humans, physical activity recently has been found to attenuate the negative relationship between central adiposity and cognitive function in healthy community-dwelling adults (Dore et al. 2008).

It is important to remember that decrements in cognitive behavior are not universally seen in obese individuals. It is most probable that it is not obesity but rather its metabolic consequences which are responsible for obesity-related alterations in brain functioning. As yet, no studies have compared cognitive functioning in metabolically healthy obese individuals and metabolically unhealthy individuals. However, this type of research is critical for helping to determine the factors mediating differences in cognitive behavior between lean and obese individuals.

Finally, while certainly important, research on obesity, brain function, and cognition could perpetuate the ongoing negative stereotypes of obesity. It is possible that this negative stereotype contributes, at least in part, to the cognitive deficits observed in some obese individuals. These stereotypes can lead to isolation, depression, and low self-esteem and can create a burden for individuals from the stereotyped groups which can undermine their performance on cognitive tasks (Schmadar et al. 2008). It is thus important for future studies to find ways to integrate the biological and social factors as potential mediators of the effects of obesity on behavior.

REFERENCES

Abbatecola, A. M., F. Lattanzio, L. Spazzafumo, A. M. Molinari, M. Cioffi, R. Canonico, L. DiCiocclo, and G. Paolisso. 2010. Adiposity predicts cognitive decline in older persons with diabetes: a 2 year follow-up. *Plos ONE* 5: e10333, doi: 10.137/journal.pone.0010333.

Anan, F., R. Masaki, T. Shimomua, M. Fujiki, Y. Umeno N. Eshima, T. Saikawa, and H. Yoshimatus. 2010. Abdominal visceral fat accumulation is associated with hippocampus volume in non-dementia patients with type 2 diabetes mellitus. *Neuroimage* 49:57–62.

Arens, R., and H. Muzumdar. 2010. Childhood obesity and obstructive sleep apnea syndrome. *J Appl Physiol* 108:436–44.

Beydoun, M. A., H. A. Beydoun, and Y. Wang, Y. 2008. Obesity and central obesity as risk factors for incident dementia and its subtypes: a systematic review and meta-analysis. *Obes Rev* 9:204–18.

Bissels, G-J., B. Bravenboer, and G. H. Willem. 2004. Glucose, insulin, and the brain: modulation of cognition and synaptic plasticity in health and disease: A preface. *Eur J Pharmacol* 490:1–4.

Boeka, A. G., and K. L. Lokken. 2008. Neuropsychological performance of a clinical sample of extremely obese individuals. *Arch Clin Neuropsych* 23:467–74.

Botero, D., and J. I. Wolfsdorf. 2005. Diabetes mellitus in children and adolescents. *Arch Med Res* 36:281–90.

Braak, H., and E. Braak. 1991. Demonstration of amyloid deposits and neurofibrillary changes in whole brain sections. *Brain Pathol.* 1:213–6.

Brogna, A., D. Hevey, G. O. O'Callaghan, R. Yoder, and D. O'Shea. 2011. Impaired decision making among morbidly obese adults. *J Psychosom Res* 70:189–96.

Buettner, R., J. Scholmerich, and L. C. Bolleimer. 2007. High-fat diets: modeling the metabolic disorders of human obesity in rodents. *Obesity* 15:798–808.

Canning, H., and J. Mayer. 1966. Obesity: its possible effect on college acceptance. *New Engl J Med* 20:352–4.

Centers for Disease Control and Prevention, Division of Nutrition, Physical Activity and Obesity, National Center for Chronic Disease Prevention and Health Promotion. 2010. Overweight and obesity: childhood overweight and obesity. www.cdc.gov/nccdphp/dnpao/

Convit, A. 2005. Links between cognitive impairment in insulin resistance: an explanatory model. *Neurobiol Aging* 26S:S31–5.

Convit, A., O. T. Wolf, C. Tarshish, and M. J. de Leon. 2003. Reduced glucose tolerance is associated with poor memory performance and hippocampal atrophy among normal elderly. *Proc Natl Acad Sci USA* 100:2019–22.

Cosgrove, R. G., C. Arroyo, J. C. Warren, and J. Zhang. 2009. Impaired cognitive functioning in overweight children and adolescents. *Agro Food Industry Hi Tech* 20:48–51.

Cournot, M., J. C. Marquie, D. Ansiau, C. Martinaud, H. Fonds, J. Ferrieres, and J. B. Ruidavets. 2006. Relation between body mass index and cognitive function in healthy middle-aged men and women. *Neurology* 67:1208–14.

Craft, S. 2009. The role of metabolic disorders in Alzheimer Disease and vascular dementia. *Arch Neurol* 66:300–5.

Cserjesi, R., D. Molnar, O. Luminet, and L. Lenard. 2007. Is there a relationship between obesity and mental flexibility in children? *Appetite* 49:675–8.

Datar, A., R. Sturm, and J. L. Magnabosco. 2004. Childhood overweight and academic performance: national study of kindergartners and first-graders. *Obes Res* 12:58–68.

Debette, S., A. Beiser, U. Hoffmann, C. DeCarli, C. J. O'Donnell, J. M. Massaro, R. Au, J. J. Himali, P. A. Wolf, C. S. Fox, and S. Seshadri. 2010. Visceral fat is associated with lower brain volume in healthy middle-aged adults. *Ann Neurol* 68:136–44.

D'Hooge, R., and P. P. De Deyn. 2001. Applications of the Morris water maze in the study of learning and memory. *Brain Res Rev* 36:60–90.

Dore, G. A., M. F. Elias, M. A. Robbins, M. M. Budge, and P. K. Elias. 2008. Relation between central adiposity and cognitive function in the Maine-Syracuse Study: attenuation by physical activity. *Ann Behav Med* 35:341–50.

Duncan, G. E. 2006. Prevalence of diabetes and impaired fasting blood glucose levels among U.S. adolescents: National Health and Nutrition Examination Survey, 1999–2002. *Arch Pediatr Adol Med* 160:523–8.

Elias, M. F., P. K. Elias, L. M. Sullivan, P. A. Wolf, and R. B. D'Agostino. 2003. Lower cognitive function in the presence of obesity and hypertension: The Framingham Heart Study. *Int J Obes* 27:260–8.

Elias M. F., P. K. Elias, L. J. Sullivan, P. A. Wolf, and R. B. D'Agostino. 2005. Obesity, diabetes and cognitive deficit: the Framingham Heart Study. *Neurobiol Aging* 26S: S11–S16.

Erickson, K. I., M. W. Voss, R. S. Prakash, C. Basak, A. Szabo, L. Chaddock, J. S. Kim, S. Heo, H. Alves, S. M. White. T. R. Wojcicki, E. Mailey, V. J. Vieira, S. A. Martin, B. D. Pence, J. A. Woods, E. McAuley, and A. F. Kramer. 2011. Exercise training increases size of hippocampus and improves memory. *Proc Nat Acad Sci USA* 108:3017–22.

Farr, S. A., K. A. Yamada, D. A. Butterfield, H. M. Abdul, L. Xu, N. E. Miller, W. A. Banks, and J. E. Morley. 2008. Obesity and hypertriglyceridemia produce cognitive impairment. *Endocrinology* 149:2628–36.

Finer, N. 2011. Medical consequences of obesity. *Medicine* 39:18–23.

Flegal, M. M., M. D. Carroll, C. L. Ogden, and L. R. Curtin. 2010. Prevalence and trends in obesity among US adults, 1999–2008. *J Amer Med Assoc* 303:235–41.

Furukawa, S., T. Fujita, M. Shimabukuro, M. Iwaki, Y. Yamada, and Y. Nakajima. 2004. Increased oxidative stress in obesity and its impact on metabolic syndrome. *J Clin Invest* 114:1752–61.

Gazdzinski, S., J. Kornak, M. W. Weiner, D. J. Meyerhoff, and D. R. Nat. 2008. Body mass index and magnetic resonance markers of brain integrity in adults. *Ann Neurol* 63:652–7.

Gazdzinski, S., R. Millin, L. G. Kaiser, T. C., Durazzo, S. G. Mueller, M. W. Weiner, and D. J. Meyerhoff. 2010. BMI and neuronal integrity in healthy, cognitively normal elderly: a proton magnetic resonance spectroscopy study. *Obesity* 18:743–8.

Goldbart, A. D., B. W. Row, L. Kheirandish-Gozal, Y. Cheny, K. R. Brittian, and D. Gozal. 2006. High fat/refined carbohydrate diet enhances the susceptibility to spatial learning deficits in rats exposed to intermittent hypoxia. *Brain Res* 1090:190–6.

Gonzales, M. M., T. Taurmi, S. C. Miles, Tanaka, H., F. Shah, and A. P. Haley. 2010. Insulin sensitivity as a mediator of the relationship between BMI and working memory-related brain activation. *Obesity* 18:2131–7.

Gorospe, E. C., and J. K. Dave. 2007. The risk of dementia with increased body mass index. *Age Aging* 36:23–9.

Greenwood, C. E., and G. Winocur. 1990. High-fat diets, insulin resistance and declining cognitive function. *Neurobiol Aging* 26S:S42–S45.

Gunstad, J., R. H. Paul, R. A. Cohen, D. F. Tate, M. B. Spitznagel, and E. Gordon. 2007. Elevated body mass index is associated with executive dysfunction in otherwise healthy adults. *Comp Psychiat* 48:57–61.

Gustafson, D., E. Rothenberg, K. Blennow, B. Steen, and I. Skoog. 2003. An 18-year follow-up of overweight and risk of Alzheimer disease. *Arch Intern Med* 163:1524–8.

Gustafson, D., L. Lissner, C. Bengtsson, C. Bjorkelund, and I. Skoog. 2004. A 24-year follow-up of body mass index and cerebral atrophy. *Neurology* 63:1876–81.

Gustafson, D. R., K. Backman, M. Waern, S. Ostling, X. Guo, P. Zandi, M. M. Mielke, C. Bengtsson, and I. Skoog. 2009. Adiposity indicators and dementia over 32 years in Sweden. *Neurology* 73:1559–66.

Hassing, L. B., A. K. Dahl, V. Thorvaldsson, S. Berg, M. Gatz, N. L. Pedersen, and B. Johansson. 2009. Overweight in midlife and risk of dementia: a 40-year follow-up study. *Int J Obesity* 33:893–8.

Hebert, L. E., P. A. Scherr, J. L. Bienias, D. A. Bennett, and D. A. Evans. 2003. Alzheimer's Disease in the US population. *Arch Neurol* 60:1119–22.

Hoyer, S. 2003. Memory function and brain glucose metabolism. *Pharmacopsychiatry* 36:62–7.

Hwang, L., C. Wang, T. Li, S. Chang, L. Lin, C. Chen, C. Chen, K. Liang, I. Ho, W. Yang, and L. Chiou. 2010. Sex differences in high-fat diet-induced obesity, metabolic alterations and learning and synaptic plasticity deficits in mice. *Obesity* 18:463–9.

Jagust, W., H. Harvey, D. Munga, and M. Haan. 2005. Central obesity and the aging brain. *Arch Neurol* 62:1545–8.

Jurdak, N., and R. B. Kanarek. 2009. Sucrose-induced obesity impairs novel object recognition learning in young rats. *Physiol Behav* 96:1–5.

Jurdak, N., and R. B. Kanarek. 2010. The effects of twelve weeks of sucrose feeding on spatial learning and memory and brain-derived neurotrophic factor in young female rats. Abstract. *Society for Neuroscience* 509:6.

Jurdak, N., A. Lichtenstein, and R. B. Kanarek. 2008. Diet-induced obesity impairs spatial learning in young male rats. *Nutr Neurosci* 11:1–6.

Kanarek, R. B., and E. Hirsch. 1977. Dietary-induced overeating in experimental animals. *Fed Proc* 36:154–8.

Kanoski, S. E., and T. L. Davidson. 2010. Different patterns of memory impairments accompany short- and longer-term maintenance on a high-energy diet. *J Exp Psychol Anim Behav Process* 36:313–9.

Kanoski, S. E., and T. L. Davidson. 2011. Western diet consumption and cognitive impairment: links to hippocampal dysfunction and obesity. *Physiol Behav* 103:59–68.

Kanoski, S. E., R. L. Meisel, A. J. Mullins, and T. L. Davidson. 2007. The effects of energy-rich diets on discrimination reversal learning and on BDNF in the hippocampus and prefrontal cortex of rat. *Behav Brain Res* 182:57–66.

Kanaya, A. M., K. Lindquist, T. B. Harris, L. Launer, C. Rosano, S. Satterfield, and K. Yaffe. 2009. Total and regional adiposity and cognitive change in older adults. *Arch Neurol* 66:329–35.

Karelis, A. D. 2008. Metabolically healthy but obese individuals. *Lancet* 372:1281–3.

Li Y, Q. Dai, J. C. Jackson, and J. Zhang. 2008. Overweight is associated with decreased cognitive functioning among children and adolescents. *Obesity* 16:1809–15.

Lindqvist, A., P. Mohapel, B. Bouter, H. Frielingsdorf, D. Pizzo, P. Brundin, and C. Erlanson-Albertsson. 2006. High-fat diet impairs hippocampal neurogenesis in male rats. *Eur J Neurol* 13:1385–8.

Linnarsson, S., A. Bjorklund, and P. Ernfors. 1997. Learning deficit in BDNF mutant mice. *Eur J Neurosci* 9:2581–7.

Lokken, K. L., A. G. Boeka, H. M. Austin, J. Gunstad, and C. M. Harmon. 2009. Evidence of executive dysfunction in extremely obese adolescents: A pilot study. *Surg Obes Relat Dis*, 5:547–52.

Luchsinger, A. J. 2010. Type 2 diabetes and related conditions relation to dementia: an opportunity for prevention? *J Alzheimers Dis* 20:723–36.

Luchsinger, A. J., and D. R. Gustafson. 2009. Adiposity and Alzheimer's disease. *Curr Opin Clin Nutr Metabol Care* 12:15–21.

McNay, E. C., C. T. Ong, R. J. McCrimmon, J. Cresswell, J. S. Bogan, and R. S. Sherwin. 2010. Hippocampal memory processes are modulated by insulin and high-fat-induced insulin resistances. *Neurbiol Learn Mem* 93:546–53.

Messier, C., M. Tsiakas, M. Gagnon, and A. Desrochers. 2010. Effect of age and glucoregulation on cognitive performance. *J Clin Exp Neuropsychol* 32:809–21.

Mizuno, M., Yamada, K., Olariu, A., Nawa, and H. Nabeshim. 2000. Involvement of brain-derived neurotrophic factors in spatial memory formation and maintenance in a radial arm maze in rats. *J Neurosci* 20:7116–21.

Molteni, R., R. J. Bernard, Z. Ying, C. K. Roberts, and F. Gomez-Pinilla. 2002. A high fat, refined sugar diet reduces brain-derived neurotrophic factor, neuronal plasticity, and learning. *Neurosci* 112:803–14.

Molteni, R., A. Wu, S. Vaynman, Z. Ying, R. J. Barnard, and F. Gómez-Pinilla. 2004. Exercise reverses the harmful effects of consumption of a high-fat diet on synaptic and behavioral plasticity associated to the action of brain-derived neurotrophic factor. *Neurosci* 123:429–40.

Morris, R. 1984. Development of a water-maze procedure for studying spatial learning in the rat. *J Neurosci Methods* 11:47–60.

Morrison, C. D., P. P. Pistell, D. K. Ingram, W. D. Johnson, Y. Liu, S. O. Feranandez-Kim, C. L. White, M. N. Purpera, R. M. Uranga, A. J. Bruce-Keller, and J. N. Keller. 2010. High fat diet increases hippocampal oxidative stress and cognitive impairment in aged mice: implications for decreased NrF2 signaling. *J. Neurochem.* 114:1581–9.

Murray A. J., N. S. Knight, L. E. Cochlin, S. McAleese, R. M. J. Deacon, N. P. Rawlins, and K. Clarke. 2009. Deterioration of physical performance and cognitive function in rats with short-term high-fat feeding. *FASEB J* 23:4353–60.

Naderali, E. K., S. H. Ratcliffe, and M. C. Dale. 2009. Obesity and Alzheimer's disease: a link between body weight and cognitive function in old age. *Am J Alzheimer's Dis.* 24:445–8.

Pannacciulli, N., A. Del Parigi, K. Chen, D. S. Le, N. T. E. M. Reiman, and P. A. Tataranni. 2006. Brain abnormalities in human obesity: a voxel-based morphometric study. *Neuroimage* 31:1419–25.

Park C. R., R. J. Seeley, S. Craft, and S. C. Woods. 2000, Intracerebroventricular insulin enhances memory in a passive-avoidance task. *Physiol Behav* 68:509–14.

Park H. R., M. Park, J. Choi, K-Y. Park, H. Y. Chung, and J. Lee. 2010. A high-fat diet impairs neurogenesis: involvement of lipid peroxidation and brain-derived neurotrophic factor. *Neurosci. Lett* 482:235–9.

Pathan, A., A. Gaikwas, B. Visanad, and P. Romarao. 2008. Rosiglitazone attenuates the cognitive deficits induced by high fat feeding in rats. *Eur J Pharmacol* 589:176–9.

Perry, R. J., and J. R. Hodges. 1999. Attention and executive deficits in Alzheimer's disease. A critical review. *Brain* 122:383–404.

Pistell, P. J., C. D. Morrison, S. Gupta, A. G. Knight, J. N. Keller, D. K. Ingram, and A. J. Bruce-Keller. 2010. Cognitive impairment following high fat diet consumption is associated with brain inflammation. *J Neuroimmunol* 219:25–32.

Profenno, L. A., A. P. Porsteinsson, and S. V. Faraone, S. V. 2010. Meta-analysis of Alzheimer's disease risk with obesity, diabetes and related disorders. *Biol Psychiat* 67:505–12.

Puhl, R. M., and J. D. Latner. 2007. Stigma, obesity, and the health of the nation's children. *Psychol Bull* 133:557–80.

Raji, C. A., A. J. Ho, N. N. Parikshak, J. T. Becker, O. L. Lopez, L. H. Kuller, X. Hua, A. D. Leow, A. W. Toga, and P. M. Thompson. 2010. Brain structure and obesity. *Human Brain Mapping* 31:353–64.

Ramaswamy R., M. Mirochna, and L. C. Perlmutter. 2010. The negative association of BMI with classroom effort in elementary school children. *J Child Health Care* 14:161–9.

Reagan, L. P. 2007. Insulin signaling effects on memory and mood. *Curr Opin Pharmacol* 7:633–7.

Reijmer, Y., D. E. van den Berg, C. Ruis, L. J. Kappelle, and G. J. Biessels. 2010. Cognitive dysfunction in patients with type 2 diabetes. *Diab-Metabol Res Rev.* 26:507–19.

Roberts, C. K., B. Freed, and W. J. McCarthy. 2010. Low aerobic fitness and obesity are associated with lower standardized test scores in children. *J Pediatr*. 156:711–8, 718.e1.

Ross, A. P., T. J., Bartness, J. G. Mielke, and M. B. Parent. 2009. A high fructose diet impairs spatial memory in male rats. *Neurobiol Learn Mem* 92:410–6.

Ryan C., M. Freed, J. Rood, A. Cobitz, B. Waterhouse, and M. Strachan. 2006. Improving metabolic control leads to better working memory in adults with type 2 diabetes. *Diab Care* 29:345–51.

Saland, J. M. 2007. Update on the metabolic syndrome in children. *Curr Opin Pediatr* 19: 183–91.

Schmader, T., M. Johns, and C. Forbes. 2008. An integrated process model of stereotype threat on performance. *Psychol Rev* 115:336–56.

Shimamura, A. P. 2010. Hierarchical relational binding in the medial temporal lobe: the strong get stronger. *Hippocampus* 20:1206–16.

Shoelson, S. E., L. Herrero, and A. Naaz. 2007. Obesity, inflammation, and insulin resistance. *Gastroenterology* 132:2169–80.

Stefan, N., K. Kantartzis, J. Machann, F. Schick, C. Thamer, K. Rittig, B. Balletshofer, F. Machicao, A. Fritsche, and H-U. Haring. 2008. Indentification and characterization of metabolically benign obesity in humans. *Arch Intern Med* 168:1609–16.

Stranahan, A. M., E. D. Norman, K. Lee, R. G. Cutler, R. S. Telljohann, J. M. Egan, and M. P. Mattson. 2008. Diet-induced insulin resistance impairs hippocampal synaptic plasticity and cognition in middle-aged rats. *Hippocampus* 18:1085–8.

Taki, Y., S. Kinomura, K. Sato, K. Inoue, R. Goto, K. Okada, S. Uchida, R. Kawashima, and H. Fukuda. 2008. Relationship between body mass index and gray matter volume in 1,428 healthy individuals. *Obesity* 16:119–24.

Uranga, R. M., A. J. Bruce-Keller, C. D. Morrison, S. O. Ferandex-Kim, P. J. Ebenezer, L. Zhang, K. Dasuri, and J. N. Keller, J. N. 2010. Intersection between metabolic dysfunction, high fat diet consumption, and brain aging. *J Neurochem* 114:344–61.

Valladolid-Acebes, I., P. Stucchi, V. Cano, M. S. Fernandez-Alfonso, B. Merino, M. Gil-Ortega, A. Fole, L. Morales, M. Ruiz-Gayo, and N. Del Olmo. 2011. High-fat diets impair spatial learning in the radial-arm maze in mice. *Neurobiol Learn Mem* 95: 80–5.

Velho, S., F. Paccaud, G. Waeber, P. Vollenweider, and P. Marques-Vidal. 2010. Metabolically healthy obesity: different prevalences using different criteria. *Eur J Clin Nutr* 64:1043–51.

Verdejo-Garcia, A., M. Perez-Exposito, J. Schmidt-Rio-Valle, M. J. Fernandez-Serrano, and C. Campoy. 2010. Selective alterations within executive functions in adolescents with excess weight. *Obesity* 18:1572–8.

Volkow, N. D., G. J. Wang, F. Telang, J. S. Fowler, R. Z. Goldstein, N. Alia-Klein, J. Logan, C. Wong, P. K. Thanos, Y. Ma, and K. Pradhan. 2009. Inverse association between BMI and prefrontal metabolic activity in healthy adults. *Obesity* 12:60–5.

Vorhees, C. V., and M. T. Williams. 2006. Morris water maze: procedures for assessing spatial and related forms of learning and memory. *Nat Protoc* 1:848–58.

Vos, M. B., and J. Welsh J. 2010. Childhood obesity: update on predisposing factors and prevention strategies. *Curr Gastroenterol Rep* 12:280–7.

Ward, M. A., C. M. Carlsson, M. A. Trivedi, M. A. Sager, and S. C. Johnson. 2005. The effect of body mass index on global brain volume in middle-aged adults: a cross-sectional study. *BMC Neurology* 5:23 doi:10.1186/1471-2377-5-23.

Walther, K., A. C. Birdsill, E. L. Glisky, and L. Ryan. 2010. Structural brain differences and cognitive functioning related to body mass index in older females. *Human Brain Mapping* 31:1052–64.

Whitmer, R. A., E. P. Gunderson, E. Barrett-Connor, C. P. Quesenberry, and K. Yaffe, K. 2005. Obesity in middle age and future risk of dementia: a 27-year longitudinal population based study. *BMJ* doi:10.1136/bmj.38446.466238.EO.

Whitmer, R. A., D. R. Gustafson, E. Barrett-Connor, N. M. Haan, E. P. Gunderson, and K. Yaffe. 2008. Central obesity and increased risk of dementia more than three decades later. *Neurology* 71:1057–64.

Wildna, R. P., P. Muntner, K. Reynolds, A. P. McGinn, S. Rajpathak, J. Wylie-Rosett, and M. R. Sowers. 2008. The obese without cardiometabolic risk factor clustering and the normal weight with cardiometabolic risk factor clustering. *Arch Intern Med* 168: 1617–24.

Winocur, G., and C. E. Greenwood. 1999. The effects of high fat diets and environmental influences on cognitive performance in rats. *Behav Brain Res* 101:153–9.

Winocur, G., and C. E. Greenwood. 2005. Studies of the effects of high fat diets on cognitive function in a rat model. *Neurobiol Aging* 26:46–9.

World Health Organization. 2009. Obesity and overweight. World Health Organization. www.who.int/dietphysicalactivity/publications/facts/en/

World Health Organization. 2011. Obesity and overweight Fact Sheet No. 311. World Health Organization. www.who.int/mediacentre/factsheets/fs311/en/

Wu, A., R. Molteni, Z. Ying, and F. Gomez-Pinilla. 2003. A saturated-fat diet aggravates the outcome of traumatic brain injury on hippocampal plasticity and cognitive function by reducing brain-derived neurotrophic factors. *Neuroscience* 119:365–75.

Yau, P. L., D. C. Javier, C. M. Ryan, W. H. Tsu, B. A. Ardekani, S. Ten, and A. Convit. 2010. Preliminary evidence for brain complications in obese adolescents with type 2 diabetes mellitus. *Diabetologia* 53:2298–306.

Yu, H., X. Bi, W. Ma, L. He, L. Yuan, J. Feng, and R. Xiao. 2010. Long-term effects of high lipid and high energy diet on serum lipid, brain fatty acid composition, and memory and learning ability in mice. *Int J Dev Neurosci* 28:271–6.

Zhang, L., A. J. Bruce-Keller, K. Dasuri, A. T. Nguyen, Y. Liu, and J. N. Keller. 2009. Diet-induced metabolic disturbances as modulators of brain homeostasis. *Biochim Biophys Acta* 1792:417–22.

8 Dietary Supplements for Weight Loss

Igho Onakpoya and Edzard Ernst

CONTENTS

Introduction 154
What Are Dietary Supplements? 154
Prevalence and Sales Figures 154
Reasons for Use 156
Mechanism of Action 156
Efficacy 158
  Guar Gum 158
  Chromium Picolinate 158
  Ephedrine 159
  Citrus aurantium (Bitter Orange) 159
  Conjugated Linoleic Acid 159
  Calcium 160
  Glucomannan 160
  Chitosan 160
  Camellia sinensis (Green Tea) 160
  Garcinia Extracts (Hydroxycitric Acid) 160
  Pyruvate 161
  Psyllium 161
  Hydroxymethylbutyrate 161
  Pausinytalia yohimbe (Yohimbe) 161
Safety 162
  Guar Gum 163
  Chromium Picolinate 163
  Ephedrine 163
  Citrus aurantium (Bitter Orange) 163
  Conjugated Linoleic Acid 163
  Glucomannan 164
  Chitosan 164
  Camellia sinensis (Green Tea) 164
  Garcinia Extracts (Hydroxycitric Acid) 164
  Hydroxymethylbutyrate 164
  Yohimbe 164
  Pyruvate 165

Conclusion 165
Acknowledgment 165
References 165

INTRODUCTION

Overweight and obesity are important risk factors for a wide range of serious diseases and have thus become topical issues of immense concern (Ogden et al. 2007). The prevalence of these conditions has increased remarkably, and measures put in place seem to have not, to date, successfully addressed the epidemic.

Overweight and obesity are defined as an excess in body weight. These excesses are usually defined in terms of the Body Mass Index (BMI) in kg/m^2. Based on the World Health Organization (WHO) criteria, a BMI range between 25 and 29 kg/m^2 is defined as overweight, while a BMI $\geq 30 kg/m^2$ is defined as obesity. Presently, there are over 1.5 billion individuals overweight worldwide, and over 400 million of these are clinically obese (WHO 2006). The problem has increased the burden on health care systems, costing the United States as much as $117 billion in 2000 (Wadden et al. 2003).

Overweight and obesity are usually due to genetic, environmental, and lifestyle factors (Wadden et al. 2003). A variety of strategies aimed at preventing or reducing body weight are currently being employed. These involve not only the use of conventional, but also complementary and alternative treatments, all aimed at reducing body weight. Chief amongst these complementary and alternative treatments is the use of dietary supplements.

WHAT ARE DIETARY SUPPLEMENTS?

Dietary supplements are nutritional substances, usually for oral use. The U.S. Food and Drug Administration (FDA) defines dietary supplements as foods, rather than as drugs (Pillitterri et al. 2008). Such supplements usually contain one or more dietary ingredients. In the United States and many other countries, dietary supplements can be marketed without FDA approval and without scientific proof of safety or efficacy. The use of dietary supplements is common, and for some indications, food supplements seem to be gradually replacing prescription medications (Marinac et al. 2007). Dietary supplements for weight loss (DSFWLs) include herbal supplements and substances used in conventional medicine to treat conditions other than overweight and obesity. The latter category includes drugs such as calcium, niacin, as well as chromium and other trace elements. A plethora of dietary supplements marketed as slimming (weight loss) aids can be purchased in many stores or on the internet. Table 8.1 lists the most commonly used dietary supplements.

PREVALENCE AND SALES FIGURES

It is difficult to give exact prevalence figures regarding the usage of DSFWLs. Since supplements can easily be purchased over the counter (OTC) and on the internet, monitoring the usage amongst consumers can be challenging. A U.S. survey (Blanck

TABLE 8.1
Most Commonly Used Dietary Supplements and Reasons Why Individuals Use Them

Dietary Supplement	Reasons for Use of Dietary Supplements for Weight Loss
Camellia sinensis (Green tea) … most common	• Easy availability
Conjugated linoleic acid (CLA)	• Inflated claims of efficacy
Chitosan	• Need to avoid dieting and exercise
Chromium picolinate	• Perception that "natural equals safe"
Garcinia cambogia (hydroxycitric acid)	• No need to consult with a health professional
Citrus aurantium (bitter orange)	• Social perception of overweight and obesity
Guarana	• A desire to lose weight fast
Caffeine	
Ginseng	
Dandelion	
Pyruvate	
Guar gum	
Glucomannan	
L-carnitine	
Ephedra/ephedrine	
Yerba mate	
Psyllium	

et al. 2007) suggested that over 15 percent of adults were estimated to have ever used a DSFWL. The likelihood of purchasing such supplements increased with increasing BMI. Another U.S. survey (Pillitteri et al. 2008) indicated that about 45 percent of women and 20 percent of men had used a DSFWL. Both surveys suggested that the use of DSFWLs was much more common in females than in males.

The sale of DSFWLs has increased steadily in the past few years (Marinac et al. 2007). In 2005, U.S. sales of DSFWLs topped \$1.6 billion (Pillitteri et al. 2008), while sales in Western Europe surpassed \$1.4 billion in 2009 (Euromonitor International 2009). The growth of the DSFWL industry is clearly associated with the increased prevalence of overweight and obesity (Pillitteri et al. 2008).

The DSFWL industry has also suffered some setbacks. This is largely due to some inflated, misleading, or incorrect claims about the effectiveness and safety of these preparations. These have led to court cases that generated adverse publicity, as well as guidance to avoid some of these supplements by regulatory agencies. In 2009, a marketer of DSFWLs was ordered to pay the United States \$37 million for making false claims about its product in a book (Federal Trade Commission 2009). In 2009, Iovate Health Sciences (a DSWFL producer) had to withdraw its Hydroxycut product (which contained hydroxycitric acid) from the market after the FDA warned its consumers about the risk of some serious adverse effects associated with the consumption of the supplement (Nutrition Business Journal 2009a). Hydroxycut was reputed to be the top-selling slimming pill when this warning was issued by the FDA, and the sales of DSFWLs declined sharply after publication of this report.

REASONS FOR USE

Dietary interventions are an important part of the treatment of obesity and overweight. Patients generally find dietary modification less demanding than regular exercise (Wadden et al. 2003), particularly if this does not entail dieting but simply taking a DSFWL. Other main reasons why patients use dietary supplements (see also Table 8.1) include the following:

1. *Availability of such supplements*: Some individuals who are intent on losing weight feel embarrassed to be questioned by health care professionals and thus resort to self-treatment with DSFWLs (Saper et al. 2004). Such supplements can easily be obtained via the internet, health shops, pharmacies, supermarkets, and so on.
2. *Unjustified therapeutic claims*: Manufacturers of DSFWLs tend to promise rapid weight loss without the hassle of dietary restriction or exercise (Saper et al. 2004). Obese patients are often desperate and vulnerable to such claims and usually fail to critically question the evidence.
3. *Conviction that "natural supplements" are safer*: "Natural" is frequently equated with "risk-free." Unfortunately, the popular press regularly contributes to this dangerous misconception. In a survey of 3,500 adults conducted in the United States in 2005–2006 (Pillitteri et al. 2008), about 33 percent of respondents believed that dietary weight loss supplements were safer than prescription medications.

MECHANISM OF ACTION

Various mechanisms of action have been proposed for DSFWLs. Some DSFWLs have also been purported to act via more than one mechanism (Table 8.2). The postulated mechanisms of actions include the following:

1. *Increasing energy expenditure*: Producers of some DSFWLs have claimed that use of such products results in increased energy expenditure, thereby resulting in a reduction of weight. Supplements in this category include *Ephedra*, bitter orange, guarana, caffeine, and yerba mate (Saper et al. 2004). The action is allegedly achieved by stimulation of the sympathetic system. Of course, this could result in adverse events involving the cardiovascular, nervous, and gastrointestinal systems. *Ephedra*-containing DSFWLs have been reported to cause about a two- to fourfold increase in psychiatric, autonomic, gastrointestinal, and cardiovascular adverse events, and were withdrawn from the U.S. market after regulatory action by the FDA. Bitter orange is thought to contain synephrine, which is also sympathomimetic and could theoretically cause similar adverse events.

TABLE 8.2
Proposed Mechanisms of Action of Common Weight Loss Dietary Supplements

Dietary Supplement	Proposed Mechanism of Action
Guar gum	Thought to increase the viscosity of the bowel content, thereby producing a feeling of postprandial fullness, which could in turn decrease appetite
Chromium picolinate	Believed to increase the basal metabolic rate and insulin sensitivity, thereby causing a reduction in body weight
Ephedra/ephedrine	Postulated to cause weight reduction via its thermogenetic effect
Citrus aurantium	Believed to contain synephrine (a phenylephrine analogue) which has a sympathomimetic effect
Conjugated linoleic acid	Reduction in adipocyte size, modifying adipocyte differentiation, stimulating apoptotic mechanisms, and regulating lipid metabolism
Calcium	Suppressing calcitriol levels and decreased adipose tissue intracellular Ca^{2+}, resulting in coordinated increase in lipid utilization and decrease in lipogenesis
Glucomannan	Thought to prolong gastric emptying time, and thereby increasing satiety and reducing body weight
Chitosan	Believed to cause weight loss by binding to negatively charged lipids, hence reducing their gastrointestinal uptake … results from animal studies
Camellia sinensis (green tea)	Postulated to induce weight loss by stimulating fat oxidation and energy expenditure
Hydroxycitric acid	Inhibition of citrate lyase, pancreatic amylase and intestinal alpha-glucosidase, and suppression of appetite
Pyruvate	Increased metabolic rates in muscle tissue
Yerba mate	Increased rate of energy expenditure
Dandelion	Increased elimination of water from the body
Psyllium	Increased satiety

2. *Modulating carbohydrate metabolism*: Producers of some DSFWLs such as chromium, hydroxycitric acid, and ginseng have claimed they modulate carbohydrate metabolism (Saper et al. 2004). It is claimed that chromium decreases the basal metabolic rate and insulin sensitivity (Pittler and Ernst 2004), and hydroxycitric acid inhibits pancreatic alpha-amylase and intestinal alpha-glucosidase and increases insulin sensitivity (Yamada et al. 2007).
3. *Increased satiety*: Another mechanism through which DSFWLs are postulated to cause weight loss is by decreasing the absorption of carbohydrates from the gastrointestinal system. This is achieved by increasing the viscosity of the bowel contents, thereby slowing down the rate of movement of food through the alimentary tract, theoretically increasing satiety (Pittler and Ernst 2004). Some other supplements such as hydroxycitric acid are also believed to act on the brain's satiety center via effects on neurotransmitters such as serotonin (Toromanyan et al. 2007).
4. *Increased fat oxidation*: It has been suggested that other DSFWLs, such as green tea and pyruvate, increase fatty acid oxidation (Saper et al. 2004). This is proposed to increase fat loss, thereby leading to a reduction in body weight.

5. *Reduction of fat synthesis*: Hydroxycitric acid is a commonly marketed DSFWL which supposedly acts via this mechanism. It is thought that ingestion of this supplement could lead to the inhibition of the mitochondrial enzyme citrate lyase (Saper et al. 2004). This in turn could lead to a decreased acetyl coenzyme A, thereby resulting in decreased synthesis of fatty acids in theory.
6. *Decreased dietary fat absorption*: This is another mechanism advocated by some authors as a mechanism by which some DSFWLs cause weight reduction. Supplements in this group include chitosan and are believed to bind to negatively charged lipids, thereby reducing their intestinal uptake.
7. *Suppression of calcitriol levels*: Calcium supplements are claimed to act via this mechanism. Suppressing calcitriol levels leads to decreased intracellular Ca^{2+}, supposedly resulting in increased lipid utilization and lipid synthesis (Zemel et al. 2005).
8. *Increase water elimination*: Some DSFWLs such as dandelion are thought to possess diuretic effect (Saper et al. 2004). By increasing the amount of water eliminated from the gastrointestinal tract, such substances are believed to have a role in reducing body weight.

The mechanism of action of other DSFWLs such as laminaria and blue green algae is not specified. St. John's wort (*Hypericum perforatum*), a commonly used herbal antidepressant, is sometimes also promoted for weight loss (Saper et al. 2004). Its mechanism of action in this context is unclear.

EFFICACY

A number of randomized controlled trials (RCTs) testing DSFWLs in overweight and obesity have been conducted. Table 8.3 summarizes the key details from systematic reviews of some of these supplements.

Guar Gum

Pittler and Ernst (2001) systematically reviewed the efficacy for guar gum on weight loss from twenty RCTs including 366 participants. They found guar gum not to be efficacious for reducing body weight. Findings from sensitivity analyses were consistent with this result. The authors reported that the included studies had small sample sizes and some were of low methodological quality.

Chromium Picolinate

Pittler et al. (2003) evaluated the efficacy of chromium picolinate for weight loss in a systematic review of seventeen RCTs with 961 participants. They found it to have a "relatively small" effect in reducing weight. Sensitivity analyses revealed a significant effect of chromium picolinate on weight reduction in obese patients. The quality of the included studies was variable, and none of them achieved the maximum score

TABLE 8.3
Effects of Dietary Supplements on Body Weight: Summary of Findings from Clinical Trials

Dietary Supplement	Conclusions Drawn from Analyses of Randomized Controlled Trials
Guar gum	Not effective for reducing body weight
Chromium picolinate	Small effect on body weight reduction
Ephedra/ephedrine	May promote small losses in weight on the short term
Citrus aurantium	No evidence that it has a positive effect on weight loss
Conjugated linoleic acid (CLA)	Not enough evidence to warrant its usage as a weight loss supplement
Calcium	No statistically significant association with a reduction in body weight
Glucomannan	May beneficially affect body weight
Chitosan	Evidence not convincing enough to recommend its usage as a slimming pill
Green tea (*Camellia sinensis*)	Statistically significant effect on body weight reduction when given with caffeine, but clinical relevance is not certain
Hydroxycitric acid	Small, statistically significant effect on body weight reduction, but clinical relevance is uncertain
Pyruvate	No evidence that pyruvate supplementation causes weight loss
Psyllium	No significant effect on body weight

in the methodological quality assessment. The definition of the terms "overweight" and "obese" was also unclear in some of the included studies.

Ephedrine

Shekelle et al. (2003) examined the efficacy and safety of *Ephedra* spp. and ephedrine for weight loss in 1,451 participants from seventeen RCTs. Statistical analyses revealed a significant short-term effect on body weight. Serious risks of *Ephedra* and ephedrine intake were also identified. Some of the included primary studies were of low methodological quality, with one study having a score of 0.

Citrus aurantium (Bitter Orange)

Bent et al. (2004) reviewed the safety and efficacy of *Citrus aurantium* (bitter orange) for weight loss. Only one trial of moderate methodological quality with twenty-three participants met their inclusion criteria. The authors concluded that there is no evidence that *Citrus aurantium* is effective for weight loss. The trial included in this review did not contain only *Citrus aurantium,* which limits the conclusions that can be drawn about *C. aurantium*-specific effects.

Conjugated Linoleic Acid

Salas-Salvado et al. (2006) systematically reviewed the effects of conjugated linoleic acid (CLA) on weight loss in twenty-one clinical trials including 852 participants. In aggregate, these data suggest no significant effect of CLA on body weight. Most of the included trials lasted less than one year in duration. The methodological quality of the included studies was not evaluated.

Calcium

Trowman et al. (2006) reviewed the effects of calcium supplements on body weight in thirteen RCTs with a total of 1,127 participants. After statistical adjustment for discrepancies in the baseline characteristics of participants, there was no evidence of a significant effect of calcium supplements on body weight. The methodological quality of the included studies was not evaluated, but the authors did state that the randomization procedures in most of the included studies were "flawed."

Glucomannan

Sood et al. (2008) assessed the efficacy of glucomannan for weight reduction in a systematic review of fourteen RCTs with 531 participants. They found that it reduced weight significantly. The effect was greatest in obese patients and in parallel studies. The methodological quality of the included primary studies was not reported. Therefore, the results of this review are less reliable in our view. Any future assessment of these data should take trial quality into account.

Chitosan

Jull et al. (2008) assessed the effects of chitosan on body weight in 1,219 participants from fifteen RCTs. There was some evidence of short-term weight loss in obese and overweight individuals. The authors reported that most of the studies included were of poor methodology. Five of the included studies with the greatest impact on the meta-analytic results did not meet the allocation concealment criteria, and four of these included agents other than chitosan. These important caveats prevented the authors from making firm conclusions.

Camellia sinensis (Green Tea)

Phung et al. (2010) reviewed the effects of green tea catechins (GTC) with or without caffeine on body weight in fifteen trials with 1,243 participants. They found GTC not to be efficacious for weight reduction and weight maintenance, but that the addition of caffeine to GTC resulted in significant reduction in body weight, with the clinical significance of this reduction modest at best. The authors concluded that administration of GTC alone does not alter body composition.

Garcinia Extracts (Hydroxycitric Acid)

A number of RCTs involving the use of *Garcinia* extracts are available, but the findings from these clinical trials have not been consistent. We analyzed the results of twelve RCTs with 706 participants and found a small, statistically significant difference in body weight loss favoring hydroxycitric acid (HCA) over placebo (Onakpoya et al. 2011). The clinical relevance of this effect is uncertain, and the quality of most of the trials is poor.

Pyruvate

A few clinical trials involving the use of this supplement have been conducted. We analyzed the results of five RCTs with 150 participants, and did not find any evidence to suggest that the intake of pyruvate causes any significant weight loss when compared with placebo (Onakpoya et al. 2010).

Psyllium

Only one RCT involving the use of this supplement was identified in a previous systematic review (Pittler and Ernst 2004). The study involved 125 overweight diabetic patients and lasted six weeks. The trial investigators reported no significant differences in weight loss between psyllium and placebo, but stated that tolerance to the supplement was good.

Hydroxymethylbutyrate

A review analyzed the results from four trials in which hydroxymethylbutyrate supplement was tested (Pittler and Ernst 2004). There were significant intergroup differences in two of the RCTs with respect to fat mass loss, but the studies were not clear in reporting of blinding issues.

Pausinytalia yohimbe (Yohimbe)

Pittler and Ernst (2004) also assessed the effects of yohimbe on weight loss in three RCTs with eighty-six participants. The results were not consistent, and the authors concluded that the effectiveness of yohimbe as a weight-reducing agent is uncertain.

The efficacy of some other weight loss supplements (e.g., St. John's wort [*Hypericum perforatum*]) is untested, and there are no data to support their role in weight loss (Saper et al. 2004). Overall, the findings from most of these trials are limited by poor methodologies, in particular inadequate randomization, small sample sizes, short duration, and wide dosage ranges. Thus, the results from analyses of these trials fail to demonstrate the efficacy or safety of these supplements as weight-reducing agents, when compared to placebo. It is therefore not plausible to recommend the use of these DSFWLs. In May 2008, GlaxoSmithKline (together with the American Dietetic Association, the Obesity Society, and Shaping America's Health) filed a petition with the FDA requesting for weight loss claims to be treated as disease claims, because there was no scientific evidence to back up the weight loss claims of the marketers of these supplements (nutraingredients-usa.com).

Most of the clinical trials involving DSFWLs also try to adjust the lifestyle factors of trial participants. These range from diet control such as diet restriction (e.g., low-fat diet, low-energy diet, and low-caloric diet) to different forms of exercise (e.g., no exercise, normal exercise regimen, moderate exercise, and resistance training). Low caloric intake alone is one of the most important components for achieving weight loss (Walden 2003). However, given the short duration of most of these trials, the extent to which these lifestyle adjustments influenced the results of the trials is frequently unclear.

SAFETY

A major concern about the use of DSFWLs is safety. Because most of these supplements have not been subjected to rigorous examination before being licensed, their safety has not been established (Pittler et al. 2005). The number of adverse events associated with the use of DSFWLs appears to be underreported as physicians' knowledge about dietary supplement safety and adverse event reporting is poor (Ashar et al. 2007).

There have been a considerable number of reports of adverse events linked to the intake of DSFWLs. Many DSFWLs contain mixtures of different active ingredients, which can make it difficult to predict potential risks associated with their use (Chan 2009). Table 8.4 lists some of the adverse reactions which have been associated with the use of these DSFWLs.

TABLE 8.4
Adverse Events Associated with Some of the Commonly Used Weight Loss Supplements

Dietary Supplement	Associated Adverse Events
Guar gum	Flatulence, heartburn, nausea
Chromium picolinate	Generally believed to be safe; risk of interaction with conventional drugs
Ephedra/ephedrine	Increased risk of cardiovascular, autonomic, gastrointestinal, and autonomic adverse events
Citrus aurantium	Risk of cardiovascular adverse events, due to synephrine content
Conjugated linoleic acid	Decrease in HDL cholesterol
Guarana	Inhibition of platelet aggregation
Glucomannan	Diarrhea, flatulence, loose stools, possibility of esophageal obstruction
Chitosan	Increased risk of gastrointestinal adverse events
Camellia sinensis (green tea)	Considered safe, but caffeine content may result in caffeine-associated side effects
Hydroxycitric acid	Increased risk of gastrointestinal adverse events
Pyruvate	Possible negative effect on blood lipid profile
Yerba mate	Sleeplessness, stomach upset
Dandelion	Dehydration, electrolyte imbalance
Psyllium	Allergy, possibility of adverse drug reactions
Licorice	Pseudoaldosteronism, which could result in hypertension and hypokalemia

Guar Gum

A number of gastrointestinal adverse events have been reportedly linked to the ingestion of guar gum. These include diarrhea and flatulence. Toxicity studies have suggested that long-term intake of this supplement does not result in any serious health consequences (McIvor et al. 1995). Guar gum has a potential to interact with drugs. It can slow the absorption of digoxin and paracetamol, as well as decrease the absorption of metformin, phenoxymethylpenicillin, and some formulations of glibenclamide (Fugh-Berman 2000).

Chromium Picolinate

Information regarding the safety of long-term use of this supplement is quite limited. The intake of chromium picolinate has been associated with nausea, vomiting, rash, dizziness, headache, loss of appetite, and unexplained bleeding (Vincent 2003). The supplement has also been reported to interact with medications such as prostaglandins and nonsteroidal anti-inflammatory drugs (NSAIDs; Vincent 2003). The intake of chromium picolinate has been linked to serious renal impairments (Wani et al. 2006) and liver problems (Lanca et al. 2002). Reports that chromium can cause DNA damage, and therefore could be carcinogenic, have been disputed (Hinninger et al. 2007).

Ephedrine

This supplement is probably the most widely known DSFWL with a variety of serious adverse events associated with its use. Adverse events include hypertension, tachycardia, palpitations, anxiety, and psychosis (Ashar et al. 2003). In a systematic review of its safety, the authors reported that individuals who take this supplement have a 2.2 to 3.6 increased odds of psychiatric, autonomic, upper gastrointestinal, and cardiovascular symptoms (Shekelle et al. 2003). The FDA banned the use of ephedrine as a weight loss supplement in 2004 because of concerns over its safety (Pittler and Ernst 2004).

Citrus aurantium (Bitter Orange)

This supplement is reported to contain synephrine, which is a sympathomimetic. Adverse events which may be associated with the use of *C. aurantium* include high blood pressure, tachycardia, and the risk of cardiovascular events.

Conjugated Linoleic Acid

Individuals taking CLA may experience mild to moderate gastrointestinal symptoms such as nausea and frequent stools (Saper et al. 2004). CLA is also reported to have effects on glucose and lipid metabolism, especially an increase in HDL cholesterol (Sala-Salvado et al. 2006), although this could be considered a beneficial effect. An isolated case of hepatotoxicity associated with its use has also been reported (Ramos et al. 2008).

Glucomannan

This supplement seems to be well tolerated, even in the long term (Vita et al. 1992). Adverse events associated with the use of glucomannan include loose stools, flatulence, diarrhea, abdominal discomfort, and esophageal obstruction (Sood et al. 2008).

Chitosan

The main adverse events associated with the use of this supplement are gastrointestinal, and include constipation, indigestion, bloating, and abdominal pain (Jull et al. 2008). It is considered to be generally safe. Some reports have suggested that chitosan could bind soluble vitamins, thus leading to hypovitaminosis in predisposed individuals.

Camellia sinensis (Green Tea)

Green tea is considered to be generally safe. Adverse events associated with green tea intake are usually due to its caffeine content, and include sleep disturbances, palpitation, and headache. There has also been a reported case of hepatitis as a result of the intake of this weight loss herb (Vanstraelen et al. 2008).

Garcinia Extracts (Hydroxycitric Acid)

There are a number of adverse events associated with the use of HCA. These include nausea, headache, and upper respiratory and gastrointestinal symptoms (Pittler and Ernst 2004). Though the supplement was considered to be generally safe (Soni et al. 2004), the FDA warned its users about the potential for serious adverse events in 2009 (Nutrition Business Journal 2009b). This resulted in its withdrawal from the market.

Hydroxymethylbutyrate

This supplement has been described as generally safe in humans. Findings from analyses of adverse events on its safety from nine human clinical studies did not suggest any significant differences in adverse events between hydroxymethylbutyrate and placebo (Nissen et al. 2000).

Yohimbe

Yohimbe is currently approved for the treatment of erectile dysfunction due to its sympathomimetic effect (Giampreti et al. 2009). Several adverse events have been reportedly associated with the use of this alpha-2 antagonist. These include sleep disturbances, nervousness, headache, and joint pain (Ernst and Pittler 2004). Intoxication with this supplement could result in malaise, vomiting, loss of consciousness, and seizures (Giampreti et al. 2009).

Pyruvate

A number of adverse events have been linked to the use of pyruvate, including diarrhea, gas, bloating, and increased LDL cholesterol.

Adverse events have also been reported from DSFWLs that have never been tested in clinical trials for weight loss. Licorice has been reported to cause pseudoaldosteronism, hypertension, and hypokalemia (Saper et al. 2004). Long-term use of dandelion could lead to dehydration and electrolyte disturbances due to its laxative effect. Psyllium has been reported to be generally safe. The safety of some other DSFWLs such as apple cider vinegar, guggul (an extract from the myrrh tree, *Commiphora mukul*), and laminaria is uncertain.

CONCLUSION

Obesity and overweight are serious health problems that are difficult to treat. The market is awash with DSFWLs claiming to be easy, effective, and safe solutions to weight loss. A comprehensive examination indicated that none of these supplements has been proven to reduce body weight safely. We therefore believe that the claims made for DSFWLs are unjustified and these compounds waste users' money.

ACKNOWLEDGMENT

Igho Onakpoya is funded by a grant from GlaxoSmithKline. The funder had no role in the preparation of this chapter. Edzard Ernst declares no potential conflict of interest.

REFERENCES

Ashar, B. H., R. G. Miller, K. J. Getz, and C. P. Pichard. 2003. A critical evaluation of internet marketing of products that contain Ephedra. *Mayo Clin Proc* 78: 944–6.

Ashar, B. H., T. N. Rice, and S. D. Sisson. 2007. Physicians' understanding of the regulation of dietary supplements. *Arch Intern Med* 167 (9): 966–9.

Bent, S., A. Padula, and Neuhaus. 2004. Safety and efficacy of *Citrus aurantium* for weight loss. *Am J Cardiol* 94 (10): 1359–61.

Blanck, H. M., M. K. Serdula, C. Gillespie, et al. 2007. Use of nonprescription dietary supplements for weight loss is common among Americans. *JADA* 107 (3): 441–7.

Chan, T. Y. 2009. Potential risks associated with the use of herbal anti-obesity products. *Drug Saf* 32 (6): 453–6.

Euromonitor International. Consumer Health. 2009. OTC Obesity 2009. http://www.euromonitor.com/OTC_Obesity

Fugh-Berman, A. 2000. Herb-drug interactions. *The Lancet* 355 (9198): 134–8.

Giampreti, A., D. Lonati, C. Locatelli, L. Rocchi, and M. T. Campailla. 2009. Acute neurotoxicity after yohimbine ingestion by a bodybuilder. *Clin Toxicol* 47 (8): 827–9.

Hininger, I., R. Benaraba, M. Osman, H. Faure, A. Marie Roussel, and R. A. Anderson. 2007. Safety of trivalent chromium complexes: no evidence for DNA damage in human HaCaT keratinocytes. *Free Radic Biol Med* 42 (12): 1759–65.

Jull, A. B., C. Ni Mhurchu, D. A. Bennett, C. A. Dunshea-Mooij, and A. Rodgers. 2008. Chitosan for overweight or obesity. *Cochrane Database Syst Rev* Issue 3. Art. No. CD003892. doi:10.1002/14651858.CD003892.pub3.

Lanca, S., A. Alves, A. I. Vieira, et al. 2002. Chromium-induced toxic hepatitis. *Eur J Intern Med* 13: 518–20.

Marinac, J. S., C. L. Buchinger, L. A. Godfrey, J. M. Wooten, C. Sun, and S. Willsie. 2007. Herbal products and dietary supplements: a survey of use, attitudes, and knowledge among older adults. *JAOA* 107 (1): 13–23.

McIvor, M. E., C. C. Cummings, and A. I. Mendeloff. 1985. Long-term ingestion of guar gum is not toxic in patients with noninsulin-dependent diabetes mellitus. *Am J Clin Nutr* 41: 891–4.

Nissen, S., R. L. Sharp, L. Panton, M. Vukovich, S. Trappe, and J. C. Fuller, Jr. 2000. b-Hydroxy-b-Methylbutyrate (HMB) supplementation in humans is safe and may decrease cardiovascular risk factors. *J Nutr* 130 (8): 1937–45.

Federal Trade Commission. 2009. Judge orders Kevin Trudeau to pay more than $37 million for false claims about weight-loss book. http://www.ftc.gov/opa/2009/01/trudeau/shtm

NUTRAIngredients-USA.com. 2008. GSK petition claims weight loss supplements have no science. http://www.nutratingredients-usa.com/Regulation

Nutrition Business Journal. 2009a. How worrisome is the hydroxycut recall for the dietary supplement industry? http://www.nutritionbusinessjournal.com/supplements/news/0505

Nutrition Business Journal. 2009b. Weight loss supplement sales plummet following Iovate's hydroxycut recall. http://www.nutritionbusinessjournal.com/supplements/0901

Ogden, C. L., S. Z. Yanovski, M. D. Carroll, and K. M. Flegal. 2007. The epidemiology of obesity. *Gastroenterology* 132 (6): 2087–2102.

Onakpoya, I., K. Hunt, B. Wider, and E. Ernst. 2010. Pyruvate supplementation for weight loss: a systematic review and meta-analysis of randomised clinical trials.

Onakpoya, I., S. K. Hung, R. Perry, B. Wider, and E. Ernst. 2011. The use of *Garcinia* extract (Hydroxycitric acid) as a weight loss supplement: a systematic review and meta-analysis of randomised clinical trials. *J Obes* (2011). doi:10.1155/2011/509038.

Phung, O. J., W. L. Baker, L. J. Matthews, M. Lanosa, A. Thorne, and C. I. Coleman. 2008. Effect of green tea catechins with or without caffeine on anthropometric measures: a systematic review and meta-analysis. *Am J Clin Nutr* 91 (1): 73–81.

Pillitteri, J. L., S. Shiffman, J. M. Rohay, A. M. Harkins, S. L. Burton, and T. A. Wadden. 2008. Use of dietary supplements for weight loss in the United States: results of a national survey. *Obesity* 16 (4): 790–6.

Pittler, M. H., and E. Ernst. 2001. Guar gum for body weight reduction: meta-analysis of randomized trials. *Am J Med* 110 (9): 724–30.

Pittler, M. H., and E. Ernst. 2004. Dietary supplements for body-weight reduction: a systematic review. *Am J Clin Nutr* 79: 529–36.

Pittler, M. H., K. Schmidt, and E. Ernst. 2005. Adverse events of herbal food supplements for body weight reduction: systematic review. *Obes Rev* 6 (2): 93–111.

Pittler, M. H., C. Stevenson, and E. Ernst. 2003. Chromium picolinate for reducing body weight: meta-analysis of randomized trials. *Int J Obes* 27 (4): 522–9.

Ramos, R., J. Mascarenhas, P. Duarte, C. Vicente, and C. Casteleiro. 2009. Conjugated linoleic acid-induced toxic hepatitis: first case report. *Dig Dis Sci* 54 (5): 1141–3.

Salas-Salvadó, J., F. Márquez-Sandoval, and M. Bulló. 2006. Conjugated linoleic acid intake in humans: a systematic review focusing on its effects on body composition, glucose, and lipid metabolism. *Crit Rev Food Sci Nutr* 46 (6): 479–88.

Saper, R. B., D. M. Eisenberg, and R. S. Phillips. 2004. Common dietary supplements for weight loss. *Am Fam Physician* 70 (9): 1731–8.

Shekelle, P. G., M. L. Hardy, M. C. Morton, et al. 2003. Efficacy and safety of ephedra and ephedrine for weight loss and athletic performance: a meta-analysis. *JAMA* 289 (12): 1537–45.

Soni, M. G., G. A. Burdock, H, G, Preuss, S. J. Stohs, S. E. Ohia, and D. Bagchi. 2004. Safety assessment of (-)-hydroxycitric acid and Super CitriMax®, a novel calcium/potassium salt. *Food Chem Toxicol* 42: 1513–29.

Sood, N., W. L. Baker, and C. I. Coleman. 2008. Effect of glucomannan on plasma lipid and glucose concentrations, body weight, and blood pressure: a systematic review and meta-analysis. *Am J Clin Nutr* 88 (4): 1167–75.
Toromanyan, E., G. Aslanyan, E. Amroyan, E. Gabrielyan, and A. Panossian. 2007. Efficacy of Slim339® in reducing body weight of overweight and obese human subjects. *Phytother Res* 21 (12): 1177–81.
Trowman, R., J. C. Dumville, S. Hahn, and D. J. Torgerson. 2006. A systematic review of the effects of calcium supplementation on body weight. *Br J Nutr* 95 (6): 1203–11.
Vanstraelen, S., J. Rahier, and A. P. Geubel. 2008. Jaundice as a misadventure of a green tea (*Camellia sinensis*) lover: a case report. *Acta Gastroenterol Belg* 71 (4): 409–12.
Vincent, J. B. 2003. The potential value and toxicity of chromium picolinate as a nutritional supplement, weight loss agent and muscle development agent. *Sports Med* 33 (3): 213–30.
Vita, P. M., A. Restelli, P. Caspani, and R. Klinger. 1992. Chronic use of glucomannan in the dietary treatment of severe obesity. *Minever Med* 83 (3): 135–9.
Wadden, T. A., B. G. McGuckin, R. A. Rothman, and S. L. Sargent. 2003. Lifestyle modification in the management of obesity. *J Gastro Surg* 7 (4): 452–63.
Wani, S., C. Wescamp, J. Marple, and L. Spry. 2006. Acute tubular necrosis associated with chromium picolinate-containing dietary supplement. *Ann Pharmacother* 40 (3): 563–6.
World Health Organization. 2006. Obesity and Overweight. Fact Sheet No. 311, September. http://www.who.int/mediacentre/factsheet
Yamada, T., H. Hida, and Y. Yamada. 2007. Chemistry, physiological properties and microbial production of hydroxycitric acid. *Appl Microbiol Biotechnol* 75 (5): 977–82.
Zemel, M. B., J. Richards, S. Mathis, A. Milstead, L. Gebhardt, and E. Silva. 2005. Dairy augmentation of total and central fat loss in obese subjects. *Int J Obes* 29 (4): 391–7.

9 Sweet Taste Preferences and Cravings in Gestational Diabetes Mellitus (GDM)

Implications for Diet and Medical Management

Beverly J. Tepper, Lisa M. Belzer, John C. Smulian, and Shou-En Lu

CONTENTS

Introduction 170
Risks and Consequences of GDM 170
Endocrine and Metabolic Profiles in Healthy Pregnancy and GDM 171
  Healthy Pregnancy 171
  GDM Pregnancy 171
Medical Management of GDM and the Need for New Nutritional Strategies 172
Sweet Taste Changes and Diet in Type 2 Diabetes 172
Changes in Sweet Taste and Cravings in Healthy Pregnancy 173
  Food Cravings in Healthy Pregnancy 174
  Changes in Taste Perception and Preference in Healthy Pregnancy 174
Sweet Taste Changes and Appetite in GDM: Empirical Studies 174
  Liking for Sweet Taste and Glucose Intolerance in GDM 175
  Liking for Sweet-Fat Dairy Drinks in Mild GDM 175
  A Role for the Gestational Hormones in Sweet Taste Changes in Mild GDM 178
  Sweet Cravings and Dietary Intake in Mild GDM 179
Conclusions and Future Research Needs 182
Summary and Implications for Clinical Practice 183
References 184

INTRODUCTION

Gestational diabetes mellitus (GDM) is a form of carbohydrate intolerance of variable severity that is first recognized during pregnancy (Metzger et al. 2007). Its metabolic profile and management are similar to those of diabetes mellitus type 2 (type 2 DM) (American Diabetes Association, 2010b). GDM is associated with a variety of obstetric complications and significant perinatal morbidity (Metzger et al. 2007). Women with previous GDM are also at increased risk of developing insulin resistance and type 2 DM later in life (Verma et al. 2002). As many as 50 percent of women with a GDM pregnancy will develop type 2 diabetes within five years of the index pregnancy (Kim et al. 2002). Thus, GDM can be considered a sentinel event for the future development of type 2 DM, which currently affects 17 million Americans and is predicted to rise to 29 million cases by 2050 (Winer and Sowers, 2004). Recent projections indicate that prevention and control of GDM with subsequent lifestyle changes could substantially improve the health of women and save millions of dollars in health care costs (Winer and Sowers, 2004). Currently, there is no consensus on what the optimum nutritional strategy should be for the management of GDM.

RISKS AND CONSEQUENCES OF GDM

GDM affects 3–8 percent of all pregnancies in the United States (Metzger et al. 2007). Prevalence rates are higher in African American (4.6 percent), Asian (8.6 percent), and Hispanic women (5.1 percent) as compared to Non-Hispanic whites (3.4 percent) (Dabelea et al. 2005). The extent to which the excess risk in minority groups reflects genetic differences or lifestyle remains uncertain (Shaat et al. 2004). Alarmingly, prevalence rates in all groups of women have doubled in less than a decade (Dabelea et al. 2005). This trend parallels the dramatic rise in obesity in the general population. Currently, greater than 30 percent of American women of child-bearing age are obese (Ogden et a., 2007), and obesity prior to pregnancy increases the risk of GDM more than fivefold (Chu et al. 2007). In clinical practice, most women who develop GDM are overweight prior to pregnancy (Metzger et al. 2007). Other risk factors for GDM are age of the mother, parity, and family history of diabetes. GDM can also recur in subsequent pregnancies with frequencies ranging from 30 to 84 percent (Kim et al. 2007).

GDM is associated with disruptions in maternal glucose metabolism, including hyperinsulinemia, insulin resistance, and decreased oral glucose tolerance (Catalano et al. 2003). Women with GDM also have a disproportionately high risk of preeclampsia and cesarean delivery (Yogev and Langer, 2008; Yogev et al. 2004). The infant of a GDM mother is also at significant risk for metabolic abnormalities, macrosomia (large for gestational age [LGA]), and respiratory distress at birth (Yogev and Langer 2008). Shoulder dystocia (the inability of the infant's shoulder to pass through the birth canal) is an obstetric emergency at delivery, and is one of the most common birth complications of LGA infants (Fadl et al. 2010). LGA infants also exhibit hyperinsulinemia and elevated leptin at birth (Westgate et al. 2006) which persist into childhood (Malee et al. 2002). Additionally, such children are at increased risk for childhood obesity (Vohr and Boney 2008) and type 2 diabetes in adulthood (Pettitt and Jovanovic 2001). Thus, GDM in the mother has consequences for her progeny that can persist into subsequent generations.

ENDOCRINE AND METABOLIC PROFILES IN HEALTHY PREGNANCY AND GDM

Healthy Pregnancy

Pregnancy is a unique metabolic state designed to maximize growth and development of the fetus (King 2006). Early pregnancy is characterized by increased maternal fat storage, decreased energy expenditure, and delayed glucose clearance from the blood. These events are timed to meet the increasing demand for glucose and amino acids by the fetus. Levels of the gestational hormones (estrogen, progesterone, and human placental lactogen) as well as insulin and cortisol rise sharply during pregnancy and play a major role in these events. Insulin needs to increase fourfold during pregnancy such that insulin release plateaus before thirty weeks' gestation age. Despite insulin resistance, women experiencing a healthy pregnancy maintain normal glucose tolerance. During late pregnancy, maternal fat stores begin to fall and are mobilized to support the energy needs of the rapidly growing fetus. The hormonal or metabolic events that trigger these changes are not well described but probably reflect ongoing insulin resistance, which promotes lipolysis and the subsequent loss of adipose stores (King, 2006).

Leptin levels also increase during pregnancy, peak in mid- to late pregnancy, and decrease to baseline at delivery (Domali and Messinis 2002; Henson and Castracane 2006). Since the pattern of change in circulating leptin parallels the process of maternal fat accumulation and mobilization in pregnancy, leptin may be a marker for these processes (Henson and Castracane 2006). The hormonal milieu of pregnancy could increase the production of leptin, contributing to leptin resistance and weight gain. Leptin also plays a variety of roles in reproductive function, including implantation of the egg and fetal bone and lung development (Henson and Castracane 2006).

GDM Pregnancy

Gestational diabetes occurs when a woman has limited pancreatic β-cell capacity and an inability to enhance insulin secretion in response to the metabolic effects of the gestational hormones (Metzger et al. 2007). The exact pathophysiology is unknown but most likely involves a combination of pronounced insulin resistance at peripheral tissues and inadequate β-cell insulin secretion. Gestational hormones rise rapidly and insulin demand begins to outstrip secretory capacity, usually in mid-pregnancy. In most clinical settings, universal screening for GDM is typically done at 24–28 weeks gestational age to allow for the timing of these changes (American Diabetes Association 2010a).

Leptin is also of critical interest in GDM, although its role is poorly understood. Leptin is normally secreted in direct proportion to the size of the adipose mass (Benoit et al. 2004). Thus, plasma leptin is higher in women who are overweight at conception and also rises more sharply in women who gain more weight during pregnancy (Butte 2000; Highman et al. 1998). However, studies have shown higher, lower, or no change in leptin in GDM in comparison to women without GDM (Festa et al. 1999; Kautzky-Willer et al. 2001; Lappas et al. 2005; Maghbooli et al. 2007; Simmons and Breie, 2002).

MEDICAL MANAGEMENT OF GDM AND THE NEED FOR NEW NUTRITIONAL STRATEGIES

Diet is the cornerstone in the medical management of GDM, and nutritional counseling is typically initiated immediately following diagnosis. The primary goals of nutritional management of GDM are similar to those of type 2 diabetes, including the control of body weight (or limiting excess weight gain in the case of the pregnant diabetic) and the reduction of glucose intolerance (American Diabetes Association 2007). Dietary guidance for both diseases includes monitoring overall energy intake, balancing macronutrient selection, and encouraging the consumption of whole grains, fruits, and vegetables. In the past, simple sugars were restricted in the diabetic diet. Under current guidelines, there is a greater focus on controlling the total amount of carbohydrate in meals rather than the source of the carbohydrate (i.e., complex carbohydrates vs. simple sugars). Other than the recommendations to space meals and snacks across the day and to limit excess calories (e.g., soft drinks and sweets) (American Diabetes Association 2009), there is no specific and uniform guidance for women with GDM.

A pregnancy complicated by GDM is highly stressful, and compliance with current diet recommendations is low (Langer 2002). In fact, 30–60 percent of women with GDM are unable to control their symptoms by diet alone (Langer 1999). In these cases, insulin or an oral hypoglycemic medication (e.g., Metformin or Glyburide) may be added to the regimen. To be effective, nutritional counseling must take into consideration personal food preferences and eating habits (Armstrong et al. 1991). Translating nutrient guidelines into individualized meal plans is a significant challenge for the health care professional. Data are lacking on food preferences and appetite changes in women with GDM. Obtaining a better understanding of these changes is a fundamental first step toward improving diet strategies for this disease.

SWEET TASTE CHANGES AND DIET IN TYPE 2 DIABETES

As stated earlier, GDM and type 2 diabetes share a common pathophysiology, and nutritional guidance for both diseases is similar. Since changes in sweet taste in type 2 diabetes may provide insights for understanding these same alterations in GDM, we begin with a brief review of the literature on sweet taste in type 2 diabetes.

Why study changes in sweet taste in diabetes? The symptoms of what we now know as uncontrolled diabetes mellitus have been recognized since antiquity as excessive thirst and urination, muscle wasting, and the production of "sweet" urine (King and Rubin 2003). As early as the eighteenth century, it was known that excess sugar in the blood (later identified as glucose) was the source of sweetness in diabetic urine. Pioneering work in the late nineteenth and early twentieth centuries demonstrated that diabetes was a disease of the endocrine pancreas (King and Rubin 2003). Later experiments characterized the disease by its major features: peripheral insulin resistance and an inability to transport glucose into cells for energy metabolism (Lin and Sun 2010). Thus, the notion that diabetes interferes with the body's ability to recognize glucose and may be associated with an increased desire for sweet taste is based on long historical precedent.

The first laboratory studies demonstrating that taste ability, especially sensitivity to sweet taste, is reduced in type 2 diabetes were conducted in the 1960s (Schelling et al. 1965). Further study of this question revealed that taste thresholds (i.e., the ability to distinguish weak taste solutions from plain water) are elevated for certain simple sugars (e.g., glucose and sucrose) which could alter the perception and palatability of sweet foods and beverages for individuals with type 2 diabetes (Abbasi 1981; Gondivkar et al. 2009; Lawson et al. 1979; Perros et al. 1996; Schelling et al. 1965)

Despite the attractiveness of this theory, studies have shown only small differences (Dye and Koziatek 1981) or no differences in hedonic ratings for sweet-tasting solutions between individuals with and without type 2 diabetes (Lawson et al. 1979; Tepper et al. 1996). In contrast, Laitenin and colleagues (1991) reported that in persons with newly diagnosed type 2 diabetes, hedonic ratings for sweetened fruit juices declined during the first three months of intensive diet therapy that was successful in improving metabolic control.

Tepper et al. (1996) reported no differences in pleasantness ratings for fruit drinks sweetened with sucrose, fructose, or aspartame between individuals with well-controlled type 2 diabetes and age-matched controls. Nevertheless, we observed positive correlations between liking ratings for the fruit beverages and reported dietary intake of sweet foods ($r = 0.46–0.55$; $p \leq 0.05$) among individuals with diabetes, but not among control subjects. Thus, for persons with diabetes, those who preferred the sweeter beverages in laboratory tests also consumed more sweetness in the diet, primarily as sweetened beverages. To our knowledge, these are the first empirical data linking hedonic reactions to sweet taste with habitual dietary behavior in type 2 diabetes. The observation that some but not all individuals with diabetes showed exaggerated taste and dietary preferences for sweet taste suggests that there is individual variation in sweet taste palatability and the motivation to seek out and consume sweets within the diabetic population. Understanding these individual differences and their underlying mechanisms could help to clarify the role of sweet taste in the diabetic diet.

CHANGES IN SWEET TASTE AND CRAVINGS IN HEALTHY PREGNANCY

Pregnancy is thought to have unique effects on taste and appetite. Since women with GDM are glucose intolerant and pregnant, it is important to distinguish the general effects of pregnancy on taste and appetite changes during gestation from those attributable to GDM. The following sections review taste changes and cravings in healthy pregnancy.

Food Cravings in Healthy Pregnancy

Food cravings are very common in pregnancy with as many as 85 percent of women reporting food cravings at some point during pregnancy (Pope et al. 1992). Often these cravings are for sweet or sweet-fat foods such as fruit, fruit juice, ice cream, cake, and pies (Hook 1978; Pope et al. 1992; Worthington-Roberts et al. 1989). Such cravings can influence dietary intake. For example, pregnant adolescents who reported craving sweet foods consumed more sugar and 400 kcal/day more than those without cravings (Pope et al. 1992). Food cravings are generally reported to be most frequent and intense during the second trimester and to decline thereafter (Pope et al. 1992). Sweet foods are generally preferred during the second trimester, whereas salty foods and starches are preferred during the third trimester (Bowen, 1992). It is important to point out, however, that most studies on food cravings in pregnancy have generally been retrospective in nature, depending on memory of food behavior in the past, which may be less reliable (Hook 1978; Worthington-Roberts et al. 1989). Rodin and Radke-Sharpe (1991) conducted one of the few prospective studies on food cravings in pregnancy reported in the literature. In contrast to other studies, they found that the number of pregnant women with food cravings during the second trimester was higher than the number of nonpregnant women with cravings, at the same time point in the study. However, the overall incidence of cravings did not change in pregnant women across gestation.

Changes in Taste Perception and Preference in Healthy Pregnancy

Studies have investigated changes in perception and preference for the basic tastes during pregnancy (Brown and Toma 1986; Dippel and Elias 1980). Results suggest a general decrease in gustatory function early in pregnancy (Kolble et al. 2001; Kuga et al. 2002; Ochsenbein-Kolble et al. 2005), but the influence of this change on taste preference is less clear. One study demonstrated a lower preference for sucrose in pregnant women as compared to nonpregnant women, but no change in sweet preference across pregnancy stages (Dippel and Elias 1980). Other studies did not observe changes in sweet taste preference in pregnancy (Brown and Toma 1986; Duffy et al. 1998; Kolble et al. 2001; Ochsenbein-Kolble et al. 2005). The most common finding reported was for increased pleasantness for salty taste during the second trimester (Brown and Toma 1986; Duffy et al. 1998; Ochsenbein-Kolble et al. 2005). Thus, the literature as a whole provides more convincing evidence for changes in salty taste during healthy pregnancy than for changes in sweet taste.

SWEET TASTE CHANGES AND APPETITE IN GDM: EMPIRICAL STUDIES

Despite the accumulating knowledge on sweet taste changes in type 2 diabetes, no study had addressed this question in GDM. In the mid-1990s, we began investigating sweet taste changes in GDM. The guiding hypothesis for our work is that women with GDM show increased preference and craving for sweet foods that might alter their food habits, and could ultimately threaten their dietary adherence. Two studies were conducted to test this hypothesis. In an initial study, Tepper and Seldner (1999) examined women with GDM at a single time point during pregnancy (28–32 weeks gestation) and at twelve weeks postpartum. Positive results from this study suggested that we follow women with GDM across pregnancy and into the early postpartum period to characterize the time course of taste and diet changes in this population. Thus, Belzer (2008) conducted a prospective study to examine changes in taste ratings, food cravings, and diet parameters in pregnant women who developed GDM during gestation. We also measured plasma profiles of selected gestational and metabolic hormones in order to correlate these patterns with the taste and diet changes. The following sections highlight the major findings from these studies.

Liking for Sweet Taste and Glucose Intolerance in GDM

Tepper and Seldner (1999) studied three groups of women: pregnant women with GDM (n = 25), pregnant women without GDM (n = 30), and nonpregnant control women (n = 19). Pregnant women entered the study at the time of their diagnostic screening test for GDM (28–32 weeks gestational age), and most of the women with a diagnosis of GDM participated in the study prior to their referral to nutritional therapy for diabetes.

This study examined the relationships between plasma glucose one hour after a 50 g oral glucose load (i.e., during the clinical screening test) and two measures of sweet taste preference: hedonic ratings for glucose solutions (0.01 M–1.0 M), and reported intakes of simple sugars as fruit and fruit juice. As shown in Figure 9.1, plasma glucose concentrations after oral glucose in women with GDM were positively correlated with both liking for the glucose solutions as well as reported dietary intake of simple sugars as fruit and fruit juice. That is, women with GDM who had more severe glucose intolerance also had higher preferences for glucose and higher reported intakes of sweet carbohydrates than women with less severe glucose intolerance. These relationships were not observed in women without GDM. It is noteworthy that a small number of women with GDM with higher plasma glucose (> 230 mg/dl or 12.8 mmol/L), who later required insulin to control their disease gave the highest liking ratings for glucose and consumed more fruit and fruit juice than the rest of the group with GDM. These data suggest that the degree of glucose intolerance in GDM might contribute to higher taste and dietary preferences for sweet foods in this population.

Liking for Sweet-Fat Dairy Drinks in Mild GDM

Since sweet-fat beverages and snacks represent another source of sweetness and calories in the diabetic diet, determining if women with GDM experience changes in perception and acceptance of these types of foods may be highly relevant for understanding dietary behavior in this population. Belzer et al. (Belzer 2008; Belzer et al. 2009, 2010) recruited pregnant women receiving prenatal care from the outpatient clinic of an urban hospital. The study population was predominantly Latina. Three groups of women were studied using a prospective design: pregnant women with normal glucose tolerance (NGT group; n = 93), pregnant women who developed GDM

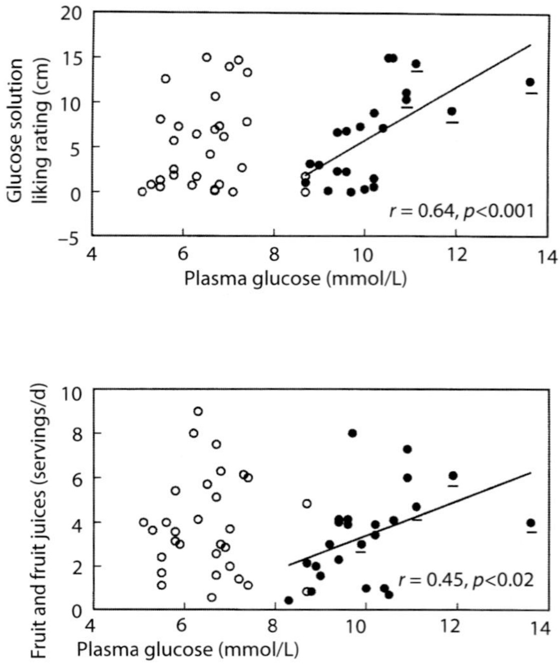

FIGURE 9.1 Correlation between plasma glucose concentrations 1 hr after a 50-g glucose load, and liking ratings for glucose solutions and daily servings of fruit/fruit juice for women with (●) and without (○) GDM. There were significant correlations for women with GDM, but not for women without GDM. Underlined data points identify four women with more severe GDM who received insulin treatment after the taste studies were conducted. Reproduced from Tepper & Seldner (1999) with permission from the American Society of Nutrition.

during the course of the study (GDM group; $n = 15$), and nonpregnant control women (CONT group; $n = 19$) who were attending the clinic for other reasons. Pregnant women were studied at 16–20 weeks, 24–28 weeks, and 34–38 weeks gestation, and at 6–10 weeks after delivery; nonpregnant women were tested at equivalent time intervals. A high percentage (~14 percent) of the women we studied developed GDM, which was confirmed by diagnostic testing at 24–28 weeks gestational age. Women in the GDM group received nutritional counseling for diabetes soon after their diagnosis; a diabetic exchange plan was used (American Diabetes Association). The NGT group received standard nutritional guidance for pregnancy. Importantly, all the women with GDM whom we studied had a mild form of the disease and were able to maintain metabolic control to term by diet alone; none received insulin or other medications for diabetes management. Further clinical details of the study can be found in Belzer et al. (2009).

The study examined intensity and liking ratings for strawberry-flavored dairy drinks varying in sucrose concentration at each test session. Figure 9.2 shows that liking ratings for the sweetness, creaminess, and flavor of the moderately sweetened

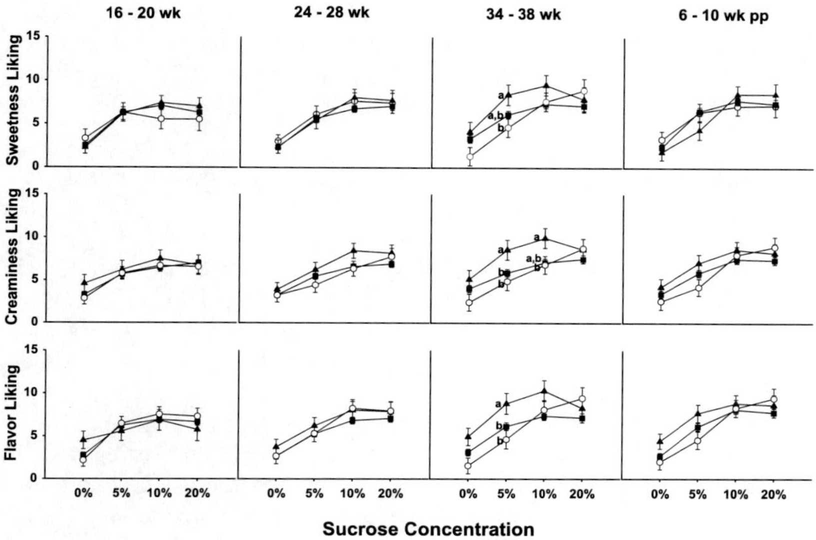

FIGURE 9.2 Sweetness, creaminess and flavor liking ratings of sucrose-sweetened dairy drinks. Drinks in women with GDM (▲), women with NGT (■), and nonpregnant controls (○). At 34–38 weeks of gestation, women with GDM showed higher ratings for sweetness liking of 5% sucrose drink, when compared to nonpregnant controls, and higher creaminess liking ratings for 10% sucrose drink when compared to women with NGT. Also, women with GDM showed higher creaminess and flavor liking of 5% sucrose drink as compared to both women with NGT and nonpregnant women. Reproduced from Belzer et al. (2009) with permission from Oxford University Press.

dairy drinks were higher in the GDM group in comparison to the other groups during late pregnancy (34–38 weeks gestation), but not at other time points in the study.

These findings provide several insights into the sweet taste responses of women with GDM. First, the GDM group gave higher liking ratings to the dairy drinks at 34–38 weeks gestation but not at 24–28 weeks gestation when their diabetes was emerging and their glucose intolerance was more severe. This suggests that enhanced palatability for sweet-fat mixtures is a feature of late pregnancy in GDM, and is not necessarily associated with the onset of diabetic symptoms. As described shortly, cravings for sweet foods also peaked at 34–38 weeks gestation, suggesting that late pregnancy is a high-risk time for the development of exaggerated reactions to sweet-fat foods in GDM.

Second, the GDM group liked the moderately sweetened (5 percent and 10 percent sucrose) dairy drinks more than the other groups did, but the women with GDM did not give higher liking ratings to the dairy drinks with the highest sweetness content (20 percent sucrose). Thus, GDM appears to increase the palatability of stimuli within the midrange of sweetness intensity, but not at more extreme sucrose concentrations. This observation is relevant for understanding everyday eating behavior since commercial dairy drinks contain ~7–10 percent added sweetener.

Finally, we also collected intensity ratings for all the attributes of dairy drinks, and no differences were found in the perception of the samples among the groups (data not shown). This finding implies that enhanced sweet taste palatability in GDM primarily reflects a shift in hedonic valence rather than a reaction to diminished perception of sweet taste. In this regard, GDM is distinct from type 2 diabetes in which the ability to perceive sweet taste may be compromised.

A Role for the Gestational Hormones in Sweet Taste Changes in Mild GDM

A major objective of the study by Belzer et al. (2008) was to track changes in the gestational hormones across pregnancy in women with GDM and NGT, and to relate these changes to alterations in sweet taste. Figure 9.3 shows changes in fasting plasma insulin, leptin, and cortisol at each time point of the study. Consistent with the literature in healthy pregnancy (Boden, 1996), we found that fasting levels of all three hormones rose to reach a peak in late pregnancy in women with NGT and fell to baseline after delivery. In women with mild GDM, fasting insulin was only modestly elevated at 24–28 weeks gestational age (consistent with their mild diabetic symptoms). However, leptin tended to rise earlier in women with GDM, and coincided with the development of insulin resistance. Cortisol rose across pregnancy stages in both groups of pregnant women but was not further elevated in the GDM group relative to the NGT group.

We questioned whether rising levels of these hormones, especially leptin, were associated with changes in sweet taste in women with GDM. Studies in animals have shown that leptin acts on taste receptor cells where it modulates behavioral and neurological responses to sweet taste (Kawai et al. 2000; Yoshida et al. 2006). Moreover, hyperleptinemia in animal models of diabetes is associated with exaggerated neural (chorda tympani) responses to sugars (Kawai et al. 2000). Although there are no data on sweet taste and leptin in diabetic humans, studies in nondiabetic individuals show that plasma leptin is highly correlated with recognition threshold for sweet taste (Nakamura et al. 2008). Thus, human data suggest a role for leptin in sweet taste sensitivity, although its effects on liking are presently unknown.

Belzer et al. (2009) also examined relationships between the sensory ratings for the sweetened dairy drinks and fasting plasma values for each of these hormones. Separate regression equations were calculated for each subject group, and repeated for each time point of the study. Glucose solutions were also evaluated in this study, and regression equations were calculated for these data as well. Results showed a positive association between fasting leptin and the sweetness liking ratings of the 10 percent sucrose-sweetened dairy drink ($R^2 = 0.42$, $p = 0.0166$) in the GDM group at 24–28 weeks gestational age. The GDM group also showed a positive association between fasting insulin and sweetness liking ratings of the glucose solutions at this same time point ($R^2 = 0.63$, $p = 0.0037$). These relationships were not observed in the other groups of women, or in women with GDM at other time points in the study. Due to the limited scope of these findings, we do not consider these associations to be of major importance in mild GDM. We note, however, that these associations were observed at an earlier stage of pregnancy than the group differences in liking ratings of the dairy drinks (Figure 9.2), suggesting different underlying mechanisms for these effects.

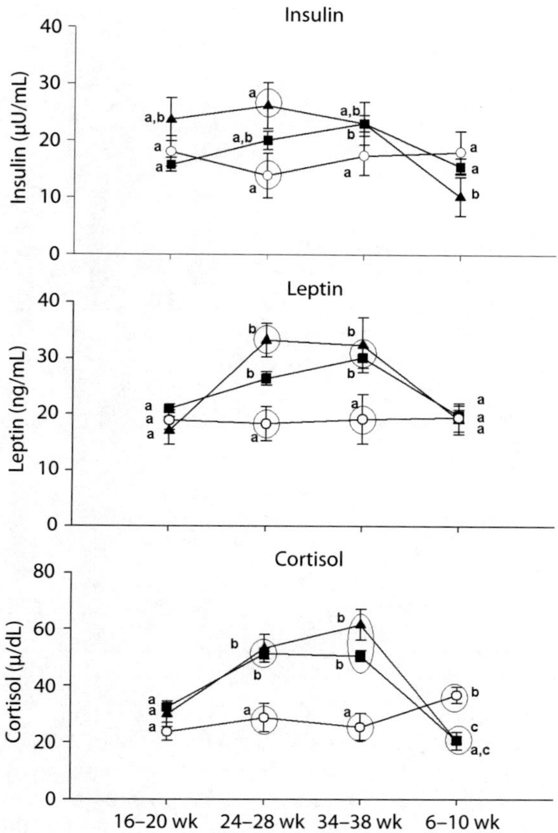

FIGURE 9.3 Fasting serum insulin, leptin and cortisol (mean + SE) across pregnancy and postpartum in women with GDM (▲), women with NGT (■) and nonpregnant controls (○). Mixed model analysis revealed a main effect of group on leptin and cortisol ($p = 0.0024$ and $p < 0.0001$, respectively). A main effect of gestational age was found for leptin and cortisol ($p < 0.001$), but not for insulin. Within-group effects across sessions are indicated by lowercase letters. Values with different lowercase letters are significantly different. Between-group effects were tested with ANOVA at each session. Circled values are significantly different from each other at each session. The significance level for all comparisons was set at $p = 0.017$, after adjusting for the Bonferroni correction.

Sweet Cravings and Dietary Intake in Mild GDM

We also tracked food cravings (via questionnaire) and collected dietary information by food frequency questionnaire at all time points of the study (Belzer 2008; Belzer et al. 2010). Cravings for sweet foods were reported by 55 percent of the NGT group, but only 40 percent of the GDM group. Nevertheless, cravings for sweet foods peaked at 34–38 weeks gestation in women with GDM at twice the mean frequency of the NGT group (12.1 + 2.8 vs. 5.9 + 1.0 times/wk) (see Figure 9.4). Thus, although

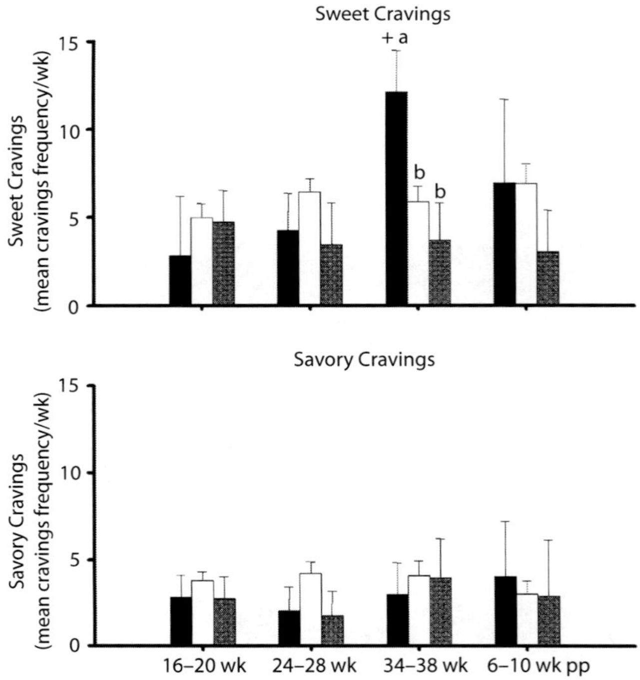

FIGURE 9.4 Mean frequency of sweet and savory cravings across sessions. At 34–38 weeks gestation, means without a common letter differ. [+] Higher mean frequency of sweet cravings in the GDM group at 34–38 weeks gestation as compared to 24–28 weeks gestation at $p < 0.017$ following Bonferroni correction. Reproduced from Belzer et al. (2010) with permission from Elsevier Ltd. Women with GDM (■); women with NGT (□); nonpregnant controls (■).

a surprisingly small percentage of women with GDM reported sweet cravings, this subgroup experienced sweet cravings almost twice daily during late pregnancy. Due to the small sample size of this subgroup, further characterization these cravings was not possible. Future studies will need to determine the specific types of sweets that are craved (e.g., snacks, sweet drinks, or fruit) and if the intensity of sweet cravings was related to "giving in" to the cravings. We also do not know why some of the women with GDM whom we studied seemed to be more vulnerable to sweet cravings when all the women with GDM achieved a similar level of good diabetic control. Finally, it would be important to know if frequent sweet cravings are a source of emotional distress for these women and hence a barrier to dietary adherence.

Belzer and coworkers (Belzer, 2008; Belzer et al. 2010) also analyzed nutrient intakes as well as consumption patterns for sweet foods. Figure 9.5 shows that the intake of sweet foods and beverages, including fruit and fruit juice, peaked in the NGT group at midpregnancy (24–28 weeks gestation) and declined thereafter. This finding agrees with published reports in healthy pregnancy (Pope et al. 1992). However, this peak was absent at 24–28 weeks gestation in women with GDM. Moreover, despite a higher reported frequency of sweet cravings at 34–38 weeks gestation (see Figure 9.4)

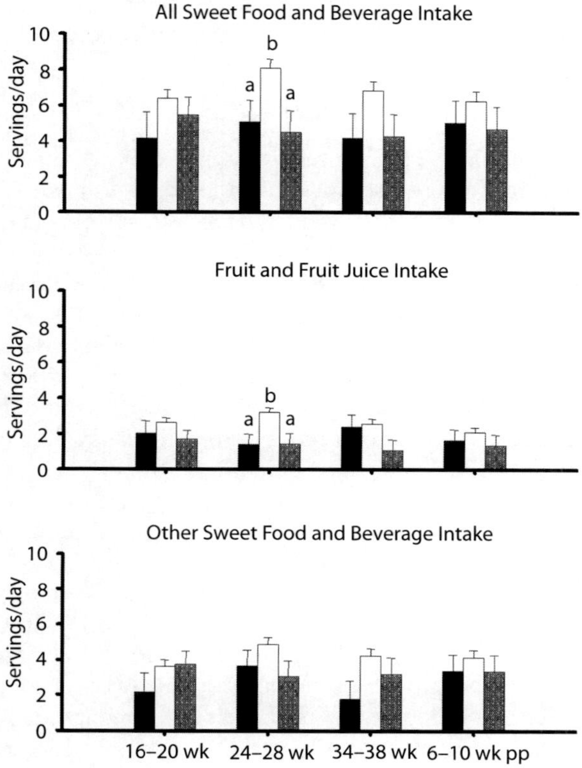

FIGURE 9.5 Reported daily sweet food and beverage intakes across sessions. At 24–28 weeks gestation, means without a common letter differ at $p < 0.017$ following Bonferroni correction. Reproduced from Belzer et al. (2010) with permission from Elsevier Ltd. Women with GDM (■); women with NGT (□); nonpregnant controls (■).

in a subset of the women with GDM, the GDM group as a whole did not increase its consumption of sweet foods at any time point during pregnancy.

With respect to nutrient intake, we (Belzer 2008; Belzer et al. 2010) also showed that women with GDM consumed less than the recommended 2,000 kcal/day (for women who are overweight prior to conception; American College of Obstetrics and Gynecology 2001), and comparatively less carbohydrate and more fat than women in the NGT group. The GDM group also consumed marginally less sugar (as percent energy) as compared to the NGT group, but this difference failed to reach statistical significance. Although dietary reporting bias cannot be ruled out, key pregnancy outcomes indicated that the intakes of women with GDM were plausible. Women with GDM gained less weight during pregnancy than women with NGT, fetal birth weight was comparable in both groups, and all women with GDM delivered healthy infants (Belzer et al. 2009). Thus, in contrast to the general obstetric population with GDM, the women with mild GDM whom we studied adhered to their diets and successfully managed their disease by diet alone. These studies need to be repeated in a larger cohort of women with more severe glucose intolerance to fully ascertain the impact of sweet taste changes on dietary behavior in this population.

CONCLUSIONS AND FUTURE RESEARCH NEEDS

Altered taste sensation in GDM is a relatively new area of research inquiry, and our studies to date raise more questions than they answer about the role of these changes in dietary behavior in these women. However, several general conclusions can be drawn from our present work that may generate new hypotheses and guide future investigations in this field. Our studies suggest that changes in sweet taste in GDM are fundamentally different in time course, frequency, and origin than those observed in healthy pregnancy. Women with NGT showed increased intake of sweet foods in midpregnancy that declined as pregnancy progressed to term, a pattern which is typical in healthy pregnancy. In contrast, women with GDM showed no changes in taste or appetite for sweet foods at midpregnancy. Rather, women with GDM showed increased preference and cravings for sweet foods during late pregnancy. The mechanisms underlying this late-pregnancy rise in palatability of sweet foods in GDM are presently unknown. However, we speculate that both metabolic and psychosocial factors play a role in these responses.

From a metabolic perspective, it seems likely that glucose intolerance, which is intrinsic to GDM, contributes to increased preference and cravings for sweets as pregnancy progresses to term. This is supported, in part, by our findings showing that healthy pregnant women who do not develop glucose intolerance late in pregnancy also do not experience increased preference and cravings for sweets (although they are mildly insulin resistant at that time). Furthermore, the persistence of glucose intolerance late in the gestational period may be a critical factor in the development of taste changes and sweet cravings in GDM. This is based on our findings showing that exaggerated responses to sweets in the GDM group occurred at 34–38 weeks gestation, but not at 24–28 weeks gestation, when their disease was untreated and their symptoms were more severe. Thus, late pregnancy appears to be a time of increased risk for exaggerated responses to sweet taste in GDM.

We observed increased preference and cravings for sweet foods in the GDM group at the time when the diabetic diet was in effect. One explanation for these findings is that the diet restrictions may have led to psychological feelings of deprivation in women with GDM which triggered their sweet cravings. This explanation seems unlikely in light of studies in the general population showing that when dietary compliance is high, energy-restricted or therapeutic diets *decrease* rather than increase food cravings (Barnard et al. 2009; Harvey et al. 1993; Martin et al. 2006). However, one study reported that an energy-restricted but varied diet of typical foods led to more food cravings (Gilhooly et al. 2007). The relationship between diet restrictions and food cravings in GDM has not been studied. However, findings in women with polycystic ovary syndrome, a condition in which insulin resistance is also a key feature, may have relevance to GDM. One study showed that carbohydrate cravings fell by 50 percent following diet therapy with a low glycemic load diet (Herriot et al. 2008). Thus, we conclude that although the diet restrictions may have contributed

FIGURE 9.6 (*See color insert.*) Model depicting the influence of negative diet perceptions on the relationship between sweet food cravings and dietary adherence in GDM.

to sweet cravings in women with GDM, they are probably not the driving force for sweet cravings in this disease.

Previous studies have stressed the importance of negative diet perceptions (i.e., the perceived difficulty and perceived barriers to following the diet) on the risk of developing type 2 diabetes in women with previous GDM (Jones et al. 2009). However, no studies have examined the influence of negative diet perceptions on adherence to diet therapy and overall metabolic control in pregnant women with active GDM. According to our model illustrated in Figure 9.6, poor glycemic control in GDM could generate more frequent cravings for sweets. Whether or not a woman acts on these cravings may be mediated by the presence of negative diet perceptions. A woman with negative diet perceptions may find sweet cravings more difficult to ignore, which could impact her food choices and undermine her efforts to follow the diet. Poor dietary adherence could also exacerbate negative diet perceptions, leading to a vicious cycle of noncompliance. Future research should focus on identifying such women and developing novel strategies to intervene in this pathway.

SUMMARY AND IMPLICATIONS FOR CLINICAL PRACTICE

The optimum diet for GDM continues to be the subject of ongoing debate (Uplinger, 2009). Current trends in nutritional therapy favor a reduction in the total carbohydrate content of the diet to 30–55 percent of energy to minimize postprandial hyperglycemia (Reader, 2007). Low glycemic index (GI) diets that are high in legumes and whole, unprocessed grains have been used with some success in GDM (Moses et al. 2009), but large clinical trials establishing their efficacy are lacking. We point out that the sweetness content of reduced-carbohydrate and low-GI diets is low due to restrictions on the use of carbohydrate sweeteners (sucrose, fructose) as well as the consumption of fruit and/or fruit juice (for those women who do not tolerate these foods well). Adherence to these diets could be difficult to achieve for women with GDM who crave sweets. In the United States, several nonnutritive sweeteners are approved for use by pregnant women (acesulfame potassium, aspartame, neotame, and sucralose, although aspartame is restricted in women with phenylketonuria) (Uplinger 2009). Although these sweeteners are considered safe for use in pregnancy, they are not broadly endorsed by the medical nutrition community for pregnant women. The introduction of novel sweeteners derived from natural products may change this perception. One example is rebaudioside A, an extract from the leaf of the stevia plant, that is Generally Recognized as Safe (GRAS) and approved by the U.S. Food and Drug Administration (FDA; 2008) for use in specific products.

There are other food-based approaches that could aid dietary compliance in GDM. Whole-grain, breakfast cereals, cereal bars, and snacks containing slowly digestible starch may be useful for limiting postprandial glucose excursions in individuals with type 2 diabetes (Lehmann and Robin, 2007), and may be useful in GDM as well. Another avenue of approach is the use of probiotics, specific strains of microorganisms that are commonly found in fermented dairy products such as yogurt. Probiotics modify the composition of intestinal microflora to regulate inflammatory pathways involved in insulin resistance and glucose homeostasis, among other mechanisms (Shi et al. 2006; Shoelson et al. 2007). Recent prospective studies in pregnancy have shown that probiotic supplements (consumed as capsules) have positive effects on maternal glucose homeostasis and reduce the incidence of GDM (Laitinen et al. 2009; Luoto et al. 2010). Whether probiotic-supplemented foods would be efficacious in GDM is not known but warrants further attention.

Better medical nutrition therapies are critically needed for GDM. Nutrition educators should be aware that heightened palatability for sweet foods may present a particular challenge for some women with GDM and should be considered when designing personalized diet plans for such women.

REFERENCES

Abbasi, A. A. 1981. Diagnostic and therapeutic significance of taste impairment. *Geriatrics* 36: 73–8.

American College of Obstetrics and Gynecology. 2001. Clinical management guidelines for obstetrician-gynecologists. *ACOG Practice Bulletin* (30).

American Diabetes Association. 2007. Nutrition recommendations and interventions for diabetes: a position statement of the American Diabetes Association. *Diabetes Care* 30:S48-S65.

American Diabetes Association. 2009. Meal planning exchange lists. www.diabetes.org/nutrition-and-recipes/nutrition/exchangelist.jsp

American Diabetes Association. 2010a. Diagnosis and classification of diabetes mellitus. *Diabetes Care* 33 (suppl. 1): S62–69.

American Diabetes Association. 2010b. Standards of medical care in diabetes: 2010. *Diabetes Care*, 33 (suppl. 1): S11–61.

Armstrong, C. L., L. P. Brown, R. York, D. Robbins, and A. Swank. 1991. From diagnosis to home management: nutritional considerations for women with gestational diabetes. *Diabetes Educ* 17 (6): 455–9.

Barnard, N. D., L. Gloede, J. Cohen, D. J. Jenkins, G. Turner-McGrievy, A. A. Green, et al. 2009. A low-fat vegan diet elicits greater macronutrient changes, but is comparable in adherence and acceptability, compared with a more conventional diabetes diet among individuals with type 2 diabetes. *J Am Diet Assoc* 109 (2): 263–72.

Belzer, L. M. 2008. Taste and endocrine factors in women with gestational diabetes mellitus. Doctoral dissertation. Rutgers, The State University of New Jersey.

Belzer, L. M., J. C. Smulian, S-E. Lu, and B. J. Tepper. 2009. Changes in sweet taste across pregnancy in mild gestational diabetes mellitus: relationship to endocrine factors. *Chem Senses* 34 (7): 595–605.

Belzer, L. M., J. C. Smulian, S-E. Lu, and B. J. Tepper. 2010. Food cravings and intake of sweet foods in healthy pregnancy and mild gestational diabetes mellitus: a prospective study. *Appetite* 55: 609–615.

Benoit, S. C., D. J. Clegg, R. J. Seeley, and S. C. Woods. 2004. Insulin and leptin as adiposity signals. *Recent Prog Horm Res* 59: 267–85.

Boden, G. 1996. Fuel metabolism in pregnancy and in gestational diabetes mellitus. *Obstet Gynecol Clin North Am* 23: 1–10.

Bowen, D. J. 1992. Taste and food preference changes across the course of pregnancy. *Appetite* 19 (3): 233–42.

Brown, J. E., and R. B. Toma. 1986. Taste changes during pregnancy. *Am J Clin Nutr* 43: 414–18.

Butte, N. F. 2000. Carbohydrate and lipid metabolism in pregnancy: normal compared with gestational diabetes mellitus. *Am J Clin Nutr* 71 (5): 1256S-61S.

Catalano, P. M., J. P. Kirwan, S. Haugel-de Mouzon, and J. King. 2003. Gestational diabetes and insulin resistance: role in short- and long-term implications for mother and fetus. *J Nutr* 133 (5 suppl. 2): 1674S-83S.

Chu, S. Y., W. M. Callaghan, S. Y. Kim, C. H. Schmid, J. Lau, L. J., England, et al. 2007. Maternal obesity and risk of gestational diabetes mellitus. *Diabetes Care* 30 (8): 2070–6.

Dabelea, D., J. K. Snell-Bergeon, C. L. Hartsfield, K. J. Bischoff, R. F.Hamman, and R. S. McDuffie. 2005. Increasing prevalence of gestational diabetes mellitus (GDM) over time and by birth cohort: Kaiser Permanente of Colorado GDM Screening Program. *Diabetes Care* 28 (3): 579–84.

Dippel, R. L., and J. W. Elias. 1980. Preferences for sweet in relationship to use of oral contraceptives and pregnancy. *Horm Behav* 14: 1–6.

Domali, E., and I. E. Messinis. 2002. Leptin in pregnancy. *J Matern Fetal Neonatal Med* 12 (4): 222–30.

Duffy, V. B., L. M. Bartoshuk, R. Striegel-Moore, and J. Rodin. 1998. Taste changes across pregnancy. *Ann N Y Acad Sci* 855: 805–9.

Dye, C. J., and D. A. Koziatek. 1981. Age and diabetes effects on threshold and hedonic perception of sucrose solutions. *J Gerontol* 36 (3): 310–15.

Fadl, H. E., I. K. Ostlund, A. F.Magnuson, and U. S. Hanson. 2010. Maternal and neonatal outcomes and time trends of gestational diabetes mellitus in Sweden from 1991 to 2003. *Diabet Med* 27 (4): 436–41.

Festa, A., N. Shnawa, W. Krugluger, P. Hopmeier, G. Schernthaner, and S. M. Haffner. 1999. Relative hypoleptinaemia in women with mild gestational diabetes mellitus. *Diabet Med* 16 (8): 656–62.

Gilhooly, C. H., S. K. Das, J. K. Golden, M. A. McCrory, G. E. Dallal, E. Saltzman, et al. 2007. Food cravings and energy regulation: the characteristics of craved foods and their relationship with eating behaviors and weight change during 6 months of dietary energy restriction. *Int J Obes (Lond)* 31 (12): 1849–58.

Gondivkar, S. M., A. Indurkar, S. Degwekar, and R. Bhowate. 2009. Evaluation of gustatory function in patients with diabetes mellitus type 2. *Oral Surg Oral Med Oral Pathol Oral Radiol Endod* 108 (6): 876–80.

Harvey, J., R. R. Wing, and M. Mullen. 1993. Effects on food cravings of a very low calorie diet or a balanced, low calorie diet. *Appetite* 21 (2): 105–15.

Henson, M. C., and V. D. Castracane. 2006. Leptin in pregnancy: an update. *Biol Reprod* 74: 218–29.

Herriot, A. M., S. Whitcroft, and Y. Jeanes. 2008. An retrospective audit of patients with polycystic ovary syndrome: the effects of a reduced glycaemic load diet. *J Hum Nutr Diet* 21 (4): 337–45.

Highman, T. J., J. E. Friedman, L. P. Huston, and W. W. Wong. 1998. Longitudinal changes in maternal serum leptin concentrations, body composition, and resting metabolic rate in pregnancy. *Am J Obstet Gynecol* 178: 1010–15.

Hook, E. B. 1978. Dietary cravings and aversions during pregnancy. *Am J Clin Nutr* 1: 1355–62.

Jones, E. J., C. C. Roche, and S. J. Appel. 2009. A review of the health beliefs and lifestyle behaviors of women with previous gestational diabetes. *J Obstet Gynecol Neonatal Nurs* 38 (5): 516–26.

Kautzky-Willer, A., G. Pacini, A. Tura, et al. 2001. Increased plasma leptin in gestational diabetes. *Diabetologia* 44: 164–72.

Kawai, K., K. Sugimoto, K. Nakashima, H. Miura, and Y. Ninomiy. 2000. Leptin as a modulator of sweet taste sensitivities in mice. *Proc Natl Acad Sci USA* 97 (20): 11044–9.

Kim, C., D. K. Berger, and S. Chamany. 2007. Recurrence of gestational diabetes mellitus: a systematic review. *Diabetes Care* 30 (5): 1314–9.

Kim, C., K. M. Newton, and R. H. Knopp. 2002. Gestational diabetes and the incidence of type 2 diabetes: a systematic review. *Diabetes Care* 25 (10): 1862–8.

King, J. C. 2006. Maternal obesity, metabolism, and pregnancy outcomes. *Annu Rev Nutr* 26: 271–91.

King, K. M., and G. Rubin. 2003. A history of diabetes: from antiquity to discovering insulin. *Br J Nurs* 12 (18): 1091–5.

Kolble, N., T. Hummel, R. von Mering, A. Huch, and R. Huch. 2001. Gustatory and olfactory function in the first trimester of pregnancy. *Eur J Obstet Gynecol Reprod Biol* 99 (2): 179–83.

Kuga, M., M. Ikeda, K. Suzuki, and S. Takeuchi. 2002. Changes in gustatory sense during pregnancy. *Acta Otolaryngol Suppl* (546): 146–53.

Laitinen, J. H., H. M. Tuorila, and M. I. J.Uusitupa. 1991. Changes in hedonic reponses to sweet and fat in recently diagnosed non-insulin-dependent diabetic patients during diet therapy. *Eur J Clin Nutr* 45: 393–400.

Laitinen, K., T. Poussa, and E. Isolauri. 2009. Probiotics and dietary counselling contribute to glucose regulation during and after pregnancy: a randomised controlled trial. *Br J Nutr* 101 (11): 1679–87.

Langer, O. 1999. Roadmap for the management of gestational diabetes. *Fetal and Maternal Medicine Review* 11: 1–5.

Langer, O. 2002. When diet fails: insulin and oral hypoglycemic agents as alternatives for the management of gestational diabetes mellitus. *J Matern Fetal Neonatal Med* 11 (4): 218–25.

Lappas, M., K. Yee, M. Permezel, and G. E. Rice. 2005. Release and regulation of leptin, resistin and adiponectin from human placenta, fetal membranes, and maternal adipose tissue and skeletal muscle from normal and gestational diabetes mellitus-complicated pregnancies. *J Endocrinol* 186: 457–65.

Lawson, W. B., A. Zeidler, and A. Rubenstein. 1979. Taste detection and preference in diabetics and their relatives. *Psychosom Med* 41 (3): 219–27.

Lehmann, U., and F. Robin. 2007. Slowly digestible starch—its structure and health implications: a review. *Trends in Food Science and Technology* 18: 346–55.

Lin, Y., and Z. Sun. 2010. Current views on type 2 diabetes. *J Endocrinol* 204 (1): 1–11.

Luoto, R., K. Laitinen, M. Nermes, and E. Isolauri. 2010. Impact of maternal probiotic-supplemented dietary counselling on pregnancy outcome and prenatal and postnatal growth: a double-blind, placebo-controlled study. *Br J Nutr* 103 (12): 1792–9.

Maghbooli, Z., A. Hossein-Nezhad, M. Rahmani, A. R. Shafaei, and B. Larijani. 2007. Relationship between leptin concentration and insulin resistance. *Horm Metab Res* 39 (12): 903–7.

Malee, M. P., A. Verma, G. Messerlian, R. Tucker, and B. R. Vohr. 2002. Association between maternal and child leptin levels 9 years after pregnancy complicated by gestational diabetes. *Horm Metab Res* 34 (4): 212–16.

Martin, C. K., P. M. O'Neil, and L. Pawlow. 2006. Changes in food cravings during low-calorie and very-low-calorie diets. *Obesity (Silver Spring)* 14 (1): 115–21.

Metzger, B. E., T. A. Buchanan, D. R. Coustan, A. de Leiva, D. b. Dunger, D. R. Hadden, et al. 2007. Summary and recommendations of the Fifth International Workshop-Conference on Gestational Diabetes Mellitus. *Diabetes Care* 30 (suppl. 2): S251–60.

Moses, R. G., M. Barker, M. Winter, P. Petocz, and J. C. Brand-Miller, J. C. 2009. Can a low-glycemic index diet reduce the need for insulin in gestational diabetes mellitus? A randomized trial. *Diabetes Care* 32 (6): 996–1000.

Nakamura, Y., K. Sanematsu, R. Ohta, S. Shirosaki, K. Koyano, K. Nonaka, et al. 2008. Diurnal variation of human sweet taste recognition thresholds is correlated with plasma leptin levels. *Diabetes* 57 (10): 2661–5.

Ochsenbein-Kolble, N., R. von Mering, R. Zimmermann, and T. Hummel. 2005. Changes in gustatory function during the course of pregnancy and postpartum. *Bjog* 112 (12): 1636–40.

Ogden, C. L., S. Z. Yanovski, M. D. Carroll, and K. M. Flegal. 2007. The epidemiology of obesity. *Gastroenterology* 132 (6): 2087–2102.

Perros, P., T. W. MacFarlane, C. Counsell, and B. M. Frier. 1996. Altered taste sensation in newly-diagnosed NIDDM. *Diabetes Care* 19 (7): 768–70.

Pettitt, D. J., and L. Jovanovic. 2001. Birth weight as a predictor of type 2 diabetes mellitus: the U-shaped curve. *Curr Diab Rep* 1 (1): 78–81.

Pope, J. E., J. D. Skinner, and B. R. Carruth. 1992. Cravings and aversions of pregnant adolescents. *J Am Diet Assoc* 92: 1479–82.

Reader, D. M. 2007. Medical nutrition therapy and lifestyle interventions. *Diabetes Care* 30 (suppl. 2): S188–93.

Rodin, J., and N. Radke-Sharpe. 1991. Changes in appetitive variables as a function of pregnancy. In *Chemical Senses: Appetite and Nutrition*, vol. 4, ed. M. I. Friedman, M. G. Tordoff and M. R. Kare, 325–40. New York: Marcel Dekker.

Schelling, J. L., L. Tetreault, L. Lasagna, and M. Davis. 1965. Abnormal taste threshold in diabetes. *Lancet* 19: 508–12.

Shaat, N., M. Ekelund, A. Lernmark, S. Ivarsson, A. Nilsson, R. Perfekt, et al. 2004. Genotypic and phenotypic differences between Arabian and Scandinavian women with gestational diabetes mellitus. *Diabetologia* 47 (5): 878–84.

Shi, H., M. V. Kokoeva, K. Inouye, I Tzameli, H, Yin, and J. S. Flier. 2006. TLR4 links innate immunity and fatty acid-induced insulin resistance. *J Clin Invest* 116 (11): 3015–25.

Shoelson, S. E., L. Herrero, and A. Naaz. 2007. Obesity, inflammation, and insulin resistance. *Gastroenterology* 132 (6): 2169–80.

Simmons, D., and B. H. Breier. 2002. Fetal overnutrition in Polynesian pregnancies and in gestational diabetes may lead to dysregulation of the adipoinsular axis in offspring. *Diabetes Care* 25: 1539–44.

Tepper, B. J., L. M. Hartfiel, and S. H. Schneider. 1996. Sweet taste and diet in type II diabetes. *Physiol Behav* 60 (1): 13–18.

Tepper, B. J., and A. C. Seldner. 1999. Sweet taste and intake of sweet foods in normal pregnancy and pregnancy complicated by gestational diabetes mellitus. *Am J Clin Nutr* 70: 277–84.

U. S. Food and Drug Administration. 2008. GRAS Assessment High Purity Steviol Glycosides. Agency Response Letter GRAS Notice No. GRN 000304. www.accessdata.fda.gov/scripts/fcn/gras_notices/grn_304.pdf

Uplinger, N. 2009. The controversy continues: nutritional management of the pregnancy complicated by diabetes. *Curr Diab Rep* 9 (4): 291–5.

Verma, A., C. M. Boney, R. Tucker, and B. R. Vohr. 2002. Insulin resistance syndrome in women with prior history of gestational diabetes mellitus. *J Clin Endocrinol Metab* 87 (7): 3227–35.

Vohr, B. R., and C. M. Boney. 2008. Gestational diabetes: the forerunner for the development of maternal and childhood obesity and metabolic syndrome? *J Matern Fetal Neonatal Med* 21 (3): 149–57.

Westgate, J. A., R. S. Lindsay, J, Beattie, N, S, Pattison, G. Gamble, L, F, Mildenhall, et al. 2006. Hyperinsulinemia in cord blood in mothers with type 2 diabetes and gestational diabetes mellitus in New Zealand. *Diabetes Care* 29 (6): 1345–50.

Winer, N., and J. R. Sowers. 2004. Epidemiology of diabetes. *J Clin Pharmacol* 44 (4): 397–405.

Worthington-Roberts, B., R. E. Little, M. D, Lambert, and R. Wu. 1989. Dietary cravings and aversions in the postpartum period. *J Am Diet Assoc* 89: 647–51.

Yogev, Y., and O. Langer. 2008. Pregnancy outcome in obese and morbidly obese gestational diabetic women. *Eur J Obstet Gynecol Reprod Biol* 137 (1): 21–26.

Yogev, Y., E. M. Xenakis, and O. Langer. 2004. The association between preeclampsia and the severity of gestational diabetes: the impact of glycemic control. *Am J Obstet Gynecol* 191 (5): 1655–60.

Yoshida, R., N. Shigemura, K. Sanematsu, K. Yasumatsu, S. Ishizuka, and Y. Ninomiya. 2006. Taste responsiveness of fungiform taste cells with action potentials. *J Neurophysiol* 96 (6): 3088–95.

10 Homocysteine, B Vitamins, and Cognitive Function

Joshua W. Miller

CONTENTS

Introduction 189
Homocysteine, B Vitamins, and One-Carbon Metabolism 190
Neurological Effects of B Vitamin Deficiencies 192
Folate Deficiency 192
Vitamin B_{12} Deficiency 193
Vitamin B_6 Deficiency 194
Homocysteine, B Vitamins, and Cognitive Function 196
Homocysteine 196
Folate 199
Vitamin B_{12} 200
Vitamin B_6 202
B Vitamin Intervention Studies 204
Conclusion 207
References 208

INTRODUCTION

The maintenance of mental alertness and robust cognitive function is an essential component of our quality of life as we age. The fastest growing portion of the human population is older adults, and world projections indicate that the number of individuals age eighty years and older will approach 370 million by the year 2050.[1] The sobering aspect of this estimate is that the current prevalence of Alzheimer's disease in adults age eighty-five years and older approaches 50 percent.[2] Many more suffer from cognitive impairments secondary to strokes and cerebrovascular disease. We are therefore in the midst of a growing epidemic that exacts a significant toll on the physical, emotional, and financial well-being of afflicted individuals, their families and caretakers, and institutional and governmental resources.

Though finding a cure for Alzheimer's disease and other dementias is an important research goal, it is also prudent to pursue strategies to prevent, slow, or ameliorate their onset. An area that has held significant hope and promise in this regard is nutrition. That overt micronutrient deficiencies can affect neurological function has long been recognized. That suboptimal levels of micronutrients in otherwise healthy older adults can influence cognitive function was first suggested by Goodwin et al. in 1983.[3] In this seminal study, cross-sectional associations were evaluated between blood levels of several vitamins and cognitive function test scores in noninstitutionalized individuals age sixty years and older. Low blood concentrations of several vitamins, including folate and vitamin B_{12}, were associated with low scores on tests of abstract thinking or short-term memory. In the ensuing twenty-five-plus years, numerous studies have evaluated the associations between micronutrients and cognitive function and risk of dementia and Alzheimer's disease. Particular attention has been paid to the B vitamins, primarily folate, vitamin B_{12}, and vitamin B_6, and the sulfur amino acid, homocysteine, the metabolism of which is dependent on these vitamins.

HOMOCYSTEINE, B VITAMINS, AND ONE-CARBON METABOLISM

"One-carbon metabolism" refers to one-carbon units, originating primarily in the amino acid serine, that are transferred between substrates and utilized for the synthesis of nucleotides for DNA and RNA, the de novo formation of methionine and cysteine, the metabolism of homocysteine, and the synthesis of S-adenosylmethionine (SAM) used as substrate in methylation reactions. Folate, vitamin B_{12}, and vitamin B_6 play essential roles in several of the reactions involved in one-carbon metabolism.

As shown in Figure 10.1, folate, in the form of tetrahydrofolate, acquires the one-carbon unit from serine in a pyridoxal-5′-phosphate (vitamin B_6)–dependent reaction to form methylenetetrahydrofolate and glycine. 5,10-Methylenetetrahydrofolate then serves as a substrate for thymidine synthesis, is converted to 10-formyltetrahydrofolate and used for the synthesis of purines (guanine and adenine), or is reduced to form 5-methyltetrahydrofolate catalyzed by the enzyme methylenetetrahydrofolate reductase, with FAD (riboflavin) serving as a cofactor.

5-Methyltetrahydrofolate serves as the methyl donor in the conversion of homocysteine to methionine, with tetrahydrofolate regenerated in the process. This reaction is catalyzed by methionine synthase (methyltetrahydrofolate-homocysteine methyltransferase), with vitamin B_{12} in the form of methylcobalamin serving as a cofactor. Methionine is then incorporated into protein or is activated by the addition of an adenosyl group from ATP to form SAM. SAM serves as the "universal methyl donor" for many essential methylation reactions, including those involving DNA, RNA, histones, membrane phospholipids, proteins, and neurotransmitters. S-adenosylhomocysteine (SAH) is formed as a product of these methylation reactions, and is subsequently metabolized to form adenosine and reform homocysteine. The equilibrium of the interconversion of SAH and homocysteine actually favors the formation of SAH in vivo, but the reaction is pulled in the forward direction by the rapid metabolism and removal of the homocysteine product.

The other metabolic fate of homocysteine is catabolism through cystathionine synthesis. Homocysteine irreversibly condenses with serine to form cystathionine, which is further metabolized to form cysteine and alpha-ketobutyrate. These two reactions are catalyzed by cystathionine β-synthase and cystathionine γ-lyase, respectively, which are both pyridoxal-5′-phosphate-dependent enzymes. Cysteine is subsequently incorporated into proteins; used as a component of the antioxidant

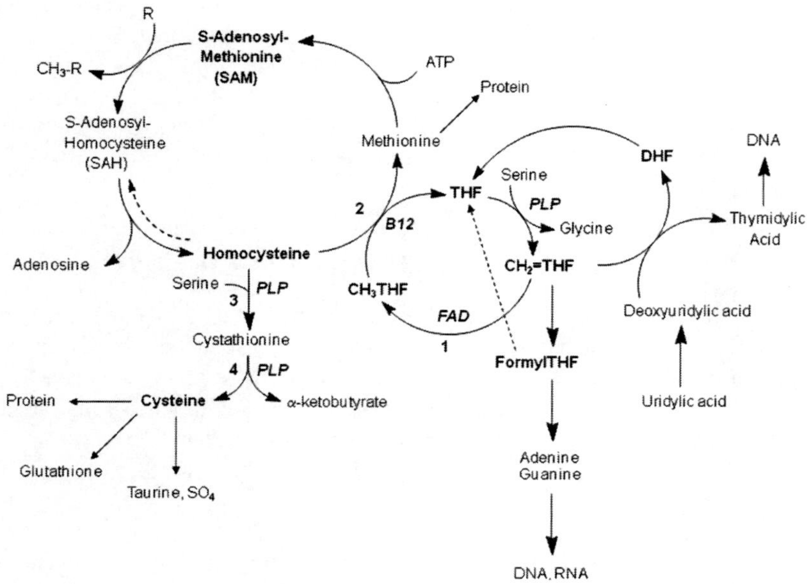

FIGURE 10.1 B vitamins, homocysteine, and one-carbon metabolism. One-carbon metabolism refers to the utilization of one-carbon units originating in the amino acid serine for (1) synthesis of nucleotides to be incorporated into DNA and RNA; (2) the metabolism of homocysteine to form methionine and the universal methyl donor, S-adenosylmethionine (SAM); and (3) the catabolism of homocysteine through cystathionine synthesis. The B vitamins—folate, vitamin B_{12}, and vitamin B_6—play important roles in these reactions, and deficiencies of these vitamins lead to inhibition of DNA and RNA synthesis, accumulation of homocysteine, and impairment of methylation reactions. One or more of these biochemical consequences may affect brain function and structure leading to neurological and cognitive impairments. Abbreviations: THF, tetrahydrofolate; CH_3, methyl; CH_2=, methylene; DHF, dihydrofolate; PLP, pyridoxal-5′-phosphate (active cofactor form of vitamin B_6); FAD, flavin adenine dinucleotide (active cofactor form of riboflavin); R, methyl acceptor (e.g., DNA, RNA, histones, phospholipids, proteins, neurotransmitters); 1, methylenetetrahydrofolate reductase; 2, methionine synthase; 3, cystathionine β-synthase; 4, cystathionine γ-lyase.

tripeptide, glutathione; or further metabolized to form taurine, inorganic sulfate, and other downstream products.

Predictably, nutritional deficiencies of folate, vitamin B_{12}, and vitamin B_6, as well as genetic defects in key enzymes, disrupt one-carbon metabolism. Blocks in the methylation and catabolism of homocysteine lead to its accumulation in cells and export into the blood (hyperhomocysteinemia) and urine (hyperhomocystinuria). In addition, intracellular levels of SAM and SAH decrease and increase, respectively. The decrease in SAM results from the block in methionine synthesis caused by folate and vitamin B_{12} deficiencies or genetic defects in methylenetetrahydrofolate reductase, methionine synthase, and related enzymes. The increase in SAH occurs because blocks in methionine synthesis and cystathionine synthesis (due to B_6 deficiency or a genetic defect in cystathionine beta-synthase) lead to intracellular accumulation of homocysteine. Diminished product removal favors the reverse reaction toward SAH synthesis, and thus SAH accumulates intracellularly. The biochemical consequence of the respective decrease and increase of SAM and SAH is impairment of methylation reactions due to reduced availability of SAM and feedback inhibition by SAH. The other major metabolic consequence of B vitamin deficiencies is impairment of DNA and RNA synthesis resulting from inhibition of thymidine, guanine, and adenine synthesis.

NEUROLOGICAL EFFECTS OF B VITAMIN DEFICIENCIES

Many of the neurological manifestations of B vitamin deficiencies, particularly folate and vitamin B_{12}, are believed to be direct or indirect consequences of disruptions in one-carbon metabolism, though other mechanisms may be involved. The observed neuropathology results from pathogenetic mechanisms that affect biochemical, physiological, and structural components of the nervous system.

Folate Deficiency

The neurological consequences of folate deficiency are many and varied. Neuropsychiatric manifestations include insomnia, irritability, fatigue, and forgetfulness.[4] There may be cognitive impairment, dementia, and depression.[5–12] Peripheral neuropathies and related disorders are also observed, including decreased or absent reflexes, diminished vibration sense, and restless leg syndrome.[5,6] Optical neuropathy and progressive vision loss have also been reported.[13]

Several potential mechanisms have been postulated to explain the effects of folate deficiency on neurological function. Mechanisms attributed to excess accumulation of homocysteine have received significant attention. Hyperhomocysteinemia is an established independent risk factor for vascular disease, including stroke and cerebrovascular disease,[14] as well as dementia and Alzheimer's disease.[15,16] These associations may be related to atherosclerotic and thrombotic damage to the vascular system that acutely (i.e., stroke and infarction) or chronically limits blood flow to the brain. Vascular dysfunction and damage may result from homocysteine-related impairment of nitric oxide-induced vasodilation, oxidative stress, endoplasmic reticulum stress, or induction of apoptosis.[17–20] In addition, homocysteine or products of its oxidative metabolism, including homocysteine sulfinic acid and homocysteic acid, may induce excitotoxicity through interactions with the N-methyl-d-aspartate (NMDA) receptor, one of the synaptic excitatory glutaminergic receptors.[21,22] Oxidative stress may also play a role as homocysteine has been shown to induce the formation of reactive oxygen species.[23–26]

It is the case, however, that homocysteine may not be a causative agent in folate deficiency-induced neurological dysfunction, but rather a marker of other factors that play a more direct role. Specifically, hyperhomocysteinemia may be a reflection of altered intracellular concentrations of SAM and SAH induced by folate deficiency. Reduced SAM and increased SAH concentrations in cerebrospinal fluid have been detected in folate-deficient patients with depression and dementia.[27,28] The implication of this is that key methylation reactions involving proteins, neurotransmitters, phospholipids, and DNA in the central nervous system may be impaired in folate deficiency.[29]

Specific protein methylation reactions that may be affected include carboxymethylation of myelin basic protein, which is essential for maintaining the structure of the myelin sheath around axons and efficient neurotransmission,[30] and carboxymethylation of protein phosphatase 2A, which is important for preventing the accumulation of phosphorylated Tau and amyloid precursor protein observed in Alzheimer's disease.[31–33]

Methylation reactions are also important in the metabolism of neurotransmitters. The conversions of both norepinephrine to epinephrine and serotonin to melatonin are SAM dependent. In addition, the first step in the catabolism of catecholamines (L-dopa, dopamine, norepinephrine, and epinephrine) is O-methylation catalyzed by catechol-O-methyltransferase (COMT). The activity of COMT is important for terminating synaptic activity induced by catecholamines and may be important for reducing the auto-oxidation of catecholamines in the presence of transition metals, such as copper.[34]

Another important methylation reaction in the brain is the conversion of phosphatidylethanolamine to phosphatidylcholine within cellular membranes. This conversion, which requires three successive methylation reactions, is necessary for maintaining lipid bilayers and membrane fluidity. This is essential for efficient receptor-mediated synaptic neurotransmission, including that involving the beta-adrenergic, muscarinic, and gamma-aminobutyric acid (GABA) systems.[35–37]

DNA methylation may also be important. Methylation of gene promoters leads to recruitment of methylated DNA-binding proteins and co-repressor complexes that turn off gene transcription.[38] One such gene is brain-derived neurotrophic factor (BDNF), which is demethylated when a neuron is depolarized, thus leading to increased BDNF expression.[39,40] Increased expression of BDNF is believed to play a role in synaptic plasticity and long-term potentiation. However, prolonged over-expression of BDNF, which could occur as a consequence of folate deficiency and reduced availability of SAM, has been associated with impaired learning and memory.[41] Other genes that may be influenced by DNA promoter methylation include nicotinic receptors, which may be involved in the pathogenesis of Alzheimer's disease; glial fibrillary acidic protein, which is important for neurogenesis; and synapsin I, which affects neurotransmitter release.[42] COMT expression is also affected by promoter methylation.[43]

Vitamin B_{12} Deficiency

The classic neurological manifestation of vitamin B_{12} deficiency is subacute combined degeneration (SACD). This is a consequence of neuronal demyelination in the central and peripheral nervous systems. Particularly vulnerable are the long tracts of the posterior and lateral columns of the spinal cord that are involved in the conduction of vibration and position sense, as well as motor function. Clinical manifestations include gait ataxia, peripheral neuropathy, cognitive disturbances, dementia, depression, and autonomic dysfunction.

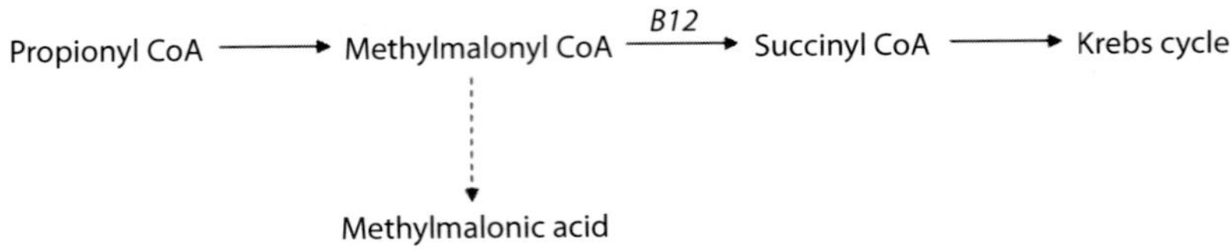

FIGURE 10.2 Vitamin B_{12} and succinyl Co-A synthesis. In addition to the conversion of homocysteine to methionine (Figure 10.1), vitamin B_{12} serves as a cofactor in the conversion of methylmalonyl CoA to succinyl CoA, which subsequently enters the Krebs cycle for the production of energy. Deficiency of vitamin B_{12} will lead to accumulation of methylmalonyl CoA and an increase in the synthesis of methylmalonic acid, normally a minor side reaction. Methylmalonic acid has been implicated in demyelination that occurs in severe vitamin B_{12} deficiency.

Because vitamin B_{12} serves as a cofactor in the folate-dependent conversion of homocysteine to methionine, many of the same putative mechanisms cited above underlying folate deficiency-associated neurological dysfunction may underlie B_{12} deficiency-associated dysfunction. Decreased levels of SAM and increased levels of SAH and homocysteine are consequences of both folate and B_{12} deficiency. However, though there are similarities, there are also differences in the physiological and clinical manifestations caused by deficiencies of the two vitamins. For example, SACD is often seen in cases of severe B_{12} deficiency, while SACD is rarely a consequence of folate deficiency. Though decreased synthesis of SAM and consequent impairment of methylation reactions are most often cited as the causes of demyelination in B_{12} deficiency,[44] two other mechanisms have been proposed that may explain the differences with folate deficiency. First, in addition to the conversion of homocysteine to methionine, vitamin B_{12}, in the form of adenosylcobalamin, serves as a cofactor in the conversion of methylmalonyl CoA to succinyl CoA (Figure 10.2), a mitochondrial pathway involved in the metabolism of odd-chain fatty acids for energy production. B_{12} deficiency leads to impairment of this pathway and accumulation of methylmalonic acid and its precursors, which have been suspected to induce demyelination.[45] Second, vitamin B_{12} deficiency causes increased levels of the inflammatory cytokine, tumor necrosis factor-alpha (TNF-alpha), in rats[46–48] and humans.[49,50] Inflammation in the brain is part of the pathogenesis of Alzheimer's disease, and a connection between B_{12} deficiency and Alzheimer's disease has long been suspected. It has been proposed that B_{12} deficiency may precipitate or contribute to Alzheimer's disease symptoms by promoting or exacerbating the inflammatory response in the brain.[51]

Vitamin B_6 Deficiency

The neurological consequences of vitamin B_6 deficiency are typically very different from those of folate and B_{12} deficiencies. Reported clinical manifestations of B_6 deficiency include ataxia, hypersensitivity to sound (hyperacousis), irritability, impaired alertness, abnormal head movements, and epileptiform convulsions.[52,53] The latter are indicated by electroencephalographic abnormalities that normalize with B_6 supplementation.

The stark differences between the neurological dysfunction induced by B_6 deficiency compared with folate or B_{12} deficiency likely stem from differences in function among the vitamins. While B_6 serves as a cofactor in three reactions involved in one-carbon and homocysteine metabolism (Figure 10.1), it also serves as a cofactor in over 140 other enzymatic reactions. Several of these reactions involve the synthesis of neurotransmitters, including dopamine (by dopa decarboxylase), serotonin (by hydroxytryptophan decarboxylase), and GABA (by glutamate decarboxylase) (Figure 10.3). The latter reaction consists of the conversion of the primary excitatory neurotransmitter, glutamate, to the main inhibitory neurotransmitter, GABA. If B_6 is lacking, the balance between the two neurotransmitters is altered, and this is believed to underlie the seizures associated with deficiency of this vitamin.[52]

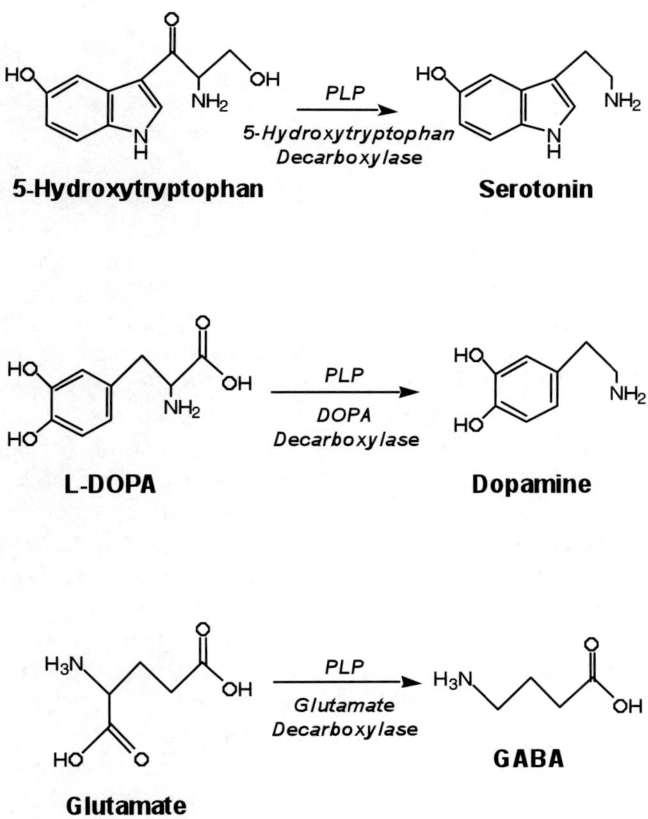

FIGURE 10.3 Vitamin B6 and neurotransmitter synthesis. Pyridoxal-5′-phosphate (PLP), the active cofactor form of vitamin B_6, is required for several decarboxylation reactions involved in the synthesis of key neurotransmitters, including serotonin, dopamine, and γ-aminobutyric acid (GABA). Vitamin B_6 deficiency may inhibit these reactions leading to disruptions in neurological functions and cognitive impairments.

HOMOCYSTEINE, B VITAMINS, AND COGNITIVE FUNCTION

The neurological consequences of B vitamin deficiencies described above are typical of overt deficiency states. However, severe deficiencies with clear neurological signs and symptoms are relatively rare. The prevalence of folate deficiency is now very low in the United States, Canada, and other countries in which folic acid fortification has been instituted to protect against neural tube birth defects.[54] Vitamin B_{12} deficiency is more common, but is primarily observed in older adults with reduced capacity to absorb B_{12} due to pernicious anemia or atrophic gastritis.[44] Vitamin B_6 deficiency is almost never encountered, except in conjunction with multinutrient deficiencies.[52] Nonetheless, it is the case that low circulating levels of each of these vitamins and elevated plasma homocysteine concentrations persist in the general population in the absence of clinical symptoms.[55,56] As Goodwin et al.[3] first suggested, such suboptimal status may affect cognitive function.

HOMOCYSTEINE

Reports of associations between homocysteine and cognitive function began to appear in the 1990s. The first was by Bell et al. in 1992[57] who found a significant inverse correlation between plasma homocysteine concentration and cognitive function in a small cohort of elderly patients hospitalized with acute depression. A subsequent study by Riggs et al.[58] found a significant inverse correlation between homocysteine and spatial copying skills in sixty-eight subjects, age fifty-eight years and older, participating in the Boston Veterans Affairs Normative Aging Study. Shortly thereafter, studies by McCaddon et al.[59] and Joosten et al.[60] revealed that homocysteine concentrations in Alzheimer's disease patients were significantly elevated compared with healthy controls of similar age.

It was not until 1998, however, that interest in the relationship between homocysteine, cognitive function, and Alzheimer's disease began to take off. In their landmark paper, Clarke et al.[15] compared 164 patients with Alzheimer's disease with 108 controls of similar age. They found that the mean homocysteine concentration in the Alzheimer's patients was significantly higher than in the controls. They also found that serum and red blood cell folate concentrations were lower in the Alzheimer's patients. Serum vitamin B_{12} concentrations were not different between the groups, nor were serum creatinine concentrations that indicate renal function, which is a major determinant of homocysteine levels in older adults.

The most convincing aspect of this study, though, was that the Alzheimer's diagnosis was verified in a large subset of the patients by postmortem histological analysis. Premortem clinical diagnosis of Alzheimer's disease is difficult, with error rates at the time of this study as high as 20–30 percent. Definitive diagnosis requires direct examination of brain tissue. In 103 of the patients who had died and were available for autopsy, histological analysis revealed that only seventy-six (74 percent) actually had neuropathological evidence of the disease. When the statistical analyses were repeated limiting the comparison to the seventy-six histologically confirmed cases, the significance of the difference in homocysteine levels between the Alzheimer's patients and the controls strengthened. In addition, the serum vitamin

B_{12} concentrations were significantly lower in the histologically confirmed cases. Secondary logistic regression analyses revealed that the odds ratio for histologically confirmed Alzheimer's disease was 4.6 for patients in the highest tertile of homocysteine (> 14.0 μmol/L) compared with those in the lowest tertile (≤ 11.0 μmol/L). Significant odds ratios of a similar magnitude were observed between the lowest and highest tertiles of serum folate and B_{12}, and all odds ratios remained significant after controlling for potentially important confounders, including age, sex, smoking, social class, and apolipoprotein E4 genotype.

The implications of these findings were twofold. First, because hyperhomocysteinemia had been implicated as an independent risk factor for vascular disease, it was consistent with the emerging hypothesis that vascular disease can contribute to the progression or even the pathogenesis of Alzheimer's disease. Second, it suggested that lowering of plasma homocysteine levels with B vitamin supplements could reduce the risk or slow the progression of Alzheimer's disease. However, it was recognized that the study was limited by its cross-sectional design and that causality between homocysteine, B vitamins, and Alzheimer's disease could not be evaluated. Therefore, prospective and intervention studies were needed.

A spate of cross-sectional and longitudinal studies followed that examined the associations between homocysteine and risk of Alzheimer's disease and dementia, and between homocysteine and neuropsychological test scores. The study of Seshadri et al.[16] in 2002 is one of the most cited. The study sample consisted of 1,092 subjects (mean age = seventy-six years) without dementia from the Framingham Study cohort. After a median follow-up period of eight years, the cumulative incidence of both overall dementia and Alzheimer's disease was about twofold higher in those subjects in the highest quartile of age-adjusted baseline homocysteine values compared with those in the three lowest quartiles combined. Moreover, baseline homocysteine was associated with decline in Mini-Mental State Examination (MMSE) scores after four years of follow-up. The MMSE represents an estimate of global cognitive functioning.[61]

In a follow-up study, this time using the Framingham offspring cohort, the influence of age on the relationship between homocysteine and cognitive function was assessed.[62] In this study, 2069 individuals without history of stroke or dementia were divided into three age groups (40–49 years, 50–59 years, and 60–82 years). Cross-sectional associations were then assessed between homocysteine concentrations and various tests of cognitive function. Significant associations between homocysteine and cognitive function were only observed in the oldest age group. The significance of this finding was that early intervention to lower homocysteine levels, prior to the development of overt cognitive impairment or dementia, could be an effective prevention strategy.

Another informative cohort with respect to homocysteine has been followed in the Sacramento Area Latino Study on Aging (SALSA).[63] The Framingham cohort and the Framingham offspring cohort cited above are predominantly Caucasian and of relatively high educational achievement. In contrast, the SALSA cohort (baseline n = 1,789) is all Hispanic/Latino and on average has fewer years of formal education. In addition, the Framingham cohort data were collected on individuals prior to the institution of folic acid fortification in the United States in 1998.[64] The SALSA cohort was recruited after the start of folic acid fortification. Folic acid fortification has had a significant effect on plasma homocysteine levels in the general population.[54] Nonetheless, elevated homocysteine levels persisted in the SALSA cohort, and baseline homocysteine levels were inversely correlated with global cognitive function scores, as well as several subdomains of cognitive function, including tests of long-term memory, attention, and visuospatial skills. In a follow-up study, incident cognitive impairment (defined as a combination of incident dementia and incident cognitive impairment but not dementia) over a 4.5-year period was associated with baseline homocysteine concentration.[65] Thus, the relationship between homocysteine and cognitive function/dementia was observed in a cohort of different ethnicity and educational background than those in the Framingham studies, and in a cohort that had been exposed to folic acid fortification.

In summary, most studies indicate that elevated homocysteine is a consistent risk marker for cognitive impairment, Alzheimer's disease, and dementia (Table 10.1), and is observed across populations with different demographic characteristics.

TABLE 10.1
Associations between Homocysteine and Cognitive Function: Selected Studies

Study Design	Main Findings	Reference
Cross-sectional, case-control	Inverse correlation between homocysteine and Mini-Mental State Examination (MMSE) in older adults with acute depression (mean age 73 years)	Bell et al.[57]
Cross-sectional, cohort	Inverse correlation between homocysteine and spatial copying skills (mean age 66 years)	Riggs et al.[58]
Cross-sectional, case-control	Homocysteine is elevated in Alzheimer's patients compared with healthy controls (median age 79 years)	McCaddon et al.[59]
Cross-sectional, case-control	Homocysteine is elevated in Alzheimer's patients compared with nondemented hospitalized and nonhospitalized healthy controls (mean age 83 years)	Joosten et al.[60]
Cross-sectional, case-control	Homocysteine is elevated in histologically confirmed Alzheimer's patients compared with healthy controls (mean age 77 years)	Clarke et al.[15]
Longitudinal, cohort	Baseline homocysteine in the upper quartile is associated with increased relative risk for dementia and Alzheimer's disease after eight years (mean age 76 years)	Seshadri et al.[16]
Cross-sectional, cohort	Homocysteine is correlated with various cognitive function tests in 60–82 years age group, but not in 40–49 years and 50–59 years age groups	Elias et al.[62]
Cross-sectional, cohort	Homocysteine is correlated with several domains of cognitive function in Latinos exposed to folic acid fortification (median age 70 years)	Miller et al.[63]
Longitudinal, cohort	Baseline homocysteine predicts incident cognitive impairment after 4.5 years in Latinos exposed to folic acid fortification	Haan et al.[65]

Folate

Because folate is involved directly in homocysteine metabolism and is thus a determinant of plasma homocysteine concentrations, many studies assessing the relationships between homocysteine and cognitive function, Alzheimer's disease, and dementia have also included evaluations of the role of folate. Both cross-sectional and longitudinal studies have verified that low folate intake or status is a risk marker, as might be expected. Less well recognized is that homocysteine and folate typically have independent associations. For example, in the study by Clarke et al.,[15] the association between homocysteine and Alzheimer's disease remained significant after controlling for folate, while the association between folate and Alzheimer's disease was attenuated after controlling for homocysteine. Similarly, in the Framingham study of incident Alzheimer's disease and dementia,[16] controlling for folate did not affect the association with homocysteine, while folate itself was not correlated with incident Alzheimer's disease or dementia.

In contrast, a different picture was observed in the SALSA cohort.[63,65,66] In the baseline cross-sectional analyses, folate was associated with dementia independent of homocysteine. Moreover, folate was associated with a different domain of cognitive function than homocysteine; both were correlated with global cognitive function scores, but folate was associated with short-term memory, while homocysteine was associated with long-term memory, attention, and visuospatial skills, as cited above. Thus, while it is a determinant of homocysteine levels, folate may affect the brain via mechanisms that are independent of homocysteine. These may include effects of low folate on neurotransmitter synthesis and metabolism through reduced levels of SAM and increased levels of SAH in the brain.

In contrast to the baseline cross-sectional analyses, however, longitudinal analysis of the SALSA study sample produced a contrary observation; baseline folate was not a significant predictor of incident cognitive impairment.[65] A potential explanation for these discrepant observations within the same cohort is that folic acid fortification, which was fully instituted in the United States in January 1998, just prior to baseline recruitment of the SALSA cohort (begun in February 1998), may have protected against incident cognitive impairment associated with low folate status. The fact that significant cross-sectional associations between folate and cognitive function/dementia were observed at baseline may be a reflection of the influence of low folate status prior to the initiation of folic acid fortification.

Another important consideration for assessing the influence of folate on cognitive function is that folate deficiency can contribute to clinical depression.[9–12] In addition, the efficacy of antidepressant medications can be influenced by folate status and enhanced by folate supplements.[9–12] A potential biochemical mechanism for this association is related to SAM. Oral SAM supplements have been shown to have antidepressant properties.[67,68] Accordingly, it is hypothesized that folate deficiency may induce depressive symptoms by reducing SAM levels in the brain. This is relevant because depressive symptoms are a strong determinant of impaired cognitive function in older adults.[69–71]

TABLE 10.2
Associations between Folate and Cognitive Function: Selected Studies

Study Design	Main Findings	Reference
Cross-sectional, case-control	Folate is lower in histologically confirmed Alzheimer's patients compared with healthy controls (mean age 77 years)	Clarke et al.[15]
Longitudinal, cohort	Baseline folate is not associated with increased relative risk for dementia and Alzheimer's disease after eight years (mean age 76 years)	Seshadri et al.[16]
Cross-sectional, cohort	Folate is lower in demented individuals compared with nondemented individuals, and is associated with global cognitive function and short-term memory scores in Latinos exposed to folic acid fortification (median age 70 years)	Ramos et al.[66]
Longitudinal, cohort	Baseline folate does not predict incident cognitive impairment after 4.5 years in Latinos exposed to folic acid fortification (median age 70 years)	Haan et al.[65]

In summary, like elevated homocysteine, low folate is often associated with cognitive function (Table 10.2). The effect of low folate on cognitive function may be mediated by both homocysteine-dependent and homocysteine-independent mechanisms.

Vitamin B_{12}

In contrast to folate and homocysteine, the putative relationship between vitamin B_{12} and cognitive function has been more difficult to ascertain. This in part stems from the heterogeneity of symptoms that is observed among individuals with B_{12} deficiency, and historical difficulty in accurately assessing B_{12} status. This was clearly demonstrated in a landmark 1988 paper by Lindenbaum et al.[72] Typically, B_{12} deficiency is suspected when there is macrocytic anemia. However, in a detailed examination of 141 consecutive cases of neuropsychiatric symptoms induced by B_{12} deficiency, Lindenbaum et al.[72] observed that 28 percent did not exhibit any hematological abnormalities. Moreover, two of these patients had circulating B_{12} concentrations that would not have been considered deficient (> 150 pmol/L), and sixteen had B_{12} concentrations within an indeterminate range (75–150 pmol/L). Nonetheless, neuropsychiatric symptoms in all the deficient patients who did not exhibit hematological abnormalities improved with B_{12} therapy. These improvements in neuropsychiatric symptoms were accompanied by decreases in circulating homocysteine and methylmalonic acid, the two primary metabolic markers of low B_{12} status.

The legacy of the Lindenbaum et al.[72] study is twofold. First is the demonstration that neuropsychiatric symptoms are often observed in the absence of anemia in B_{12} deficiency. Second is that B_{12} deficiency can be difficult to diagnose if one relies solely on circulating vitamin B_{12} concentrations. Though the Goodwin et al. study[3] did find an association between B_{12} and cognitive function scores, often in subsequent studies no such association was found.[73,74] However, associations have been observed using other analytes that putatively have better predictive value than total serum B_{12}. These include methylmalonic acid and homocysteine, as well as holotranscobalamin (holoTC).

Circulating vitamin B_{12} is bound to two transport proteins, haptocorrin (formerly known as transcobalamin I and transcobalamin III) and transcobalamin (formerly known as transcobalamin II). Measurement of total plasma or serum B_{12} represents the amount of B_{12} bound to both transport proteins together. However, it is transcobalamin that is responsible for delivering B_{12} from the site of absorption in the ileum to all the tissues of the body. Haptocorrin delivers B_{12} only to the liver. Thus, measurement of the amount of total serum B_{12} bound to transcobalamin (i.e., holoTC) is thought to represent a better indicator of overall B_{12} status than measurement of total serum B_{12}. Within the last decade, reliable assays for holoTC have become available[75,76] that have made possible the assessment of associations between holoTC and cognitive function.

Demonstration of the potential utility of holoTC and methylmalonic acid in assessing the association between B_{12} status and cognitive function was provided by Clarke et al.[77] in 2007. Participants in the Oxford Healthy Aging Project (age ≥ 65 years) were followed over a ten-year period and assessed for change in cognitive function as indicated by the MMSE. In multivariate analyses, cognitive decline was inversely correlated with holoTC and directly correlated with methylmalonic acid concentrations. Furthermore, it was estimated that a doubling of holoTC concentration would be associated with a 30 percent slowing of cognitive decline, while a doubling of methylmalonic acid would be associated with a > 50 percent acceleration of cognitive decline. In contrast, no significant association was found between total serum B_{12} and rate of cognitive decline. These results are consistent with other cross-sectional studies.[78,79]

Another growing trend in the assessment of B_{12} status that may further refine detection of B_{12} deficiency–related cognitive impairment is to combine the results of more than one analyte. Strategies have been proposed that incorporate the measurement of total serum B_{12} in combination with homocysteine, methylmalonic acid, or holoTC.[80–82] Examples of the potential utility of this approach come from the aforementioned SALSA study in which homocysteine was found to predict incidence of clinical cognitive impairment over a 4.5-year follow-up period.[65] Further analysis revealed that it was elevated homocysteine associated with low total serum B_{12} that, at least in part, was driving the association between elevated homocysteine and incident cognitive impairment.[65] In addition, it was determined in a separate analysis of the SALSA cohort that a low ratio of holoTC to total serum B_{12} was associated with lower cognitive function scores in those subjects with concomitant elevations in depressive symptoms.[83]

Another emerging issue with respect to vitamin B_{12} status and cognitive function is the possibility that excess folic acid intake may exacerbate vitamin B_{12} deficiency. In a pair of provocative studies published in 2007,[84,85] cross-sectional associations between B_{12} and folate status and cognitive function, anemia, and circulating concentrations of homocysteine and methylamonic acid were assessed in older adults who participated in the National Health and Nutrition Examination Survey (NHANES) from 1999 to 2002. Importantly, the data were collected after the institution of folic acid fortification in the United States. Not surprisingly, significant associations between low B_{12} status and increased risk of cognitive impairment and anemia were observed. What was surprising was that the associations were stronger for subjects with both low B_{12} status and elevated plasma folate concentrations (> 59 nmol/L) than for those with low B_{12} status and normal folate concentrations (< 59 nmol/L). Moreover, the combination of low B_{12} and elevated folate was associated with higher homocysteine and methylmalonic acid levels than the combination of low B_{12} and normal folate. The associations with homocysteine and methylmalonic acid were subsequently confirmed in the aforementioned SALSA cohort.[86] Also in the SALSA cohort, the combination of low B_{12} and high folate was associated with lower holoTC concentrations.[86]

Taken together, these findings suggest that excess folic acid intake, which would be expected in populations exposed to both folic acid fortification and folic acid in supplements, exacerbates B_{12} deficiency. Two possible mechanisms have been proposed to explain this putative effect. The first hypothesis is that folic acid causes the irreversible oxidation of vitamin B_{12} within cells.[85] There is no empirical evidence to support this hypothesis, though there is biological precedent. The anesthetic gas, nitrous oxide, is known to irreversibly oxidize B_{12} in vivo, leading to elevations in circulating homocysteine and methylmalonic acid as well as neurological damage. The second hypothesis is that excess folic acid intake leads to excess accumulation of dihydrofolate within cells. Dihydrofolate may then interfere with other important folate-dependent reactions that are necessary for maintaining neurological function and the production of red blood cells.[87] This hypothesis also has little empirical support. Moreover, it must be noted that the studies associating the combination of low B_{12} and high folate with more severe B_{12} deficiency were cross-sectional in design, and it was not possible to discern cause and effect. An alternate explanation for the findings that does not invoke an effect of excess folic acid on B_{12} function has been proposed.[88]

In summary, low B_{12} status is almost certainly an important determinant of cognitive function in older adults (Table 10.3). Our ability to detect these associations is maturing as new strategies are employed to refine the assessment of B_{12} status.

VITAMIN B_6

In contrast to homocysteine, folate, and vitamin B_{12}, fewer studies have evaluated the associations between vitamin B_6 and cognitive function and dementia. This is likely due to the relative rarity of overt vitamin B_6 deficiency as a clinical problem and because assays for measurement of blood levels of vitamin B_6 are not as readily available to clinicians as are those for homocysteine, folate, and B_{12}. Nonetheless, there is some literature relating plasma levels of pyridoxal-5′-phosphate (PLP), the active cofactor form of vitamin B_6, to cognitive function and dementia. As for the other B vitamins, the findings have been mixed.

In the study by Goodwin et al.,[3] significant associations between plasma PLP and cognitive function scores were not observed. In contrast, Riggs et al.[58] in 1996 found significant direct associations between PLP and two measures of short-term memory in the Boston Veterans Affairs Normative Aging Study cited above. In a follow-up study of the same cohort, Tucker et al.[89] in 2005 assessed the association

TABLE 10.3
Associations between Vitamin B_{12} and Cognitive Function: Selected Studies

Study Design	Main Findings	Reference
Case series, cohort	Low B_{12} is often associated with neuropsychiatric symptoms in the absence of anemia (age range 17–88 years)	Lindenbaum et al.[72]
Cross-sectional, case-control	B_{12} is lower in histologically confirmed Alzheimer's patients compared with healthy controls (mean age 77 years)	Clarke et al.[15]
Longitudinal, cohort	Baseline holoTC and methylmalonic acid, but not total B_{12}, predict cognitive decline over ten years (mean age 76 years)	Clarke et al.[77]
Longitudinal, cohort	Low B_{12} with elevated homocysteine predicts incident cognitive impairment after 4.5 years in Latinos exposed to folic acid fortification (median age 70 years)	Haan et al.[65]
Cross-sectional, cohort	Ratio of holoTC to total B_{12} is associated with cognitive function scores in older adults with elevated depressive symptoms (median age 70 years)	Garrod et al.[83]
Cross-sectional, cohort	Risk of cognitive impairment is greater in individuals with low B_{12} and high folate compared with individuals with low B_{12} and low folate (mean age 70 years)	Morris et al.[84]

between PLP and decline in spatial copying skills over a three-year follow-up period. Baseline PLP concentrations were directly correlated with spatial copying skills after three years, but the association was largely attenuated after controlling for confounding factors, including homocysteine, folate, and B_{12}.

Similar observations were made by Kado et al.[90] in 2005. Both cross-sectional and longitudinal associations between PLP and a composite score of tests of language, memory, conceptualization, and visuospatial ability were assessed in a cohort of older adults (age 70–79 years) participating in the McArthur Studies of Successful Aging. At baseline, the relative risk for poor cognitive performance in subjects within the lowest quartile of plasma PLP concentrations was elevated compared with subjects in the highest quartile of PLP. However, the association was largely attenuated after controlling for confounding variables, including homocysteine, B_{12}, and folate. Moreover, no association between PLP and cognitive decline over a seven-year period was observed.

Contrary to these findings are those made by Elias et al.[91] in a cohort of older adults (age 62 ± 13 years) participating in the Maine-Syracuse Longitudinal Study. PLP concentrations were directly associated with several subdomains of cognitive function, as well as a global composite cognitive function score. These associations remained strong after controlling for several confounding factors, including cardiovascular disease and related risk factors. In contrast, though folate also correlated with the same subdomains of cognitive function, the associations with folate were largely attenuated after controlling for cardiovascular disease and related risk factors.

Only a few studies have examined the relationship between B_6 status and Alzheimer's disease. In a case control study, Miller et al.[92] found a higher prevalence of low-plasma PLP (< 25 nmol/L) in Alzheimer's patients compared with controls of similar age. However, in the study by Seshadri et al.[16] cited above in which

TABLE 10.4
Associations between Vitamin B_6 and Cognitive Function: Selected Studies

Study Design	Main Findings	Reference
Cross-sectional, cohort	PLP is correlated with short-term memory (mean age 66 years)	Riggs et al.[58]
Longitudinal, cohort	Baseline PLP is correlated with spatial copying skills after three years (mean age 66 years)	Tucker et al.[89]
Cross-sectional, longitudinal, cohort	Baseline relative risk for poor cognitive performance is elevated in individuals in the lowest quartile of PLP; no association observed with cognitive decline over seven years (mean age 74 years)	Kado et al.[90]
Cross-sectional, cohort	PLP is associated with multiple domains of cognitive function (median age 62 years)	Elias et al.[91]
Cross-sectional, case-control	The prevalence of low PLP is greater in Alzheimer's disease patients compared with healthy controls (mean age 79 years)	Miller et al.[92]
Longitudinal, cohort	Baseline PLP is not associated with increased relative risk for dementia and Alzheimer's disease after eight years (mean age 76 years)	Seshadri et al.[16]
Longitudinal, cohort	Baseline PLP is not associated with risk for Alzheimer's disease after five years (mean age 77 years)	Luchsinger et al.[93]

a longitudinal association between baseline homocysteine and incident Alzheimer's disease was observed, baseline PLP plasma levels were not associated with subsequent risk of the disease. A similar finding was made by Luchsinger et al.[93] in a study of participants in the Washington Heights–Inwood Columbia Aging Project.

Taken together, there is only limited support for an association between vitamin B_6 status and cognitive function or dementia (Table 10.4).

B Vitamin Intervention Studies

Cross-sectional and longitudinal associations notwithstanding, the question of greater significance is whether B vitamin supplements improve cognitive function or protect against cognitive decline. Results of randomized clinical trials of B vitamins, either alone or in combination, have produced mixed results, with many showing no discernable benefit on cognitive function or risk of dementia (Table 10.5). However, there are studies that have shown beneficial effects, and a review of these trials reveals several important considerations.

The first consideration is the extent of cognitive impairment of those to be supplemented. In a multicenter randomized control trial of high-dose B vitamin supplements (5 mg/d of folic acid, 25 mg/d of vitamin B_6, and 1 mg/d of vitamin B_{12}), Aisen et al.[94] assessed the effect of B vitamin supplements on change in cognitive performance in patients with mild to moderate Alzheimer's disease. Despite a significant reduction in homocysteine levels in the treatment group compared with the placebo control group, no benefit of improved cognitive function or slowed cognitive decline, as indicated by the Alzheimer's Disease Assessment Scale,[95] was observed in those patients who received the B vitamins over an eighteeen-month period. In considering

TABLE 10.5
B Vitamin Intervention and Cognitive Function: Selected Studies

Study Design	Main Findings	Reference
Randomized clinical trial (folic acid 5.0 mg/d, B_{12} 1.0 mg/d, B_6 25 mg/d)	High dose B vitamin supplements for 1.5 years lowered homocysteine, but did not slow cognitive decline in patients with mild to moderate Alzheimer's disease (mean age 76 years)	Aisen et al.[94]
Randomized clinical trial (folic acid 0.8 mg/d)	Folic acid supplements for 3 years lowered homocysteine and improved global cognitive function, memory, and information-processing speed in community-dwelling older adults without Alzheimer's disease or dementia (mean age 60 years)	Durga et al.[96]
Randomized clinical trial (folic acid 0.8 mg/d, B_{12} 0.5 mg/d, B_6 20 mg/d)	High dose B vitamin supplements for 2 years lowered homocysteine and slowed the rate of brain atrophy in older adults with mild cognitive impairment (mean age 77 years)	Smith et al.[97]
Randomized clinical trial (folic acid 1.0 mg/d, B_{12} 0.5 mg/d, B_6 10 mg/d)	High dose B vitamin supplements for 2 years lowered homocysteine, but did not have a positive effect on various domains of cognitive function (mean age 74 years)	McMahon et al.[98]

this negative finding, one must keep in mind that Alzheimer's disease is caused by a specific neuropathological process that has profound effects on the structure and function of the brain. It is unlikely that B vitamin supplements can significantly reverse the structural damage caused by the accumulation of amyloid plaques and neurofibrillary tangles that are the neuropathological hallmarks of Alzheimer's disease. In addition, the length of the intervention period (eighteen months) may not have been sufficient in duration to achieve a beneficial response if one could be achieved with B vitamins in this context. Therefore, it is perhaps not surprising that these patients with overt signs of Alzheimer's disease did not benefit from B vitamin supplements.

In contrast to the study by Aisen et al.[94] is a randomized clinical trial by Durga et al.[96] This study consisted of folic acid supplements (0.8 mg/day) or placebo given to community-dwelling older adults (age 50–70 years) without overt signs of Alzheimer's disease or dementia. After a three-year intervention period, improvements in global cognitive function, memory, and information-processing speed were observed in the supplemented group compared with the placebo controls. This finding is not inconsistent with the notion that B vitamin supplements may be beneficial before or in the absence of significant structural damage caused by the Alzheimer's disease process.

The second consideration is the initial status of the individuals at the start of the intervention. In the trial by Aisen et al.,[94] both the treatment and placebo groups had baseline homocysteine and B vitamin levels within normal ranges. Because there was no benefit of the B vitamin supplements, even though there was lowering of the plasma homocysteine concentrations, this suggests that such supplements may not benefit those who are not deficient to begin with. In support of this contention is the study by Durga et al.[96] A main inclusion criterion for this study was an elevated homocysteine concentration (> 13 μmol/L) (normal range: 4–10 μmol/L). Thus, the participants in the Durga et al. trial had baseline evidence of inadequate folate status and thus were more likely to benefit from folic acid supplements.

A more recent randomized clinical trial by Smith et al.[97] extends this conclusion by providing compelling evidence that B vitamin supplements and lowering of homocysteine not only protect against cognitive decline but also attenuate atrophy of the brain. Older adults (age ≥ 70 years) with mild cognitive impairment (but without a diagnosis of Alzheimer's disease or other form of dementia) were given either high-dose B vitamin supplements (0.8 mg/day folic acid, 0.5 mg/day vitamin B_{12}, and 20 mg/day vitamin B_6) or placebo for twenty-four months. In this study, the primary outcome was change in brain volume as measured by magnetic resonance imaging (MRI), with cognitive function scores serving as a secondary outcome. In those receiving the B vitamin supplements, the mean rate of brain atrophy was 0.76 percent per year compared with 1.08 percent per year for the placebo group. Moreover, the attenuation of the rate of brain atrophy achieved with the B vitamin supplements was greatest in those subjects with the highest homocysteine levels at baseline, while cognitive function scores at the end of the twenty-four-month trial were lowest in those who had experienced the greatest amount of brain atrophy.

The third consideration is the dose of folate in supplements, particularly in populations already exposed to folic acid fortification. In light of the potential adverse effects of excess folic acid in exacerbating vitamin B_{12} deficiency, there is some concern that excess folate intake (with or without B_{12} supplements) in individuals with pernicious anemia or other B_{12} malabsorption syndromes might increase risk of cognitive impairment. Evidence consistent with this possibility comes from the randomized clinical trial by McMahon et al.[98] in which 276 healthy older adults (age ≥ 65 years) received either supplements of folate (1.0 mg/day), B_{12} (0.5 mg/day), and B_6 (10 mg/day) or placebo for twenty-four months. All subjects had an elevated plasma homocysteine concentration at baseline (> 13 μmol/L). After twenty-four months, plasma homocysteine was significantly lower and plasma folate and B_{12} were significantly higher in the supplemented group compared with the placebo group. Nonetheless, for global cognitive function, as well as subdomain tests of memory, learning, verbal fluency, semantic fluency, information-processing speed, and reasoning ability, cognitive performance was not better in the supplemented group compared with placebo. Moreover, for one subdomain, information-processing speed, the supplemented group performed significantly worse than the placebo group after twenty-four months. In addition, using a combined score of all the cognitive domain tests employed in the study, the overall cognitive function was borderline lower in the supplemented group ($p = 0.05$). The authors concluded that B vitamin supplements were not beneficial, and were perhaps harmful, though chance could not be excluded as an explanation for the findings.

It has been suggested, however, that the high dose of folate in the supplements (1 mg/day), which is equivalent to the upper tolerable limit for the vitamin,[99] may have exacerbated cognitive impairment associated with low B_{12} status.[100]

Though plasma B_{12} concentrations increased significantly in the supplemented group, a subset of the group may not have benefited from the B_{12} supplements due to undetected B_{12} malabsorption. Neurological function in these subjects may have been exacerbated by the high-dose folate supplements. This would not have been detected in the statistical analysis because group means, and not individual values, were analyzed. Additional support for a potential adverse effect of excess folate intake on cognitive function in older adults was provided by Morris et al.,[101] who found that cognitive decline in a cohort of older adults in Chicago over a six-year follow-up period was accelerated in those subjects with high folate intake.

CONCLUSION

It is clear that overt B vitamin deficiencies are detrimental to neurological function. The relationship between B vitamins and cognitive function in older adults without overt clinical signs of deficiency also has significant epidemiological support. The strongest and most consistent associations are between elevated homocysteine and poor cognitive performance and cognitive decline. This likely represents the particular utility of homocysteine as a functional indicator of the efficiency of one-carbon metabolism. In contrast, low B vitamin status in the absence of an elevated homocysteine concentration may indicate that for a given individual, his or her B vitamin status is sufficient to maintain metabolic efficiency. This in turn would suggest that any cognitive impairment may be due to other factors and B vitamin supplements may not be beneficial. This is supported by the few intervention studies that have demonstrated a benefit of B vitamin supplements on cognitive function.[96,97]

For patients and their physicians, the take-home messages are as follows:

1. B vitamins may be beneficial for maintaining cognitive function, but only under certain conditions. These conditions include metabolic evidence of low B vitamin status (i.e., elevated homocysteine and/or methylmalonic acid) and no clinical evidence of overt Alzheimer's disease or dementia. Early intervention may be the key, which suggests that screening for elevated homocysteine and low B vitamin status in midlife (years or decades prior to the typical age of onset of Alzheimer's disease and dementia) could be critical for identifying those who may benefit from B vitamin supplements.
2. The dose of B vitamins may be important, particularly for folate. Very high doses of folate are not typically necessary to achieve normalization of circulating homocysteine concentrations. Lower doses of folate may be sufficient and less likely to exacerbate B_{12} deficiency, if this is proven to occur. This is particularly important in populations already exposed to folic acid fortification in which additional folate supplements lead to very high intakes of the vitamin.
3. Before beginning B vitamin supplements, the B_{12} status of older adults should be investigated. Importantly, total serum B_{12} should not be relied upon as a reliable indicator of B_{12} status in and of itself. Other tests should

be considered in addition to total serum B_{12}, including holoTC, methylmalonic acid, and homocysteine. This will allow the identification of individuals with subtle but clinically relevant B_{12} deficiency, which in turn will inform intervention strategies for protecting against cognitive decline.

REFERENCES

1. United Nations Population Division. *World population prospects. Sex and age.* New York: United Nations Publications, 1998.
2. Evans, D. A., Funkenstein, H. H., Albert, M. S., et al. Prevalence of Alzheimer's disease in a community population of older persons: higher than previously reported. *JAMA* 1989;262:2551–6.
3. Goodwin, J. S., Goodwin, J. M., Garry, P. J. Association between nutritional status and cognitive functioning in a healthy elderly population. *JAMA* 1983;249:2917–21.
4. Herbert, V. Experimental nutritional folate deficiency in man. *Trans Assoc Am Physicians* 1962;75:307–20.
5. Botez, M. I., Cadotte, M., Beaulieu, R., et al. Neurologic disorders responsive to folic acid therapy. *CMAJ* 1976;115:217–23.
6. Botez, M. I., Fontaine, F., Botez, T., et al. Folate-responsive neurological and mental disorders: Report of 16 cases. *Eur Neurol* 1977;16:230–46.
7. Enk, C., Hougaard, K., Hippe, E. Reversible dementia and neuropathy associated with folate deficiency 16 years after partial gastrectomy. *Scand J Haematol* 1980;25:63–6.
8. Pincus, I. H., Reynolds, E. H., Glaser, G. H. Subacute combined system degeneration with folate deficiency. *JAMA* 1972;221:496–7.
9. Bottiglieri, T. Folate, vitamin B12, and neuropsychiatric disorders. *Nutr Rev* 1996; 54:382–90.
10. Fava, M., Borus, J. S., Alpert, J. E., Nierenberg, A. A., Rosenbaum, J. F., Bottiglieri, T. Folate, vitamin B12, and homocysteine in major depressive disorder. *Am J Psychiatry* 1997;154:426–8.
11. Alpert, J. E., Fava, M. Nutrition and depression: the role of folate. *Nutr Rev* 1997;55:145–9.
12. Reynolds, E. H. Folic acid, ageing, depression, and dementia. *BMJ* 2002;324:1512–5.
13. Golnik, K. C., Schaible, E. R. Folate-responsive optic neuropathy. *J Neuroophthalmol* 1994;14:163–9.
14. Refsum, H., Ueland, P. M., Nygard, O., Vollset, S. E. Homocysteine and cardiovascular disease. *Annu Rev Med* 1998;49:31–62.
15. Clarke, R., Smith, A. D., Jobst, K. A., Refsum, H., Sutton, L., Ueland, P. M. Folate, vitamin B12, and serum total homocysteine levels in confirmed Alzheimer disease. *Arch Neurol* 1998;55:1449–55.
16. Seshadri, S., Beiser, A., Selhub, J., et al. Plasma homocysteine as a risk factor for dementia and Alzheimer's disease. *New Engl J Med* 2002;346:476–83.
17. Hannibal, L., Glushchenko, A. V., Jacobsen, D. W. Folate and vascular disease: basic mechanisms. In: *Folate in Health and Disease*, 2nd ed., ed. Bailey, L. B. Boca Raton, FL: CRC Press, 2010.
18. Wilson, K. M., Lentz, S. R. Mechanisms of the atherogenic effects of elevated homocysteine in experimental models. *Semin Vasc Med* 2005;5:163–71.
19. Lentz, S. R. Mechanisms of homocysteine-induced atherothrombosis. *J Thromb Haemost* 2005;3:1646–54.

20. Cook, J. W., Taylor, L. M., Orloff, S. L., Landry, G. J., Moneta, G. L., Porter, J. M. Homocysteine and arterial disease. Experimental mechanisms. *Vascul Pharmacol* 2002;38:293–300.
21. Lipton, S. A., Kim, W. K., Choi, Y. B., et al. Neurotoxicity associated with dual actions of homocysteine at the N-methyl-D-aspartate receptor. *Proc Natl Acad Sci* 1997;94:5923–8.
22. Olney, J. W. Excitatory amino acids and neuropsychiatric disorders. *Biol Psychiatry* 1989;26:505–25.
23. White, A. R., Huang, X., Jobling, M. F., et al. Homocysteine potentiates copper- and amyloid beta peptide-mediated toxicity in primary neuronal cultures: possible risk factors in the Alzheimer's-type neurodegenerative pathways. *J Neurochem* 2001;76:1509–20.
24. Ho, P. I., Ashline, D., Dhitavat, S., et al. Folate deprivation induces neurodegeneration: roles of oxidative stress and increased homocysteine. *Neurobiol Dis* 2003;14:32–42.
25. Wyse, A. T., Zugno, A. I., Streck, E. L., et al. Inhibition of Na(+),K(+)-ATPase activity in hippocampus of rats subjected to acute administration of homocysteine is prevented by vitamins E and C treatment. *Neurochem Res* 2002;27:1685–9.
26. Matté, C., Mackedanz, V., Stefanello, F. M., et al. Chronic hyperhomocysteinemia alters antioxidant defenses and increases DNA damage in brain and blood of rats: protective effect of folic acid. *Neurochem Int* 2009;54:7–13.
27. Bottiglieri, T., Godfrey, P., Flynn, T., Carney, M. W., Toone, B. K., Reynolds, E. H. Cerebrospinal fluid S-adenosylmethionine in depression and dementia: effects of treatment with parenteral and oral S-adenosylmethionine. *J Neurol Neurosurg Psychiatry* 1990;53:1096–8.
28. Bottiglieri, T., Laundy, M., Crellin, R., Toone, B. K., Carney, M. W., Reynolds, E. H. Homocysteine, folate, methylation, and monoamine metabolism in depression. *J Neurol Neurosurg Psychiatry* 2000;69:228–32.
29. Bottiglieri, T., Reynolds, E. Folate and neurological disease: basic mechanisms. In: *Folate in Health and Disease*, 2nd ed., ed. Bailey, L. B. Boca Raton, FL: CRC Press, 2010.
30. Kim, S., Lim, I. K., Park, G. H., Paik, W. K. Biological methylation of myelin basic protein: enzymology and biological significance. *Int J Biochem Cell Biol* 1997;29:743–51.
31. Sontag, E., Nunbhakdi-Craig, V., Sontag, J. M., et al. Protein phosphatase 2A methyltransferase links homocysteine metabolism with tau and amyloid precursor protein regulation. *J Neurosci* 2007;27:2751–9.
32. Sontag, J. M., Nunbhakdi-Craig, V., Montgomery, L., Arning, E., Bottiglieri, T., Sontag, E. Folate deficiency induces in vitro and mouse brain region-specific downregulation of leucine carboxyl methyltransferase-1 and protein phosphatase 2A B(alpha) subunit expression that correlate with enhanced tau phosphorylation. *J Neurosci* 2008;28:11477–87.
33. Sontag, E., Hladik, C., Montgomery, L., et al. Downregulation of protein phosphatase 2A carboxyl methylation and methyltransferase may contribute to Alzheimer disease pathogenesis. *J Neuropathol Exp Neurol* 2004;63:1080–91.
34. Miller, J. W., Selhub, J., Joseph, J. A. Oxidative damage caused by free radicals produced during catecholamine autoxidation: protective effects of O-methylation and melatonin. *Free Radic Biol Med* 1996;21:241–9.
35. Hirata, F., Strittmatter, W. J., Axelrod, J. beta-Adrenergic receptor agonists increase phospholipid methylation, membrane fluidity, and beta-adrenergic receptor-adenylate cyclase coupling. *Proc Natl Acad Sci* 1979;76:368–72.
36. Muccioli, G., Scordamaglia, A., Bertacco, S., Di Carlo, R. Effect of S-adenosyl-L-methionine on brain muscarinic receptors of aged rats. *Eur J Pharmacol* 1992;227:293–9.

37. Di Perri, B., Calderini, G., Battistella, A., Raciti, R., Toffano, G. Phospholipid methylation increases [3H]diazepam and [3H]GABA binding in membrane preparations of rat cerebellum. *J Neurochem* 1983;41:302–8.
38. Robertson, K. D., Wolffe, A. P. DNA methylation in health and disease. *Nat Rev Genet* 2000;1:11–19.
39. Martinowich, K., Hattori, D., Wu, H., et al. DNA methylation-related chromatin remodeling in activity-dependent BDNF gene regulation. *Science* 2003;302:890–3.
40. Chen, W. G., Chang, Q., Lin, Y., et al. Derepression of BDNF transcription involves calcium-dependent phosphorylation of MeCP2. *Science* 2003;302:885–9.
41. Cunha, C., Angelucci, A., D'Antoni, A., et al. Brain-derived neurotrophic factor (BDNF) overexpression in the forebrain results in learning and memory impairments. *Neurobiol Dis* 2009;33:358–68.
42. Liu, L., van Groen, T., Kadish, I., Tollefsbol, T. O. DNA methylation impacts on learning and memory in aging. *Neurobiol Aging* 2009;30:549–60.
43. Abdolmaleky, H. M., Cheng, K. H., Faraone, S. V., et al. Hypomethylation of MB-COMT promoter is a major risk factor for schizophrenia and bipolar disorder. *Hum Mol Genet* 2006;15:3132–45.
44. Dinn, J. J., Weir, D. G., McCann, S., Reed, B., Wilson, P., Scott, J. M. Methyl group deficiency in nerve tissue: a hypothesis to explain the lesion of subacute combined degeneration. *Ir Med J* 1980;149:1–4.
45. Frenkel, E. P. Abnormal fatty acid metabolism in peripheral nerves of patients with pernicious anemia. *J Clin Invest* 1973;52:1237–45.
46. Scalabrino, G., Monzio-Compagnoni, B., Ferioli, M. E., Lorenzini, E. C., Chiodini, E., Candiani, R. Subacute combined degeneration and induction of ornithine decarboxylase in spinal cords of totally gastrectomized rats. *Lab Invest* 1990;62:297–304.
47. Scalabrino, G., Buccellato, F. R., Tredici, G., et al. Enhanced levels of biochemical markers for cobalamin deficiency in totally gastrectomized rats: uncoupling of the enhancement from the severity of spongy vacuolation in spinal cord. *Exp Neurol* 1997;144:258–65.
48. Buccellato, F. R., Miloso, M., Braga, M., et al. Myelinolytic lesions in spinal cord of cobalamin-deficient rats are TNF-alpha-mediated. *FASEB J* 1999;13:297–304.
49. Peracchi, M., Bamonti Catena, F., Pomati, M., De Franceschi, M., Scalabrino, G. Human cobalamin deficiency: alterations in serum tumour necrosis factor-alpha and epidermal growth factor. *Eur J Haematol* 2001;67:123–7.
50. Scalabrino, G., Carpo, M., Bamonti, F., et al. High tumor necrosis factor-alpha levels in cerebrospinal fluid of cobalamin-deficient patients. *Ann Neurol* 2004;56:886–90.
51. Miller, J. W. Vitamin B12 deficiency, tumor necrosis factor-alpha, and epidermal growth factor: a novel function for vitamin B12? *Nutr Rev* 2002;60:142–4.
52. Dakshinamurti, S., Dakshinamurti, K. Vitamin B6. In: *Handbook of Vitamins*, 4th ed., ed. Zempleni, J., Rucker, R. B., McCormick, D. B., Suttie, J. W. Boca Raton, FL: CRC Press, 2007.
53. Tower, D. B. Neurochemical aspects of pyridoxine metabolism and function. *Am J Clin Nutr* 1956;4:329–45.
54. Jacques, P. F., Selhub, J., Bostom, A. G., Wilson, P. W., Rosenberg, I. H. The effect of folic acid fortification on plasma folate and total homocysteine concentrations. *New Engl J Med* 1999;340:1449–54.
55. Selhub, J., Jacques, P. F., Wilson, P. W., Rush, D., Rosenberg, I. H. Vitamin status and intake as primary determinants of homocysteinemia in an elderly population. *JAMA* 1993;270:2693–8.
56. Selhub, J., Jacques, P. F., Bostom, A. G., Wilson, P. W., Rosenberg, I. H. Relationship between plasma homocysteine and vitamin status in the Framingham study population. Impact of folic acid fortification. *Public Health Rev* 2000;28:117–45.

57. Bell, I. R., Edman, J. S., Selhub, J., et al. Plasma homocysteine in vascular disease and in nonvascular dementia of depressed elderly people. *Acta Psychiatr Scand* 1992;86:386–90.
58. Riggs, K. M., Spiro, A., Tucker, K., Rush, D. Relations of vitamin B-12, vitamin B-6, folate, and homocysteine to cognitive performance in the Normative Aging Study. *Am J Clin Nutr* 1996;63:306–14.
59. McCaddon, A., Davies, G., Hudson, P., et al. Total serum homocysteine in senile dementia of the Alzheimer's type. *Int J Geriatr Psychiatry* 1998;13:235–9.
60. Joosten, E., Lesaffre, E., Riezler, R., et al. Is metabolic evidence for vitamin-B12 and folate deficiency more frequent in elderly patients with Alzheimer's disease? *J Gerontol* 1997;52:76–9.
61. Teng, E. L., Chui, H. C. The Modified Mini-Mental State (3MS) examination. *J Clin Psychiatry* 1987;48:314–8.
62. Elias, M. F., Sullivan, L. M., D'Agostino, R. B., et al. Homocysteine and cognitive performance in the Framingham offspring study: age is important. *Am J Epidemiol* 2005;162:644–53.
63. Miller, J. W., Green, R., Ramos, M. I., et al. Homocysteine and cognitive function in the Sacramento Area Latino Study on Aging (SALSA). *Am J Clin Nutr* 2003;78:441–7.
64. Food and Drug Administration. Food labeling: health claims and label statements; folate and neural tube defects. *Federal Register* 1996;61:8752–81.
65. Haan, M. N., Miller, J. W., Aiello, A. E., Whitmer, R. A., Jagust, W. J., Mungas, D. M., Allen, L. H., Green, R. Homocysteine, B vitamins and incidence of dementia and cognitive impairment: results from the Sacramento Area Latino Study on Aging. *Am J Clin Nutr* 2007;85:511–7.
66. Ramos, M. I., Allen, L. H., Mungas, D. M., Jagust, W. J., Haan, M. N., Green, R., Miller, J. W. Low folate status is associated with impaired cognitive function and dementia in the Sacramento Area Latino Study on Aging. *Am J Clin Nutr* 2005;82:1346–52.
67. Mischoulon, D., Fava, M. Role of S-adenosyl-L-methionine in the treatment of depression: a review of the evidence. *Am J Clin Nutr* 2002;76(suppl):1158S–61S.
68. Bottiglieri, T., Laundy, M., Crellin, R., Toone, B. K., Carney, M. W., Reynolds, E. H. Homocysteine, folate, methylation, and monoamine metabolism in depression. *J Neurol Neurosurg Psychiatry* 2000;69:228–32.
69. Geda, Y. E. Blowing hot and cold over depression and cognitive impairment. *Neurology* 2010;75:12–4.
70. Saczynski, J. S., Beiser, A., Seshadri, S., Auerbach, S., Wolf, P. A., Au, R. Depressive symptoms and risk of dementia: the Framingham Heart Study. *Neurology* 2010;75:35–41.
71. Dotson, V. M., Beydoun, M. A., Zonderman, A. B. Recurrent depressive symptoms and the incidence of dementia and mild cognitive impairment. *Neurology* 2010;75:27–34.
72. Lindenbaum, J., Healton, E. B., Savage, D. G., et al. Neuropsychiatric disorders caused by cobalamin deficiency in the absence of anemia or macrocytosis. *New Engl J Med* 1988;318:1720–8.
73. Kang, J. H., Irizarry, M. C., Grodstein, F. Prospective study of plasma folate, vitamin B12, and cognitive function and decline. *Epidemiology* 2006;17:650–7.
74. Campbell, A. K., Jagust, W. J., Mungas, D. M., Miller, J. W., Green, R., Haan, M. N., Allen, L. H. Low erythrocyte folate, but not plasma vitamin B-12 or homocysteine, is associated with dementia in elderly Latinos. *J Nutr Health Aging* 2005;9:39–43.
75. Ulleland, M., Eilertsen, I., Quadros, E. V., et al. Direct assay for cobalamin bound to transcobalamin (holo-transcobalamin) in serum. *Clin Chem* 2002;48:526–32.
76. Orning, L., Rian, A., Campbell, A., et al. Characterization of a monoclonal antibody with specificity for holo-transcobalamin. *Nutr Metab (Lond)* 2006;3:3.
77. Clarke, R., Birks, J., Nexo, E., et al. Low vitamin B-12 status and risk of cognitive decline in older adults. *Am J Clin Nutr* 2007;86:1384–91.

78. McCracken, C., Hudson, P., Ellis, R., McCaddon, A. Medical Research Council Cognitive Function and Ageing Study. Methylmalonic acid and cognitive function in the Medical Research Council Cognitive Function and Ageing Study. *Am J Clin Nutr* 2006;84:1406–11.
79. Lewis, M. S., Miller, L. S., Johnson, M. A., Dolce, E. B., Allen, R. H., Stabler, S. P. Elevated methylmalonic acid is related to cognitive impairment in older adults enrolled in an elderly nutrition program. *J Nutr Elder* 2005;24:47–65.
80. Miller, J. W., Garrod, M. G., Rockwood, A. L., et al. Measurement of total vitamin B12 and holotranscobalamin, singly and in combination, in screening for metabolic vitamin B12 deficiency. *Clin Chem* 2006;52:278–85.
81. Clarke, R., Refsum, H., Birks, J., et al. Screening for vitamin B-12 and folate deficiency in older persons. *Am J Clin Nutr* 2003;77:1241–7.
82. Lloyd-Wright, Z., Hvas, A. M., Moller, J., Sanders TAB, Nexo, E. Holotranscobalamin as an indicator of dietary vitamin B12 deficiency. *Clin Chem* 2003;49:2076–8.
83. Garrod, M. G., Green, R., Allen, L. H., Mungas, D. M., Jagust, W. J., Haan, M. N., Miller, J. W. The fraction of total plasma vitamin B12 bound to transcobalamin correlates with cognitive function in elderly Latinos with depressive symptoms. *Clin Chem* 54:1210–7, 2008.
84. Morris, M. S., Jacques, P. F., Rosenberg, I. H., Selhub, J. Folate and vitamin B-12 status in relation to anemia, macrocytosis, and cognitive impairment in older Americans in the age of folic acid fortification. *Am J Clin Nutr* 2007;85:193–200.
85. Selhub, J., Morris, M. S., Jacques, P. F. In vitamin B12 deficiency, higher serum folate is associated with increased total homocysteine and methylmalonic acid concentrations. *Proc Natl Acad Sci* 2007;104:19995–20000.
86. Miller, J. W., Garrod, M. G., Allen, L. H., Haan, M. N., Green, R. Metabolic evidence of vitamin B12 deficiency, including high homocysteine and methylmalonic acid and low holotranscobalamin, is more pronounced in older adults with elevated plasma folate. *Am J Clin Nutr* 2009;90:1586–92.
87. Smith, A. D., Kim, Y. I., Refsum, H. Is folic acid good for everyone? *Am J Clin Nutr* 2008;87: 517–33.
88. Berry, R. J., Carter, H. K., Yang, Q. Cognitive impairment in older Americans in the age of folic acid fortification. *Am J Clin Nutr* 2007;86: 265–7.
89. Tucker, K. L., Qiao, N., Scott, T., Rosenberg, I., Spiro A 3rd. High homocysteine and low B vitamins predict cognitive decline in aging men: the Veterans Affairs Normative Aging Study. *Am J Clin Nutr* 2005;82:627–35.
90. Kado, D. M., Karlamangla, A. S., Huang, M. H., et al. Homocysteine versus the vitamins folate, B6, and B12 as predictors of cognitive function and decline in older high-functioning adults: MacArthur Studies of Successful Aging. *Am J Med* 2005;118:161–7.
91. Elias, M. F., Robbins, M. A., Budge, M. M., et al. Homocysteine, folate, and vitamins B6 and B12 blood levels in relation to cognitive performance: the Maine-Syracuse study. *Psychosom Med* 2006;68:547–54.
92. Miller, J. W., Green, R., Mungas, D. M., Reed, B. R., Jagust, W. J. Homocysteine, vitamin B6, and vascular disease in AD patients. *Neurology* 58:1471–5, 2002.
93. Luchsinger, J. A., Tang M-X, Shea, S., Miller, J., Green, R., Mayeux, R. Plasma homocysteine levels and risk of Alzheimer's disease. *Neurology* 62:1972–6, 2004.
94. Aisen, P. S., Schneider, L. S., Sano, M., et al. High-dose B vitamin supplementation and cognitive decline in Alzheimer disease: a randomized controlled trial. *JAMA* 2008;300:1774–83.
95. Rosen, W. G., Mohs, R. C., Davis, K. L. A new rating scale for Alzheimer's disease. *Am J Psychiatry* 1984;141:1356–64.

96. Durga, J., van Boxtel, M. P., Schouten, E. G., et al. Effect of 3-year folic acid supplementation on cognitive function in older adults in the FACIT trial: a randomised, double blind, controlled trial. *Lancet* 2007;369:208–16.
97. Smith, A. D., Smith, S. M., de Jager, C. A., et al. Homocysteine-lowering by B vitamins slows the rate of accelerated brain atrophy in mild cognitive impairment: a randomized controlled trial. *PLoS One* 2010;5:e12244.
98. McMahon, J. A., Green, T. J., Skeaff, C. M., Knight, R. G., Mann, J. I., Williams, S. M. A controlled trial of homocysteine lowering and cognitive performance. *New Engl J Med.* 2006;354:2764–72.
99. Institute of Medicine. Folate. In: *Dietary reference intakes for thiamin, riboflavin, niacin, vitamin B6, folate, vitamin B12, pantothenic acid, biotin, and choline*. Washington, DC: National Academy Press, 1998.
100. Morris, M. S., Jacques, P. Folate and neurological function: epidemiological perspective. In: *Folate in Health and Disease*, 2nd ed., ed. Bailey, L. B. Boca Raton, FL: CRC Press, 2010.
101. Morris, M. C., Evans, D. A., Bienias, J. L., et al. Dietary folate and vitamin B12 intake and cognitive decline among community-dwelling older persons. *Arch Neurol* 2005;62:641–5.

11 Creatine, Brain Functioning, and Behavior

Patricia J. Allen, Kristen E. D'Anci, and Robin B. Kanarek

CONTENTS

Introduction 215
Creatine Intake, Synthesis, and Metabolism 216
Creatine and Brain-Related Disorders 218
  Creatine Deficiency 218
  Huntington's Disease 219
  Alzheimer's Disease 220
  Amyotrophic Lateral Sclerosis 220
  Parkinson's Disease 221
  Schizophrenia 222
  Psychological Stress 223
  Depression 224
  Traumatic Brain Injury 226
Creatine and Cognitive Behavior 226
Summary and Future Directions 227
References 228

INTRODUCTION

Creatine is one of the most widely used dietary supplements on the market, with a $400 million industry promoting its use. Creatine, in the form of creatine monohydrate, is popular among athletes, who use it to increase muscle mass and strength in resistance training and to improve performance in high-intensity physical activity like cycling or rowing (Bemben and Lamont 2005; Volek and Rawson 2004). While creatine is more often associated with enhancing physical performance, it is also essential for the proper functioning of the brain. There is growing evidence that creatine may be of value in the treatment of neurological conditions that are linked to dysfunctional energy metabolism such as age-related cognitive decline and neurodegenerative diseases (e.g., Parkinson's disease, Huntington's disease, and Alzheimer's disease) (Andres et al. 2008; Gualano et al. 2010). Moreover, creatine supplementation is beginning to garner attention as a complementary strategy in the treatment of psychiatric disorders such as depression and posttraumatic stress disorder (Allen et al. 2010; Amital et al. 2006; Roitman et al. 2007).

It is generally agreed that creatine up to 5 grams per day is safe and well tolerated by adults, but there is not enough evidence to make an informed recommendation in favor or against doses higher than 5 g/day (Shao and Hathcock 2006). Nonetheless, empirical and anecdotal accounts do exist that have reported mild to moderate side effects of daily creatine supplementation. The majority of these reported side effects are of gastrointestinal distress, renal dysfunction, or dehydration in study participants. Fewer have reported disturbances in mood and anxiety.

CREATINE INTAKE, SYNTHESIS, AND METABOLISM

Creatine (N-aminoiminomethyl-N-methylglycine) is an amino acid–like compound that is acquired from dietary intake of high-protein foods, such as fish, eggs, or meat. Additionally, creatine is produced endogenously from the biosynthesis of arginine, glycine, and methionine in the liver, kidney, pancreas, and, to a lesser degree, brain (Andres et al. 2008; Beard and Braissant 2010; Wyss and Kaddurah-Daouk 2000). Approximately half of an individual's daily requirement comes from dietary creatine, while the remainder is produced in the body. As creatine is synthesized endogenously, it is a nonessential nutrient. However, dietary intake of creatine, or its precursors, is necessary because the reversible enzymatic conversion of creatine to phosphocreatine produces creatinine as a byproduct, which is cleared from the body via the kidney.

De novo creatine synthesis relies upon the one-carbon metabolism cycle, with guanidinoacetate receiving a methyl group from S-adenosylmethionine to form creatine (Figure 11.1). Some estimates indicate that 70 percent of available S-adenosylmethionine is used to synthesize creatine (Beard and Braissant 2010), although these rates may be lower with high creatine intake (Wyss and Kaddurah-Daouk 2000). Creatine is actively transported into the brain via a specific creatine transporter (Beard and Braissant 2010), and dietary supplementation with creatine increases brain levels of creatine (Wyss and Wallimann 1994).

The creatine–phosphocreatine system serves as a spatial and temporal energy buffer in tissue with significant and fluctuating energy requirements, including the brain (Mudd et al. 2007; Lyoo et al. 2003; Brosnan and Brosnan 2007). Pools of cellular energy (ATP) are generated from the reaction (Phosphocreatine + ADP → Creatine + ATP), and, conversely, energy is stored in the form of phosphocreatine (Creatine + ATP → Phosphocreatine + ADP) (McLeish and Kenyon 2005; Wallimann et al. 1992). Creatine kinases, the enzymatic catalysts of these reactions, are located in tissues that consume significant amounts of energy, making this enzyme system a critical regulator of energy homeostasis (Brosnan and Brosnan 2007; Niklasson and Agren 1984; Kuzhikandathil and Molloy 1994). Of importance to neuronal function, creatine improves the survival and differentiation of dopaminergic and GABA-ergic neurons, likely by buffering against rapid depletion of energy pools (Wilken et al. 1998; Andres et al. 2005a, 2005b).

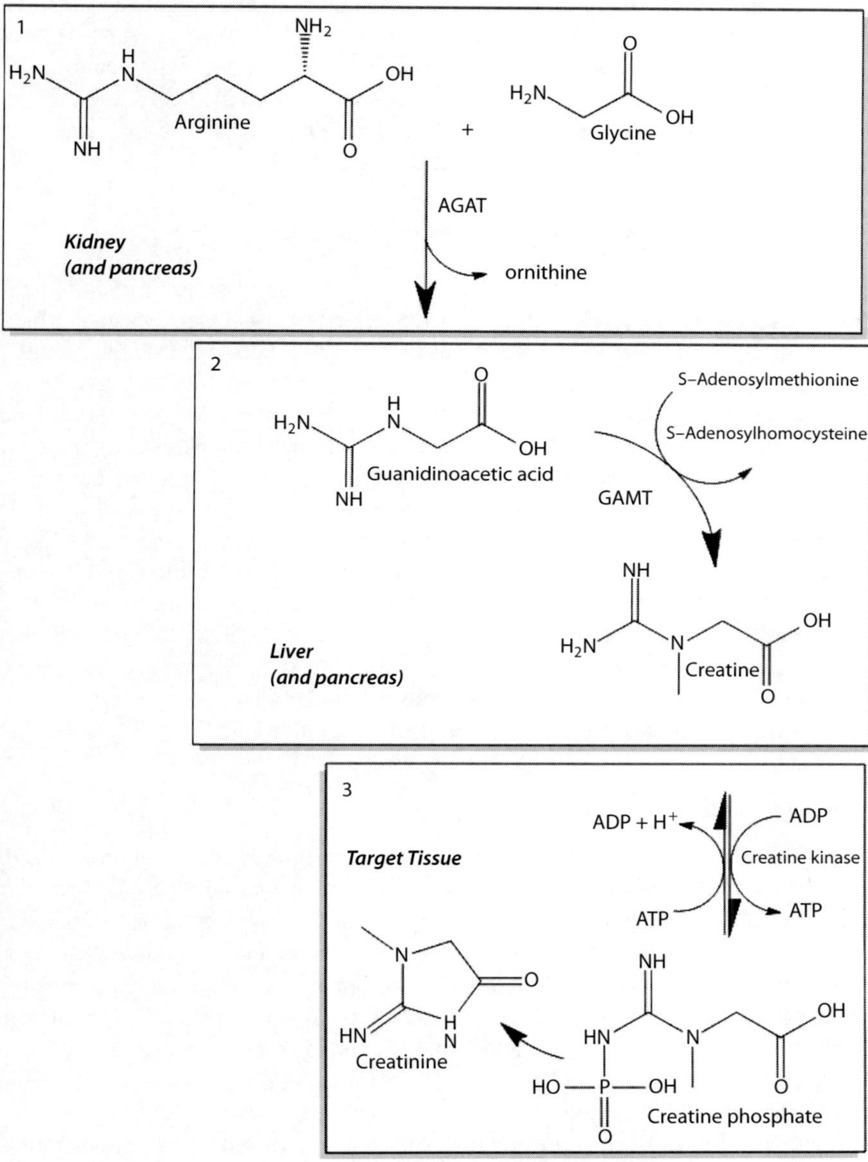

FIGURE 11.1 Biosynthesis and metabolism of creatine. Panel 1: The amino acids arginine and glycine are enzymatically converted by L-arginine:glycine amidinotransferase (AGAT) into guanidinoacetic acid (GAA) in the kidney. Panel 2: GAA is methylated by guanidinoacetate methyltransferase (GAMT) in the liver producing creatine. Panel 3: Creatine is transported to target tissue (e.g., brain or skeletal muscle) via the bloodstream. Creatine is enzymatically and reversibly converted to phosphocreatine via creatine kinase. The phosphorylation of creatine to phosphocreatine produces creatinine, which is excreted via the kidney.

Both creatine and phosphocreatine are broken down spontaneously to creatinine, which is removed from the body in the urine. The rate of loss is approximately 1.7 percent of the total body pool of creatine per day. Since over 95 percent of creatine and phosphocreatine is located in skeletal muscle, creatine losses and creatine excretion vary as a function of differences in muscle mass resulting from age, gender, and levels of daily activity (Brosnan and Brosnan 2007).

CREATINE AND BRAIN-RELATED DISORDERS

The therapeutic value of oral creatine supplementation to treat brain-related disorders is related to its ability to increase cellular energy reserves (Dechent et al. 1999; Lyoo et al. 2003; Nash et al. 1994; Ohtsuki et al. 2002; Pan and Takahashi 2007). The beneficial effects of creatine are also strongly supported by the pivotal role it plays in buffering metabolic processes to prevent energy depletion and neuronal death. Furthermore, recent work uncovering nontraditional mechanisms of action of cerebral creatine have demonstrated high potential for the use of dietary creatine to improve brain function. In humans and in animals, research has consistently shown that administration of creatine over a prolonged period of time results in measurable increases of creatine concentrations in brain and muscle tissue, which are most pronounced after four weeks (Dechent et al. 1999; Ferrante et al. 2000; Ipsiroglu et al. 2001; Perasso et al. 2003; Persky et al. 2003; Stockler et al. 1996). Although limited at present, empirical evidence on the neural and behavioral effects of creatine is steadily growing to support the safety, tolerability, and efficacy of creatine to protect the brain from neuronal damage and to improve cognitive function.

CREATINE DEFICIENCY

Evidence from studies assessing creatine deficiency in animals and humans highlights the essential role of creatine in normal cognitive function. Cerebral creatine deficiency in animals is associated with mitochondrial abnormalities (Wyss and Wallimann 1994) and loss of hippocampal mossy fiber connections (In't Zandt et al. 2004). Mice lacking either the mitochondrial or brain-type creatine kinase isoforms exhibit reductions in the size of the hippocampus, deficits in spatial learning and exploration, and poor acoustic startle reflex (Jost et al. 2002; Streijger et al. 2004). Mice lacking both isoforms (CK --/-- double knock-out mice) demonstrate a more severe physical and behavioral phenotype (Streijger et al. 2005).

Humans born with errors in creatine synthesis or an X-linked creatine transporter defect suffer from severe language and speech impairments, developmental delays, mental retardation, autistic behavior, brain atrophy, and mild epilepsy (Andres et al. 2008; Braissant and Henry 2008; Nasrallah et al. 2010; Schulze et al. 2003). In some individuals, the cognitive deficits resulting from these inborn errors can be improved, although not totally reversed, with chronic supplementation with high doses of creatine (Battini et al. 2002; Bianchi et al. 2007; Bizzi et al. 2002; Mercimek-Mahmutoglu et al. 2006; Stockler et al. 1996; Salomons et al. 2001).

Huntington's Disease

Huntington's disease is caused by an autosomal dominant mutation of a gene called huntingtin (HTT), which results in the abnormal coding of the huntingtin protein (Htt), known as mutant huntingtin protein (mHtt). Symptoms of Huntington's disease include cognitive, mental, and physical disturbances. Chorea is the classic symptom of Huntington's disease and is characterized by jerky, spontaneous, uncontrollable involuntary movements. Other movement symptoms typically develop, including loss of balance and coordination, slurred speech, jaw clenching, and difficulty walking. Cognitive impairments emerge that affect judgment, memory, attention, and emotional control, and can progress to dementia. Affected brain regions include the basal ganglia, the substantia nigra, the cortex, the hippocampus, the angular gyrus in the parietal lobe, the cerebellum, and subsets of neurons within the hypothalamus and thalamus (Ross and Tabrizi 2011).

Among all of the neurodegenerative disorders, the role of endogenous creatine and the effects of creatine supplementation on the brain have been most extensively studied in human and animal models of Huntington's disease (Ryu et al. 2005). Researchers have hypothesized that the generalized energy defects observed in Huntington's disease are directly related to abnormalities within the creatine–phosphocreatine circuit. Reduced creatine kinase is used as a biomarker to assess the progression of Huntington's disease (Kim et al. 2010). In support of this hypothesis, brain-type creatine kinase is significantly reduced in the brains of patients with Huntington's disease and mice expressing Huntington's-like lesions and behavior. The degree of reduction correlates with the severity of disease symptoms. In addition, patients with Huntington's diseases have a decreased phosphocreatine:inorganic phosphate ratio in resting muscle (Koroshetz et al. 1997).

A number of preclinical and clinical studies have evaluated the potential of creatine to treat Huntington's disease. In a rodent model of Huntington's disease, oral supplementation with 1 percent creatine for two weeks buffered ATP and phosphocreatine levels, shielded against malnate lesions, reduced 3-NP-induced lactate concentrations, and protected against 3-NP-induced oxidative injury (Matthews et al. 1998). It is hypothesized that creatine prevents the accumulation of intracellular calcium, the formation of reactive oxygen species, and ultimately damage from oxidative stress (Sestili et al. 2006). In transgenic models of Huntington's disease, a diet supplemented with 1–2 percent creatine improved survival, attenuated brain atrophy, slowed the formation of Huntingtin-positive aggregates, increased body weight, and improved motor performance (Andreassen et al. 2001; Ferrante et al. 2000) Dedeoglu and colleagues (2003) replicated these effects of creatine in R6/2 mice supplemented at different points of disease progression (at six, eight, and ten weeks of age), with the greatest benefit seen in the earliest intervention point.

Several small-scale clinical trials in patients with Huntington's disease have shown daily creatine supplementation to increase serum, muscle, and brain concentrations of creatine and to be safe and well tolerated by patients. With respect to positive outcomes, Bender and colleagues (2005) used H-MRS to show that creatine supplementation reduced levels of glutamate in the parieto-occipital cortex, which is important because excessive levels of glutamate are implicated in the pathogenesis of Huntington's disease. In patients with Huntington's disease, creatine administration lowered serum levels of 8-hydroxy-2-deoxyguanosine, which is a marker for oxidative injury to DNA (Hersch et al. 2006). In spite of these positive findings, few trials have detected significant beneficial effects on indices of cognitive, neurological, or motor function (Bender et al. 2005; Hersch et al. 2006; Tabrizi et al. 2003, 2005; Verbessem et al. 2003).

Alzheimer's Disease

Alzheimer's disease is the most common neurodegenerative disease and the most prevalent form of dementia among the elderly. Dementia is a condition characterized by a cluster of cognitive symptoms, such as diminished short-term memory, personality changes, confusion, and poor emotional control. The etiology of Alzheimer's disease remains elusive, but progressive cognitive decline is associated with significant brain atrophy and energetic dysfunction. Alzheimer's disease brains show reduced glucose metabolism; disturbances in acetylcholine neurotransmission; an accumulation of insoluble beta amyloid plaques, which are produced by a mutation in the amyloid precursor protein; and neurofibrillary tangles, which are caused by altered tau proteins found within neurons (Ertekin-Taner 2007).

Cerebral creatine and creatine kinase have been implicated as biomarkers in Alzheimer's disease. Initial studies suggested that brain-type creatine kinase levels were higher in patients suffering from Alzheimer's disease compared with controls (Thompson et al. 1980). However, this finding has been criticized because these studies did not report clinical details regarding the type of dementia studied or any details about the control groups (Court et al. 1987). More recent studies that employed more rigorous screening protocols reported decreased creatine kinase activity in the frontal lobe, hippocampus, inferior parietal lobe, and cerebellum of patients with Alzheimer's disease compared with age-matched healthy controls. These findings suggest that creatine kinase abnormalities may contribute to the abnormal metabolism, dysfunctions in neurotransmission, and neuronal loss observed in Alzheimer's disease (Aksenov et al. 1997; David et al. 1998; Hensley et al. 1995).

In support of the connection between creatine kinase activity and Alzheimer's disease pathogenesis, it has been shown that beta amyloid directly inactivates creatine kinase in vivo (Hensley et al. 1995). It is hypothesized that beta amyloid inactivates or degrades brain-type creatine kinase via oxidative stress (Burklen et al. 2006). Creatine prevents oxidative damage from the formation of reactive oxygen species through direct antioxidant activity in mammalian cell cultures. This suggests that creatine may protect against beta amyloid-induced oxidative stress in Alzheimer's disease patients through direct antioxidant actions (Sestili et al. 2006).

Amyotrophic Lateral Sclerosis

Amyotrophic lateral sclerosis (Pfefferbaum et al. 2007) is a progressive neurodegenerative disease that is characterized by the selective atrophy of motor neurons in the brain and spinal cord, which results in paralysis. ALS has been linked to antioxidant impairment, excitotoxicity from excessive glutamate, mitochondrial dysfunction, and impaired axonal transport (Wijesekera and Leigh 2009).

Abnormalities in the creatine–phosphocreatine circuit have been detected in animal models of ALS and in humans diagnosed with ALS. There are significant reductions in creatine kinase activity in the spinal tissue of genetically altered mice that are predisposed to oxidative stress (FALS mice) relative to wild-type animals (Wendt et al. 2002). In imaging studies, ALS patients have significantly lower N-acetylaspartate:creatine metabolite ratios in the motor cortex, suggesting altered creatine levels or creatine kinase activity (Vielhaber et al. 2001). Moreover, creatine deposits were detected in central regions typically affected by neurodegeneration in ALS in two out of six patients, possibly resulting from dysfunctional oxidative processes (Kastyak et al. 2010).

Early preclinical studies provide evidence for the neuroprotective effects of creatine supplementation in ALS. FALS mice supplemented with a diet containing 2 percent creatine displayed improved motor performance, increased survival rates, reversal of cholinergic deficits, and reductions in oxidative stress (Klivenyi et al. 1999; Pena-Altamira et al. 2005; Wendt et al. 2002). Other work using magnetic resonance spectroscopy showed that FALS mice maintained on a +2 percent creatine diet displayed higher brain levels of creatine (cerebellum: +25 percent; medulla: +11 percent; and cortex: +4 percent) and less N-acetyl aspartate loss in the medulla than nonsupplemented FALS mice (Choi et al. 2009). These findings suggest that the neuroprotective effects of creatine may be attributed to improved cellular energetics, antioxidant action at creatine kinase sites, reduced glutamate levels, or a combination of these effects.

The benefits of creatine observed in animal models of ALS have not translated into successful treatment of ALS. A randomized, placebo-controlled clinical trial evaluating the effects of creatine was halted after sixteen months when it was observed that creatine supplementation did not reduce the rate of decline or improve survival rates in patients with ALS compared with controls (Groeneveld et al. 2003). In another clinical trial, no benefits on any of the outcome measures, including the ALS Functional Rating Scale, were observed (Shefner et al. 2004). In a small preclinical trial, modest, acute improvements were observed in exercise performance, maximal voluntary isometric muscular contractions, and fatigue after one week of creatine supplementation in ALS patients (Mazzini et al. 2001). However, when measured six months after the initiation of supplementation, performance in patients with ALS had declined.

Parkinson's Disease

Parkinson's disease (PD) is a progressive neurodegenerative disorder that causes motor disturbances and, to a lesser degree, cognitive impairment. The cause of Parkinson's disease is connected to several genetic mutations that are associated with protein accumulation inside of cells, known as "Lewy bodies," which are inclusions that cause neurodegeneration. The prevalence of Lewy bodies increases throughout the brain as Parkinson's disease progresses, particularly affecting cells within the dopaminergic pathways between the substantia nigra and the basal ganglia (Brundin et al. 2008). Parkinson's disease is characterized by postural instability, rigidity, tremor, bradykinesia (slowing of physical movement), and akinesia (loss of physical movement). Mild cognitive symptoms can also develop in PD, affecting cognitive speed, memory, attention, and problem solving, but these symptoms are considered secondary to the disorder.

The pathogenesis of Parkinson's disease has been linked with mitochondrial dysfunction and oxidative damage (Abou-Sleiman et al. 2006). Energetic defects in Parkinson's disease are accompanied by abnormalities in the creatine–phosphocreatine circuit. For instance, serum creatine kinase was elevated in patients with Parkinson's disease, although creatine kinase levels did not correlate with age of onset, duration of disease, or symptom severity (Takubo et al. 2003).

Parkinson's-like symptoms can be induced in animals by the administration of the neurotoxin methyl-phenyl-tetrahydropyridine (MPTP). Dietary supplementation with 0.25–1 percent creatine protected against MPTP-induced loss of dopaminergic neurons in the substantia nigra of mice, possibly by buffering ATP and phosphocreatine levels or by enhancing mitochondrial function (Matthews et al. 1999). However, diets supplemented with 2 and 3 percent creatine did not exert a neuroprotective effect. In other studies, dietary creatine reduced L-DOPA-induced dyskinesia, a common side effect of therapy used to treat the symptoms of Parkinson's disease. Specifically, 2 percent creatine supplementation, when combined with L-DOPA therapy, reduced the occurrence of abnormal voluntary movement in rats with lesions to the striatum without affecting the benefits of L-DOPA to treat Parkinsonian symptoms (Valastro et al. 2009).

Emerging work has revealed that creatine may be useful in ameliorating the symptoms of Parkinson's disease. For example, a recent phase 2 clinical trial determined that 10 g/day creatine slowed the progression of the disorder, relative to placebo (NINDS NET-PD 2006) Moreover, a randomized, placebo-control trial showed that individuals with Parkinson's disease who were supplemented with creatine for a total of two years reported elevated mood and required smaller increases in the amount of pharmacotherapy needed to treat symptoms in comparison to controls (Bender et al. 2006). Patients given creatine supplements in combination with weight training showed greater improvement in strength (chest press strength, biceps curl strength, and chair rise strength) and function compared to controls (Hass et al. 2007).

Schizophrenia

Among the most consistently replicated findings in the creatine literature is the positive association between creatine kinase activity and acute, but not chronic, psychosis. Psychosis can be broadly characterized by disturbances in thought and/or loss of contact with reality. Psychosis is a defining feature of schizophrenia, a phenotypically heterogeneous disorder characterized by delusions, hallucinations, disorganized thought or speech, catatonic behavior, and/or negative symptoms. These symptoms cause significant occupational and social dysfunction.

Abnormalities in the creatine–phosphocreatine circuit in schizophrenia have generated interest in its role in the pathogenesis of the disorder. The presence of markedly elevated serum creatine kinase levels measured within the first few days of acute psychotic episodes has become so common a laboratory abnormality that it has been given the name "psychosis-associated creatine kinase-emia" (PACK). While statistics are exceedingly variable, studies have reported an average of 40–50 percent incidence of muscle-type PACK in acutely psychotic patients (Coffey et al. 1970; Faulstich et al. 1984; Gosling et al. 1972; Hermesh et al. 2002; Meltzer 1968; Martin et al. 1972; Schweid et al. 1972). Creatine kinase elevations occur to a significantly higher degree in males than females, and the pattern of creatine kinase elevation tends to repeat at the onset of subsequent acute psychotic episodes (Hermesh et al. 2002).

Soni (1976) described a negative association between serum creatine kinase levels and psychomotor activity, such that patients showed normal creatine kinase levels when displaying psychomotor depression and elevations in serum creatine kinase in excess of normal when they returned to normal psychomotor activity. In addition, antipsychotics that are clinically effective for treating psychosis patients significantly increase levels of creatine kinase, implicating the creatine–phosphocreatine circuit as a therapeutic target (Melkersson 2006; Meltzer et al. 1996). However, the utility of creatine kinase as a biomarker for acute psychosis has been called into question for at least two reasons. First, there is a vast range of variability in the detection of elevated creatine kinase levels (Tuason et al. 1974). Second, elevations in creatine kinase are not unique to psychosis. For example, elevations in creatine kinase were observed in nonpsychotic depressed patients compared to psychotic depressed patients (Segal et al. 2007), and markedly abnormal elevations in creatine kinase are also observed in chronic alcoholics (Ikeda 1977).

Schizophrenic patients who had never received antipsychotic medication have significantly lower phosphocreatine:ATP ratios in their basal ganglia compared to healthy controls (Jayakumar et al. 2010). A reduced phosphocreatine:ATP ratio corresponds with a hypermetabolic state, such that more phosphocreatine is converted to ATP to maintain energy balance within basal ganglia neurons. After one year of antipsychotic treatment, phosphocreatine:ATP ratios of the schizophrenic group normalized to levels comparable with those of healthy controls. This change in metabolic activity was associated with symptom improvement. Complementary findings of abnormalities in energy metabolism include reductions of total creatine levels in the anterior cingulate and parieto-occipital cortex of schizophrenic patients compared to bipolar patients and normal controls (Theberge et al. 2007; Ongur et al. 2009). In contrast, in a clinical trial using a randomized, double-blind cross-over design to examine the potential for creatine to buffer abnormal energy metabolism in schizophrenia, creatine supplementation was not successful in treating symptoms. These findings indicate that more research is needed to integrate the disparate results between brain chemistry and behavior in schizophrenia.

Psychological Stress

Stress has been studied extensively as a precipitating factor in the onset of psychiatric disorders. The detrimental effects of chronic stress include impairments in neuronal plasticity, cellular resiliency, and brain energy metabolism (Duman and Monteggia 2006; McEwen 2007; Sapolsky 2000). For instance, in animals, chronic stress has been shown to reduce neurogenesis, and synaptic function in the hippocampus (Duman 2004; Duman and Monteggia 2006; Pittenger and Duman 2008). Animal studies indicate that chronic stress reduces brain levels of creatine, which is in turn associated with negative effects on affective behavior (Czeh et al. 2001; Fuchs et al. 2002; Michael-Titus et al. 2008; van der Hart et al. 2002). For example, psychosocial stress occurs naturally in male tree shrews that develop social hierarchies with dominant and subordinate members. In imaging studies, subordinate tree shrews exhibited a decline in phosphocreatine and creatine (total creatine) levels, reduced hippocampal volume, and impaired neurogenesis after exposure to conflict with a dominant animal. Moreover, treatment with tricyclic antidepressants reversed the effects of stress on total creatine levels in these animals (Czeh et al. 2001; van der Hart et al. 2002).

An important gap in knowledge is whether creatine supplementation could oppose the adverse effects of stress on synaptic plasticity, cell morphology, and cell integrity in the same manner as antidepressant drugs. This hypothesis is plausible as creatine has antioxidant properties and plays a central role in buffering metabolic processes to prevent energy exhaustion and neuronal death (Brosnan and Brosnan 2007). Additionally, there is strong evidence that creatine exerts potent neuroprotective effects to combat stress via nontraditional mechanisms in the brain (Koga et al. 2005). Young chickens given creatine prior to social separation stress displayed significantly fewer stress responses, including fewer vocalizations, less spontaneous activity, and reduced plasma corticosterone concentrations. In contrast, chickens given both creatine and a $GABA_A$ antagonist displayed high levels of stress that were comparable to those of control animals. These findings indicate that creatine modulates the $GABA_A$ receptor site to attenuate effects of stress. In further support of creatine's nontraditional neuromodulatory capability, Almeida and colleagues (Almeida et al. 2006) demonstrated that creatine can be released from neurons in an excitotic, action-potential dependent manner in response to membrane depolarization—in the same fashion as classic neurotransmitters. Specifically, depolarization of the cell membrane of a neuron causes an influx of Ca^{2+} and then subsequent release of creatine. When Ca^{2+} is not present or when Na^+ channels are blocked by tetrodotoxin, creatine cannot be released from the tissue.

To date, few studies have directly evaluated the neuroprotective effects of creatine supplementation against the consequences of stress. Fewer have evaluated possible neurobiological mechanisms that could be involved, although it is clear that creatine has many effects that are not yet well defined. Both the classical and nontraditional mechanisms of creatine action remain important avenues of research that could pave the way for the development of complementary or alternative treatments for stress-related disorders, such as depression and anxiety.

Depression

Impairments in bioenergetic function within the brain, cellular resiliency, and neural plasticity have been associated with the pathogenesis of depression. For example, alterations in the creatine signal are seen in individuals with major depressive disorder (Roy et al. 2002). More specifically, changes in high-energy phosphate metabolism, particularly creatine and phosphocreatine levels, are associated with depression (McLeish and Kenyon 2005; Renshaw et al. 2001; Lyoo and Renshaw 2002; Agren and Niklasson 1988; Czeh et al. 2001; Iosifescu et al. 2008; Moore et al. 1997). Research has shown that there is a negative correlation between creatine metabolites and self-reported suicidal ideation in patients suffering from major depressive disorders (Agren and Niklasson 1988). More recently, it has been found that levels of brain creatine are inversely related to the severity of a depressive episode (Dager et al. 2004; Segal et al. 2007). Additionally, phosphorus-magnetic resonance spectroscopy has shown increased phosphocreatine and decreased ATP values in the frontal lobe and basal ganglia of depressed subjects (Iosifescu et al. 2008; Moore et al. 1997; Segal et al. 2007; Kato et al. 1992).

It has been hypothesized that creatine could be useful in preventing or treating depression. In support of this hypothesis, chronic intake of diets supplemented with creatine decreased depressive-like behavior in a dose-dependent manner in female rats in the forced swim test, an animal model of depression. However, in male rats, intake of the creatine-supplemented diets failed to reduce depressive behavior (Allen et al. 2010).

In humans, support for the potential usefulness of creatine in the treatment of depression comes from open-labeled studies demonstrating that daily oral intake of 3 to 5 grams of creatine elevated mood in depressed patients resistant to antidepressant drugs, and patients with comorbid posttraumatic stress disorder (Amital et al. 2006; Roitman et al. 2007) Moreover, in an eight-week, open-labeled study of creatine, significant improvements were observed in quality of life, mood, sleep patterns, and pain in patients with fibromyalgia (Amital et al. 2006). These improvements deteriorated four weeks after stopping creatine therapy.

An interesting point to note is the increasing number of reports suggesting that the nature of the relationship between the creatine-phosphocreatine circuit and mood may be sex dependent. Investigations using P-MRS have shown increased phosphocreatine and decreased ATP values in the frontal lobe and basal ganglia of depressed adults in vivo, though this pattern is more common in women than in men (Kato et al. 1992, 1994; Moore et al. 1997; Volz et al. 1998). In addition, female bipolar and depressed patients have lower creatine levels in the right frontal lobe compared with male patients, suggesting impaired metabolism that may result from sex differences in creatine transport function into the brain (Hamakawa et al. 1999). A recent examination of metabolite levels in unmedicated depressed patients and healthy controls, using H-MRS, revealed that phosphocreatine+creatine levels in the frontal cortex in depressed males were lower than those in healthy males, whereas phosphocreatine+creatine levels in depressed females were higher than those in healthy females (Nery et al. 2009). These observations are important as it has been widely reported that women have higher rates of depression than men, and that response to treatment in depressed women, but not men, is associated with altered energetic metabolite levels (Renshaw et al. 2001). These findings suggest that creatine may be more beneficial for treating depressions in women than in men. Currently, clinical trials are underway to evaluate the efficacy of creatine in ameliorating symptoms of depression in young women (e.g., trials NCT00851006, NCT01175616, and NCT00313417).

Traumatic Brain Injury

Traumatic brain injury (TBI) can be caused by a bump, blow, or jolt to the head, or by a penetrating head injury. TBI is characterized by primary damage which results in the disruption of neural and vascular structures, and also by secondary damage which is not the immediate consequence of the trauma, but develops within minutes, hours, or days after the injury.

Animal and human data indicate that TBI may be associated with selective changes in levels of creatine and/or phosphocreatine. For instance, there are reports of reduced NAA:creatine ratios post brain injury in rats (Casey et al. 2008) and in humans (Garnett et al. 2000; Govind et al. 2010; Signoretti et al. 2008), which may reflect changes in the creatine+phosphocreatine signal. Further, the degree of NAA:creatine reduction corresponds with the extent of injury and recovery outcome (Signoretti et al. 2008). In contrast, Babikian and colleagues (2010) found no long-term differences in brain creatine levels in the anterior and posterior corpus collosum in children who suffered moderate or severe TBI. This may indicate that elevations in creatine and phosphocreatine are transitory and possibly normalize within days following the traumatic assault.

Work in animals underscores the potential for creatine supplementation in neuroprotection and recovery from brain injury. Mice given intraperitoneal injections of creatine prior to the induction of brain damage demonstrated dose-related reductions in cortical damage (Sullivan et al. 2000). Similarly, rats fed a standard rodent diet supplemented with creatine before the induction of traumatic brain injury demonstrated reduced cortical damage in comparison with controls. Moreover, rats that consumed creatine had significantly higher mitochondrial membrane potentials, mitochondrial calcium, and ATP, and significantly reduced levels of reactive oxidative species, relative to controls. In other work, rats fed a diet supplemented with creatine before experiencing cortical contusions had significantly more sparing of cortical tissue and suppressed levels of lactate and free fatty acids than rats not given creatine (Sullivan et al. 2000). Finally, creatine supplementation can also reduce both ischemia-mediated depletion of ATP, and caspace-3-activation and cytochrome C release, which are indicators of cell damage (Scheff and Dhillon 2004).

Preliminary studies in humans suggest that creatine supplementation may be useful in the treatment of the secondary symptoms of TBI (Sakellaris et al. 2006, 2008). On a short-term basis, children and adolescents given creatine spent less time in an intensive care unit and needed to be intubated for a shorter period of time than controls not given creatine. Additionally, when examined three and six months after injury, individuals who had received creatine supplementation displayed greater improvements in cognitive functioning, self-care, sociability, and communication skills than controls (Licata and Renshaw 2010). During a six-month follow-up period, the number of children reporting headaches, dizziness, and fatigue was significantly lower in the creatine-supplemented group than in the control group (Licata and Renshaw 2010).

CREATINE AND COGNITIVE BEHAVIOR

Intake of creatine may also benefit brain function and cognitive performance in healthy individuals. Mice fed a diet enhanced with creatine exhibited better object recognition memory, more rapidly explored a novel environment, and lived longer and healthier lives than their nonsupplemented littermates (Bender et al. 2008). Further evidence of a role for creatine in memory consolidation comes from work demonstrating that direct administration of creatine into the hippocampus of male rats improved spatial learning. Interestingly, coadministration of an *N*-methyl *D*-aspartate (NMDA) receptor antagonist reversed creatine-induced enhancement of spatial learning, while coadministration of an NMDA agonist intensified spatial learning (Oliveira et al. 2008). These findings have led to the hypothesis that creatine exerts its positive effects on learning, at least in part, by modifying the activity of the NMDA receptor (Brosnan and Brosnan 2007).

Creatine may also have a positive effect on cognitive behavior in our own species, particularly in stressful situations or in conditions in which creatine availability may be less than optimal. For example, daily creatine supplementation significantly improved working memory and intelligence scores in vegetarians and vegans, who are assumed to have compromised levels of creatine due to low dietary intake of meat (Rae et al. 2003). Additionally, in young men who had undergone significant sleep deprivation paired with mild exercise, oral intake of creatine attenuated fatigue and improved performance on a complex central executive task (McMorris et al. 2006). Moreover, daily supplementation with creatine reduced the mental fatigue that occurred after a stressful math task (Watanabe et al. 2002). More recently, Hammett and colleagues (Hammett et al. 2010) found that creatine supplementation enhanced short-term memory and decreased fMRI blood oxygen–dependent responses in the primary visual cortex. Supporting the importance of stress in mediating creatine's effects on cognition, no differences in performance on a variety of neurocognitive tests were observed as a function of creatine supplementation in healthy nonvegetarian participants (Rawson et al. 2008).

Creatine may also ameliorate age-related declines in cognitive performance, as daily supplementation of creatine in older adults improved performance on tests of verbal and spatial short-term and long-term memory (McMorris et al. 2007).

SUMMARY AND FUTURE DIRECTIONS

Based on both animal and human studies, creatine represents a promising nutritional supplement for the treatment of brain disorders that are associated with energetic impairment. However, a number of questions remain with regard to how much creatine should be taken and for how long. Indeed, animal studies suggest that long-term feeding or high doses of creatine may reduce its actions with respect to protecting against the consequences of neurotoxic or traumatic injury (Matthews et al. 1999; Prass et al. 2007). Thorough pharmacokinetic and pharmacodynamic studies are needed to establish recommended guidelines for healthy adults.

Moreover, knowing the functional and clinical significance of the creatine-phosphocreatine circuit, it is essential to more fully characterize the role of this creatine in cognitive processes. For instance, if creatine has an effect on cognition, it is critical to know whether cognitive performance varies as a function of different doses of creatine or whether there is some threshold dose after which the effects of creatine begin to occur. Further experiments are required to investigate biochemical changes that may explain creatine's effects on behavior, such as the relationship between creatine and neurotransmitter action or synaptic plasticity. Moreover, there is increasing evidence that creatine monohydrate produces different behavioral effects in males and females (Allen et al. 2010), which is supported by an extensive body of literature documenting sex differences in brain creatine levels and creatine kinase activity. The significance of these observations would be greatly enhanced by additional studies that systematically evaluate possible sex differences in response to supplementation with creatine monohydrate.

In the case of neurodegenerative disorders, animal experiments have demonstrated the potential for dietary creatine to improve symptoms and slow the course of AD, HD, PD, and to a lesser extent ALS, but human studies have been less conclusive. More rigorous methodology for the evaluation of the effectiveness of creatine monohydrate should be considered in future clinical and experimental trials. In humans, this encompasses the use of more selective screening practices, increased sample size (> 100 subjects), two or more doses of creatine, doses ranging from 20 to 40 g/day, a placebo control, and multiple endpoints for a duration of at least two years. In animals, careful evaluation of the effects of varying levels of dietary creatine on different neurotransmitter systems and plasticity factors would bolster our mechanistic understanding of how brain creatine influences behavioral outcomes.

REFERENCES

Abou-Sleiman, P. M., M. M. Muqit, and N. W. Wood. 2006. Expanding insights of mitochondrial dysfunction in Parkinson's disease. *Nat Rev Neurosci* 7 (3): 207–19.

Agren, H., and F. Niklasson. 1988. Creatinine and creatine in CSF: indices of brain energy metabolism in depression. Short note. *J Neural Transm* 74 (1): 55–9.

Aksenov, M. Y., M. V. Aksenova, R. M. Payne, C. D. Smith, W. R. Markesbery, and J. M. Carney. 1997. The expression of creatine kinase isoenzymes in neocortex of patients with neurodegenerative disorders: Alzheimer's and Pick's disease. *Exp Neurol* 146 (2): 458–65.

Allen, P. J., K. E. D'Anci, R. B. Kanarek, and P. F. Renshaw. 2010. Chronic creatine supplementation alters depression-like behavior in rodents in a sex-dependent manner. *Neuropsychopharmacology* 35 (2): 534–46.

Almeida, L. S., G. S. Salomons, F. Hogenboom, C. Jakobs, and A. N. Schoffelmeer. 2006. Exocytotic release of creatine in rat brain. *Synapse* 60 (2): 118–23.

Amital, D., T. Vishne, S. Roitman, M. Kotler, and J. Levine. 2006. Open study of creatine monohydrate in treatment-resistant posttraumatic stress disorder. *J Clin Psychiatry* 67 (5): 836–7.

Amital, D., T. Vishne, A. Rubinow, and J. Levine. 2006. Observed effects of creatine monohydrate in a patient with depression and fibromyalgia. *Am J Psychiatry* 163 (10): 1840–1.

Andreassen, O. A., A. Dedeoglu, R. J. Ferrante, B. G. Jenkins, K. L. Ferrante, M. Thomas, A. Friedlich, S. E. Browne, G. Schilling, D. R. Borchelt, S. M. Hersch, C. A. Ross, and M. F. Beal. 2001. Creatine increase survival and delays motor symptoms in a transgenic animal model of Huntington's disease. *Neurobiol Dis* 8 (3): 479–91.

Andres, R. H., A. D. Ducray, A. W. Huber, A. Perez-Bouza, S. H. Krebs, U. Schlattner, R. W. Seiler, T. Wallimann, and H. R. Widmer. 2005a. Effects of creatine treatment on survival and differentiation of GABA-ergic neurons in cultured striatal tissue. *J Neurochem* 95 (1): 33–45.

Andres, R. H., A. D. Ducray, U. Schlattner, T. Wallimann, and H. R. Widmer. 2008. Functions and effects of creatine in the central nervous system. *Brain Res Bull* 76 (4): 329–43.

Andres, R. H., A. W. Huber, U. Schlattner, A. Perez-Bouza, S. H. Krebs, R. W. Seiler, T. Wallimann, and H. R. Widmer. 2005b. Effects of creatine treatment on the survival of dopaminergic neurons in cultured fetal ventral mesencephalic tissue. *Neuroscience* 133 (3): 701–13.

Babikian, T., S. D. Marion, S. Copeland, J. R. Alger, J. O'Neill, F. Cazalis, R. Mink, C. C. Giza, J. A. Vu, S. M. Hilleary, C. L. Kernan, N. Newman, and R. F. Asarnow. 2010. Metabolic levels in the corpus callosum and their structural and behavioral correlates after moderate to severe pediatric TBI. *J Neurotrauma* 27 (3): 473–81.

Battini, R., V. Leuzzi, C. Carducci, M. Tosetti, M. C. Bianchi, C. B. Item, S. Stockler-Ipsiroglu, and G. Cioni. 2002. Creatine depletion in a new case with AGAT deficiency: clinical and genetic study in a large pedigree. *Mol Genet Metab* 77 (4): 326–31.

Beard, E., and O. Braissant. 2010. Synthesis and transport of creatine in the CNS: importance for cerebral functions. *J Neurochem* 115 (2): 297–313.

Bemben, M. G., and H. S. Lamont. 2005. Creatine supplementation and exercise performance: recent findings. *Sports Med* 35 (2): 107–25.

Bender, A., D. P. Auer, T. Merl, R. Reilmann, P. Saemann, A. Yassouridis, J. Bender, A. Weindl, M. Dose, T. Gasser, and T. Klopstock. 2005. Creatine supplementation lowers brain glutamate levels in Huntington's disease. *J Neurol* 252 (1): 36–41.

Bender, A., W. Koch, M. Elstner, Y. Schombacher, J. Bender, M. Moeschl, F. Gekeler, B. Muller-Myhsok, T. Gasser, K. Tatsch, and T. Klopstock. 2006. Creatine supplementation in Parkinson disease: a placebo-controlled randomized pilot trial. *Neurology* 67 (7): 1262–4.

Bender, A., W. Samtleben, M. Elstner, and T. Klopstock. 2008. Long-term creatine supplementation is safe in aged patients with Parkinson disease. *Nutr Res* 28 (3): 172–8.

Bianchi, M. C., M. Tosetti, R. Battini, V. Leuzzi, M. G. Alessandri, C. Carducci, I. Antonozzi, and G. Cioni. 2007. Treatment monitoring of brain creatine deficiency syndromes: a 1H- and 31P-MR spectroscopy study. *AJNR Am J Neuroradiol* 28 (3): 548–54.

Bizzi, A., M. Bugiani, G. S. Salomons, D. H. Hunneman, I. Moroni, M. Estienne, U. Danesi, C. Jakobs, and G. Uziel. 2002. X-linked creatine deficiency syndrome: a novel mutation in creatine transporter gene SLC6A8. *Ann Neurol* 52 (2): 227–31.

Braissant, O., and H. Henry. 2008. AGAT, GAMT and SLC6A8 distribution in the central nervous system, in relation to creatine deficiency syndromes: A review. *J Inherit Metab Dis.*, April 4 (epub ahead of print).

Brosnan, J. T., and M. E. Brosnan. 2007. Creatine: endogenous metabolite, dietary, and therapeutic supplement. *Annu Rev Nutr* 27: 241–61.

Brundin, P., J. Y. Li, J. L. Holton, O. Lindvall, and T. Revesz. 2008. Research in motion: the enigma of Parkinson's disease pathology spread. *Nat Rev Neurosci* 9 (10): 741–5.

Burklen, T. S., U. Schlattner, R. Homayouni, K. Gough, M. Rak, A. Szeghalmi, and T. Wallimann. 2006. The creatine kinase/creatine connection to Alzheimer's disease: CK-inactivation, APP-CK complexes and focal creatine deposits. *J Biomed Biotechnol* 2006 (3): 35936.

Casey, P. A., M. C. McKenna, G. Fiskum, M. Saraswati, and C. L. Robertson. 2008. Early and sustained alterations in cerebral metabolism after traumatic brain injury in immature rats. *J Neurotrauma* 25 (6): 603–14.

Choi, J. K., E. Kustermann, A. Dedeoglu, and B. G. Jenkins. 2009. Magnetic resonance spectroscopy of regional brain metabolite markers in FALS mice and the effects of dietary creatine supplementation. *Eur J Neurosci* 30 (11): 2143–50.

Coffey, J. W., R. G. Heath, and A. F. Guschwan. 1970. Serum creatine kinase, aldolase, and copper in acute and chronic schizophrenics. *Biol Psychiatry* 2 (4): 331–9.

Court, J., N. Ferrier, H. Griffiths, B. Lauffart, R. Perry, J. Candy, A. Fairbairn, and G. Blessed. 1987. Serum creatine kinase-BB levels and cerebral cortical creatine kinase activity in senile dementia of the Alzheimer type. *J Neurol Sci* 80 (1): 111–5.

Czeh, B., T. Michaelis, T. Watanabe, J. Frahm, G. de Biurrun, M. van Kampen, A. Bartolomucci, and E. Fuchs. 2001. Stress-induced changes in cerebral metabolites, hippocampal volume, and cell proliferation are prevented by antidepressant treatment with tianeptine. *Proc Natl Acad Sci U S A* 98 (22): 12796–801.

Dager, S. R., S. D. Friedman, A. Parow, C. Demopulos, A. L. Stoll, I. K. Lyoo, D. L. Dunner, and P. F. Renshaw. 2004. Brain metabolic alterations in medication-free patients with bipolar disorder. *Arch Gen Psychiatry* 61 (5): 450–8.

David, S., M. Shoemaker, and B. E. Haley. 1998. Abnormal properties of creatine kinase in Alzheimer's disease brain: correlation of reduced enzyme activity and active site photolabeling with aberrant cytosol-membrane partitioning. *Brain Res Mol Brain Res* 54 (2): 276–87.

Dechent, P., P. J. Pouwels, B. Wilken, F. Hanefeld, and J. Frahm. 1999. Increase of total creatine in human brain after oral supplementation of creatine-monohydrate. *Am J Physiol* 277 (3 pt. 2): R698–704.

Dedeoglu, A., J. K. Kubilus, L. Yang, K. L. Ferrante, S. M. Hersch, M. F. Beal, and R. J. Ferrante. 2003. Creatine therapy provides neuroprotection after onset of clinical symptoms in Huntington's disease transgenic mice. *J Neurochem* 85 (6): 1359–67.

Duman, R. S. 2004. Depression: a case of neuronal life and death? *Biol Psychiatry* 56 (3): 140–5.

Duman, R. S., and L. M. Monteggia. 2006. A neurotrophic model for stress-related mood disorders. *Biol Psychiatry* 59 (12): 1116–27.

Ertekin-Taner, N. 2007. Genetics of Alzheimer's disease: a centennial review. *Neurol Clin* 25 (3): 611–67.

Faulstich, M. E., P. J. Brantley, and C. A. Barkemeyer. 1984. Creatine phosphokinase, the MMPI, and psychosis. *Am J Psychiatry* 141 (4): 584–6.

Ferrante, R. J., O. A. Andreassen, B. G. Jenkins, A. Dedeoglu, S. Kuemmerle, J. K. Kubilus, R. Kaddurah-Daouk, S. M. Hersch, and M. F. Beal. 2000. Neuroprotective effects of creatine in a transgenic mouse model of Huntington's disease. *J Neurosci* 20 (12): 4389–97.

Fuchs, E., B. Czeh, T. Michaelis, G. de Biurrun, T. Watanabe, and J. Frahm. 2002. Synaptic plasticity and tianeptine: structural regulation. *Eur Psychiatry* 17 (suppl. 3): 311–7.

Garnett, M. R., A. M. Blamire, B. Rajagopalan, P. Styles, and T. A. Cadoux-Hudson. 2000. Evidence for cellular damage in normal-appearing white matter correlates with injury severity in patients following traumatic brain injury: A magnetic resonance spectroscopy study. *Brain* 123 (pt. 7): 1403–9.

Gosling, R., R. J. Kerry, J. E. Orme, and G. Owen. 1972. Creatine phosphokinase activity in newly admitted psychiatric patients. *Br J Psychiatry* 121 (563): 351–5.

Govind, V., S. Gold, K. Kaliannan, G. Saigal, S. Falcone, K. L. Arheart, L. Harris, J. Jagid, and A. A. Maudsley. 2010. Whole-brain proton MR spectroscopic imaging of mild-to-moderate traumatic brain injury and correlation with neuropsychological deficits. *J Neurotrauma* 27 (3): 483–96.

Groeneveld, G. J., J. H. Veldink, I. van der Tweel, S. Kalmijn, C. Beijer, M. de Visser, J. H. Wokke, H. Franssen, and L. H. van den Berg. 2003. A randomized sequential trial of creatine in amyotrophic lateral sclerosis. *Ann Neurol* 53 (4): 437–45.

Gualano, B., G. G. Artioli, J. R. Poortmans, and A. H. Lancha Jr. 2010. Exploring the therapeutic role of creatine supplementation. *Amino Acids* 38 (1): 31–44.

Hamakawa, H., T. Kato, T. Shioiri, T. Inubushi, and N. Kato. 1999. Quantitative proton magnetic resonance spectroscopy of the bilateral frontal lobes in patients with bipolar disorder. *Psychol Med* 29 (3): 639–44.

Hammett, S. T., M. B. Wall, T. C. Edwards, and A. T. Smith. 2010. Dietary supplementation of creatine monohydrate reduces the human fMRI BOLD signal. *Neurosci Lett* 479 (3): 201–5.

Hass, C. J., M. A. Collins, and J. L. Juncos. 2007. Resistance training with creatine monohydrate improves upper-body strength in patients with Parkinson disease: a randomized trial. *Neurorehabil Neural Repair* 21 (2): 107–15.

Hensley, K., N. Hall, R. Subramaniam, P. Cole, M. Harris, M. Aksenov, M. Aksenova, S. P. Gabbita, J. F. Wu, J. M. Carney, and et al. 1995. Brain regional correspondence between Alzheimer's disease histopathology and biomarkers of protein oxidation. *J Neurochem* 65 (5): 2146–56.

Hermesh, H., D. Stein, I. Manor, T. Shechtmann, R. Blumensohn, S. Meged, R. Shiloh, Y. Benjamini, and A. Weizman. 2002. Serum creatine kinase levels in untreated hospitalized adolescents during acute psychosis. *J Am Acad Child Adolesc Psychiatry* 41 (9): 1045–53.

Hersch, S. M., S. Gevorkian, K. Marder, C. Moskowitz, A. Feigin, M. Cox, P. Como, C. Zimmerman, M. Lin, L. Zhang, A. M. Ulug, M. F. Beal, W. Matson, M. Bogdanov, E. Ebbel, A. Zaleta, Y. Kaneko, B. Jenkins, N. Hevelone, H. Zhang, H. Yu, D. Schoenfeld, R. Ferrante, and H. D. Rosas. 2006. Creatine in Huntington disease is safe, tolerable, bioavailable in brain and reduces serum 8OH2'dG. *Neurology* 66 (2): 250–2.

Ikeda, H. 1977. Serum creatine phosphokinase activity in newly admitted chronic alcoholics. *Folia Psychiatr Neurol Jpn* 31 (1): 9–16.

In't Zandt, H. J., W. K. Renema, F. Streijger, C. Jost, D. W. Klomp, F. Oerlemans, C. E. Van der Zee, B. Wieringa, and A. Heerschap. 2004. Cerebral creatine kinase deficiency influences metabolite levels and morphology in the mouse brain: a quantitative in vivo 1H and 31P magnetic resonance study. *J Neurochem* 90 (6): 1321–30.

Iosifescu, D. V., N. R. Bolo, A. A. Nierenberg, J. E. Jensen, M. Fava, and P. F. Renshaw. 2008. Brain bioenergetics and response to triiodothyronine augmentation in major depressive disorder. *Biol Psychiatry* 63 (12): 1127–34.

Ipsiroglu, O. S., C. Stromberger, J. Ilas, H. Hoger, A. Muhl, and S. Stockler-Ipsiroglu. 2001. Changes of tissue creatine concentrations upon oral supplementation of creatine-monohydrate in various animal species. *Life Sci* 69 (15): 1805–15.

Jayakumar, P. N., B. N. Gangadhar, G. Venkatasubramanian, S. Desai, L. Velayudhan, D. Subbakrishna, and M. S. Keshavan. 2010. High energy phosphate abnormalities normalize after antipsychotic treatment in schizophrenia: a longitudinal 31P MRS study of basal ganglia. *Psychiatry Res* 181 (3): 237–40.

Jost, C. R., C. E. Van Der Zee, H. J. In 't Zandt, F. Oerlemans, M. Verheij, F. Streijger, J. Fransen, A. Heerschap, A. R. Cools, and B. Wieringa. 2002. Creatine kinase B-driven energy transfer in the brain is important for habituation and spatial learning behaviour, mossy fibre field size and determination of seizure susceptibility. *Eur J Neurosci* 15 (10): 1692–706.

Kastyak, M. Z., M. Szczerbowska-Boruchowska, D. Adamek, B. Tomik, M. Lankosz, and K. M. Gough. 2010. Pigmented creatine deposits in amyotrophic lateral sclerosis central nervous system tissues identified by synchrotron Fourier transform infrared microspectroscopy and X-ray fluorescence spectromicroscopy. *Neuroscience* 166 (4): 1119–28.

Kato, T., S. Takahashi, T. Shioiri, and T. Inubushi. 1992. Brain phosphorous metabolism in depressive disorders detected by phosphorus-31 magnetic resonance spectroscopy. *J Affect Disord* 26 (4): 223–30.

Kato, T., S. Takahashi, T. Shioiri, J. Murashita, H. Hamakawa, and T. Inubushi. 1994. Reduction of brain phosphocreatine in bipolar II disorder detected by phosphorus-31 magnetic resonance spectroscopy. *J Affect Disord* 31 (2): 125–33.

Kim, J., D. J. Amante, J. P. Moody, C. K. Edgerly, O. L. Bordiuk, K. Smith, S. A. Matson, W. R. Matson, C. R. Scherzer, H. D. Rosas, S. M. Hersch, and R. J. Ferrante. 2010. Reduced creatine kinase as a central and peripheral biomarker in Huntington's disease. *Biochim Biophys Acta* 1802 (7–8): 673–81.

Klivenyi, P., R. J. Ferrante, R. T. Matthews, M. B. Bogdanov, A. M. Klein, O. A. Andreassen, G. Mueller, M. Wermer, R. Kaddurah-Daouk, and M. F. Beal. 1999. Neuroprotective effects of creatine in a transgenic animal model of amyotrophic lateral sclerosis. *Nat Med* 5 (3): 347–50.

Koga, Y., H. Takahashi, D. Oikawa, T. Tachibana, D. M. Denbow, and M. Furuse. 2005. Brain creatine functions to attenuate acute stress responses through GABAnergic system in chicks. *Neuroscience* 132 (1): 65–71.

Koroshetz, W. J., B. G. Jenkins, B. R. Rosen, and M. F. Beal. 1997. Energy metabolism defects in Huntington's disease and effects of coenzyme Q10. *Ann Neurol* 41 (2): 160–5.

Kuzhikandathil, E. V., and G. R. Molloy. 1994. Transcription of the brain creatine kinase gene in glial cells is modulated by cyclic AMP-dependent protein kinase. *J Neurosci Res* 39 (1): 70–82.

Licata, S. C., and P. F. Renshaw. 2010. Neurochemistry of drug action: insights from proton magnetic resonance spectroscopic imaging and their relevance to addiction. *Ann N Y Acad Sci* 1187: 148–71.

Lyoo, I. K., S. W. Kong, S. M. Sung, F. Hirashima, A. Parow, J. Hennen, B. M. Cohen, and P. F. Renshaw. 2003. Multinuclear magnetic resonance spectroscopy of high-energy phosphate metabolites in human brain following oral supplementation of creatine-monohydrate. *Psychiatry Res* 123 (2): 87–100.

Lyoo, I. K., and P. F. Renshaw. 2002. Magnetic resonance spectroscopy: current and future applications in psychiatric research. *Biol Psychiatry* 51 (3): 195–207.

Martin, W. A., R. E. Garey, and R. G. Heath. 1972. Cerebrospinal fluid creatine kinase in acutely psychotic patients. *J Neurol Neurosurg Psychiatry* 35 (5): 726–9.

Matthews, R. T., R. J. Ferrante, P. Klivenyi, L. Yang, A. M. Klein, G. Mueller, R. Kaddurah-Daouk, and M. F. Beal. 1999. Creatine and cyclocreatine attenuate MPTP neurotoxicity. *Exp Neurol* 157 (1): 142–9.

Matthews, R. T., L. Yang, B. G. Jenkins, R. J. Ferrante, B. R. Rosen, R. Kaddurah-Daouk, and M. F. Beal. 1998. Neuroprotective effects of creatine and cyclocreatine in animal models of Huntington's disease. *J Neurosci* 18 (1): 156–63.

Mazzini, L., C. Balzarini, R. Colombo, G. Mora, I. Pastore, R. De Ambrogio, and M. Caligari. 2001. Effects of creatine supplementation on exercise performance and muscular strength in amyotrophic lateral sclerosis: preliminary results. *J Neurol Sci* 191 (1–2): 139–44.

McEwen, B. S. 2007. Physiology and neurobiology of stress and adaptation: central role of the brain. *Physiol Rev* 87 (3): 873–904.

McLeish, M. J., and G. L. Kenyon. 2005. Relating structure to mechanism in creatine kinase. *Crit Rev Biochem Mol Biol* 40 (1): 1–20.

McMorris, T., R. C. Harris, J. Swain, J. Corbett, K. Collard, R. J. Dyson, L. Dye, C. Hodgson, and N. Draper. 2006. Effect of creatine supplementation and sleep deprivation, with mild exercise, on cognitive and psychomotor performance, mood state, and plasma concentrations of catecholamines and cortisol. *Psychopharmacology (Berl)* 185 (1): 93–103.

McMorris, T., G. Mielcarz, R. C. Harris, J. P. Swain, and A. Howard. 2007. Creatine supplementation and cognitive performance in elderly individuals. *Neuropsychol Dev Cogn B Aging Neuropsychol Cogn* 14 (5): 517–28.

Melkersson, K. 2006. Serum creatine kinase levels in chronic psychosis patients—a comparison between atypical and conventional antipsychotics. *Prog Neuropsychopharmacol Biol Psychiatry* 30 (7): 1277–82.

Meltzer, H. 1968. Creatine kinase and aldolase in serum: abnormality common to acute psychoses. *Science* 159 (821): 1368–70.

Meltzer, H. Y., P. A. Cola, and M. Parsa. 1996. Marked elevations of serum creatine kinase activity associated with antipsychotic drug treatment. *Neuropsychopharmacology* 15 (4): 395–405.

Mercimek-Mahmutoglu, S., S. Stoeckler-Ipsiroglu, A. Adami, R. Appleton, H. C. Araujo, M. Duran, R. Ensenauer, E. Fernandez-Alvarez, P. Garcia, C. Grolik, C. B. Item, V. Leuzzi, I. Marquardt, A. Muhl, R. A. Saelke-Kellermann, G. S. Salomons, A. Schulze, R. Surtees, M. S. van der Knaap, R. Vasconcelos, N. M. Verhoeven, L. Vilarinho, E. Wilichowski, and C. Jakobs. 2006. GAMT deficiency: features, treatment, and outcome in an inborn error of creatine synthesis. *Neurology* 67 (3): 480–4.

Michael-Titus, A. T., M. Albert, G. J. Michael, T. Michaelis, T. Watanabe, J. Frahm, O. Pudovkina, M. G. van der Hart, M. B. Hesselink, E. Fuchs, and B. Czeh. 2008. SONU20176289, a compound combining partial dopamine D(2) receptor agonism with specific serotonin reuptake inhibitor activity, affects neuroplasticity in an animal model for depression. *Eur J Pharmacol* 598 (1–3): 43–50.

Moore, C. M., J. D. Christensen, B. Lafer, M. Fava, and P. F. Renshaw. 1997. Lower levels of nucleoside triphosphate in the basal ganglia of depressed subjects: a phosphorus-31 magnetic resonance spectroscopy study. *Am J Psychiatry* 154 (1): 116–8.

Mudd, S. H., J. T. Brosnan, M. E. Brosnan, R. L. Jacobs, S. P. Stabler, R. H. Allen, D. E. Vance, and C. Wagner. 2007. Methyl balance and transmethylation fluxes in humans. *Am J Clin Nutr* 85 (1): 19–25.

Nash, S. R., B. Giros, S. F. Kingsmore, J. M. Rochelle, S. T. Suter, P. Gregor, M. F. Seldin, and M. G. Caron. 1994. Cloning, pharmacological characterization, and genomic localization of the human creatine transporter. *Receptors Channels* 2 (2): 165–74.

Nasrallah, F., M. Feki, and N. Kaabachi. 2010. Creatine and creatine deficiency syndromes: biochemical and clinical aspects. *Pediatr Neurol* 42 (3): 163–71.

Nery, F. G., J. A. Stanley, H. H. Chen, J. P. Hatch, M. A. Nicoletti, E. S. Monkul, K. Matsuo, S. C. Caetano, M. A. Peluso, P. Najt, and J. C. Soares. 2009. Normal metabolite levels in the left dorsolateral prefrontal cortex of unmedicated major depressive disorder patients: a single voxel (1)H spectroscopy study. *Psychiatry Res* 174 (3): 177–83.

Niklasson, F., and H. Agren. 1984. Brain energy metabolism and blood-brain barrier permeability in depressive patients: analyses of creatine, creatinine, urate, and albumin in CSF and blood. *Biol Psychiatry* 19 (8): 1183–206.

Ohtsuki, S., M. Tachikawa, H. Takanaga, H. Shimizu, M. Watanabe, K. Hosoya, and T. Terasaki. 2002. The blood-brain barrier creatine transporter is a major pathway for supplying creatine to the brain. *J Cereb Blood Flow Metab* 22 (11): 1327–35.

Oliveira, M. S., A. F. Furian, M. R. Fighera, N. G. Fiorenza, J. Ferreira, M. A. Rubin, C. F. Mello, and L. F. Royes. 2008. The involvement of the polyamines binding sites at the NMDA receptor in creatine-induced spatial learning enhancement. *Behav Brain Res* 187 (1): 200–4.

Ongur, D., A. P. Prescot, J. E. Jensen, B. M. Cohen, and P. F. Renshaw. 2009. Creatine abnormalities in schizophrenia and bipolar disorder. *Psychiatry Res* 172 (1): 44–8.

Pan, J. W., and K. Takahashi. 2007. Cerebral energetic effects of creatine supplementation in humans. *Am J Physiol Regul Integr Comp Physiol* 292 (4): R1745–50.

Pena-Altamira, E., C. Crochemore, M. Virgili, and A. Contestabile. 2005. Neurochemical correlates of differential neuroprotection by long-term dietary creatine supplementation. *Brain Res* 1058 (1–2): 183–8.

Perasso, L., A. Cupello, G. L. Lunardi, C. Principato, C. Gandolfo, and M. Balestrino. 2003. Kinetics of creatine in blood and brain after intraperitoneal injection in the rat. *Brain Res* 974 (1–2): 37–42.

Persky, A. M., M. Muller, H. Derendorf, M. Grant, G. A. Brazeau, and G. Hochhaus. 2003. Single- and multiple-dose pharmacokinetics of oral creatine. *J Clin Pharmacol* 43 (1): 29–37.

Pfefferbaum, A., E. Adalsteinsson, R. L. Bell, and E. V. Sullivan. 2007. Development and resolution of brain lesions caused by pyrithiamine- and dietary-induced thiamine deficiency and alcohol exposure in the alcohol-preferring rat: a longitudinal magnetic resonance imaging and spectroscopy study. *Neuropsychopharmacology* 32 (5): 1159–77.

Pittenger, C., and R. S. Duman. 2008. Stress, depression, and neuroplasticity: a convergence of mechanisms. *Neuropsychopharmacology* 33 (1): 88–109.

Prass, K., G. Royl, U. Lindauer, D. Freyer, D. Megow, U. Dirnagl, G. Stockler-Ipsiroglu, T. Wallimann, and J. Priller. 2007. Improved reperfusion and neuroprotection by creatine in a mouse model of stroke. *J Cereb Blood Flow Metab* 27 (3): 452–9.

Rae, C., A. L. Digney, S. R. McEwan, and T. C. Bates. 2003. Oral creatine monohydrate supplementation improves brain performance: a double-blind, placebo-controlled, cross-over trial. *Proc Biol Sci* 270 (1529): 2147–50.

Rawson, E. S., H. R. Lieberman, T. M. Walsh, S. M. Zuber, J. M. Harhart, and T. C. Matthews. 2008. Creatine supplementation does not improve cognitive function in young adults. *Physiol Behav* 95 (1–2): 130–4.

Renshaw, P. F., A. M. Parow, F. Hirashima, Y. Ke, C. M. Moore, B. Frederick Bde, M. Fava, J. Hennen, and B. M. Cohen. 2001. Multinuclear magnetic resonance spectroscopy studies of brain purines in major depression. *Am J Psychiatry* 158 (12): 2048–55.

Roitman, S., T. Green, Y. Osher, N. Karni, and J. Levine. 2007. Creatine monohydrate in resistant depression: a preliminary study. *Bipolar Disord* 9 (7): 754–8.

Ross, C. A., and S. J. Tabrizi. 2011. Huntington's disease: from molecular pathogenesis to clinical treatment. *Lancet Neurol* 10 (1): 83–98.

Roy, B. D., J. M. Bourgeois, D. J. Mahoney, and M. A. Tarnopolsky. 2002. Dietary supplementation with creatine monohydrate prevents corticosteroid-induced attenuation of growth in young rats. *Can J Physiol Pharmacol* 80 (10): 1008–14.

Ryu, H., H. D. Rosas, S. M. Hersch, and R. J. Ferrante. 2005. The therapeutic role of creatine in Huntington's disease. *Pharmacol Ther* 108 (2): 193–207.

Sakellaris, G., M. Kotsiou, M. Tamiolaki, G. Kalostos, E. Tsapaki, M. Spanaki, M. Spilioti, G. Charissis, and A. Evangeliou. 2006. Prevention of complications related to traumatic brain injury in children and adolescents with creatine administration: an open label randomized pilot study. *J Trauma* 61 (2): 322–9.

Sakellaris, G., G. Nasis, M. Kotsiou, M. Tamiolaki, G. Charissis, and A. Evangeliou. 2008. Prevention of traumatic headache, dizziness and fatigue with creatine administration. A pilot study. *Acta Paediatr* 97 (1): 31–4.

Salomons, G. S., S. J. van Dooren, N. M. Verhoeven, K. M. Cecil, W. S. Ball, T. J. Degrauw, and C. Jakobs. 2001. X-linked creatine-transporter gene (SLC6A8) defect: a new creatine-deficiency syndrome. *Am J Hum Genet* 68 (6): 1497–500.

Sapolsky, R. M. 2000. Glucocorticoids and hippocampal atrophy in neuropsychiatric disorders. *Arch Gen Psychiatry* 57 (10): 925–35.

Scheff, S. W., and H. S. Dhillon. 2004. Creatine-enhanced diet alters levels of lactate and free fatty acids after experimental brain injury. *Neurochem Res* 29 (2): 469–79.

Schulze, A., P. Bachert, H. Schlemmer, I. Harting, T. Polster, G. S. Salomons, N. M. Verhoeven, C. Jakobs, B. Fowler, G. F. Hoffmann, and E. Mayatepek. 2003. Lack of creatine in muscle and brain in an adult with GAMT deficiency. *Ann Neurol* 53 (2): 248–51.

Schweid, D. E., J. S. Steinberg, and H. S. Sudak. 1972. Creatine phosphokinase and psychosis. *Arch Gen Psychiatry* 26 (3): 263–5.

Segal, M., A. Avital, M. Drobot, A. Lukanin, A. Derevenski, S. Sandbank, and A. Weizman. 2007. Serum creatine kinase level in unmedicated nonpsychotic, psychotic, bipolar and schizoaffective depressed patients. *Eur Neuropsychopharmacol* 17 (3): 194–8.

Sestili, P., C. Martinelli, G. Bravi, G. Piccoli, R. Curci, M. Battistelli, E. Falcieri, D. Agostini, A. M. Gioacchini, and V. Stocchi. 2006. Creatine supplementation affords cytoprotection in oxidatively injured cultured mammalian cells via direct antioxidant activity. *Free Radic Biol Med* 40 (5): 837–49.

Shao, A., and J. N. Hathcock. 2006. Risk assessment for creatine monohydrate. *Regul Toxicol Pharmacol* 45 (3): 242–51.

Shefner, J. M., M. E. Cudkowicz, D. Schoenfeld, T. Conrad, J. Taft, M. Chilton, L. Urbinelli, M. Qureshi, H. Zhang, A. Pestronk, J. Caress, P. Donofrio, E. Sorenson, W. Bradley, C. Lomen-Hoerth, E. Pioro, K. Rezania, M. Ross, R. Pascuzzi, T. Heiman-Patterson, R. Tandan, H. Mitsumoto, J. Rothstein, T. Smith-Palmer, D. MacDonald, D. Burke, and Neals Consortium. 2004. A clinical trial of creatine in ALS. *Neurology* 63 (9): 1656–61.

Signoretti, S., A. Marmarou, G. A. Aygok, P. P. Fatouros, G. Portella, and R. M. Bullock. 2008. Assessment of mitochondrial impairment in traumatic brain injury using high-resolution proton magnetic resonance spectroscopy. *J Neurosurg* 108 (1): 42–52.

Soni, S. D. 1976. Serum creatine phosphokinase in acute psychosis. *Br J Psychiatry* 128: 181–3.

Stockler, S., F. Hanefeld, and J. Frahm. 1996. Creatine replacement therapy in guanidinoacetate methyltransferase deficiency, a novel inborn error of metabolism. *Lancet* 348 (9030): 789–90.

Streijger, F., C. R. Jost, F. Oerlemans, B. A. Ellenbroek, A. R. Cools, B. Wieringa, and C. E. Van der Zee. 2004. Mice lacking the UbCKmit isoform of creatine kinase reveal slower spatial learning acquisition, diminished exploration and habituation, and reduced acoustic startle reflex responses. *Mol Cell Biochem* 256–257 (1–2): 305–18.

Streijger, F., F. Oerlemans, B. A. Ellenbroek, C. R. Jost, B. Wieringa, and C. E. Van der Zee. 2005. Structural and behavioural consequences of double deficiency for creatine kinases BCK and UbCKmit. *Behav Brain Res* 157 (2): 219–34.

Sullivan, P. G., J. D. Geiger, M. P. Mattson, and S. W. Scheff. 2000. Dietary supplement creatine protects against traumatic brain injury. *Ann Neurol* 48 (5): 723–9.

Tabrizi, S. J., A. M. Blamire, D. N. Manners, B. Rajagopalan, P. Styles, A. H. Schapira, and T. T. Warner. 2003. Creatine therapy for Huntington's disease: clinical and MRS findings in a 1-year pilot study. *Neurology* 61 (1): 141–2.

Tabrizi, S. J., A. M. Blamire, D. N. Manners, B. Rajagopalan, P. Styles, A. H. Schapira, and T. T. Warner. 2005. High-dose creatine therapy for Huntington disease: a 2-year clinical and MRS study. *Neurology* 64 (9): 1655–6.

Takubo, H., S. Shimoda-Matsubayashi, and Y. Mizuno. 2003. Serum creatine kinase is elevated in patients with Parkinson's disease: a case controlled study. *Parkinsonism Relat Disord* 9 (suppl. 1): S43–6.

Theberge, J., J. E. Jensen, and L. M. Rowland. 2007. Regarding "Increased prefrontal and hippocampal glutamate concentration in schizophrenia: evidence from a magnetic resonance spectroscopy study." *Biol Psychiatry* 61 (10): 1218–9; author reply 1219–20.

Thompson, R. J., J. G. Graham, I. N. McQueen, P. A. Kynoch, and K. W. Brown. 1980. Radioimmunoassay of brain-type creatine kinase-BB isoenzyme in human tissues and in serum of patients with neurological disorders. *J Neurol Sci* 47 (2): 241–54.

Tuason, V. B., M. A. Oleshansky, and J. M. Jaranson. 1974. Creatine phosphokinase in functional psychosis. *Compr Psychiatry* 15 (5): 435–8.

Valastro, B., A. Dekundy, W. Danysz, and G. Quack. 2009. Oral creatine supplementation attenuates L-DOPA-induced dyskinesia in 6-hydroxydopamine-lesioned rats. *Behav Brain Res* 197 (1): 90–6.

Van der Hart, M. G., B. Czeh, G. de Biurrun, T. Michaelis, T. Watanabe, O. Natt, J. Frahm, and E. Fuchs. 2002. Substance P receptor antagonist and clomipramine prevent stress-induced alterations in cerebral metabolites, cytogenesis in the dentate gyrus and hippocampal volume. *Mol Psychiatry* 7 (9): 933–41.

Verbessem, P., J. Lemiere, B. O. Eijnde, S. Swinnen, L. Vanhees, M. Van Leemputte, P. Hespel, and R. Dom. 2003. Creatine supplementation in Huntington's disease: a placebo-controlled pilot trial. *Neurology* 61 (7): 925–30.

Vielhaber, S., J. Kaufmann, M. Kanowski, M. Sailer, H. Feistner, C. Tempelmann, C. E. Elger, H. J. Heinze, and W. S. Kunz. 2001. Effect of creatine supplementation on metabolite levels in ALS motor cortices. *Exp Neurol* 172 (2): 377–82.

Volek, J. S., and E. S. Rawson. 2004. Scientific basis and practical aspects of creatine supplementation for athletes. *Nutrition* 20 (7–8): 609–14.

Volz, H. P., R. Rzanny, S. Riehemann, S. May, H. Hegewald, B. Preussler, G. Hubner, W. A. Kaiser, and H. Sauer. 1998. 31P magnetic resonance spectroscopy in the frontal lobe of major depressed patients. *Eur Arch Psychiatry Clin Neurosci* 248 (6): 289–95.

Wallimann, T., M. Wyss, D. Brdiczka, K. Nicolay, and H. M. Eppenberger. 1992. Intracellular compartmentation, structure and function of creatine kinase isoenzymes in tissues with high and fluctuating energy demands: the 'phosphocreatine circuit' for cellular energy homeostasis. *Biochem J* 281 (Pt 1): 21–40.

Watanabe, A., N. Kato, and T. Kato. 2002. Effects of creatine on mental fatigue and cerebral hemoglobin oxygenation. *Neurosci Res* 42 (4): 279–85.

Wendt, S., A. Dedeoglu, O. Speer, T. Wallimann, M. F. Beal, and O. A. Andreassen. 2002. Reduced creatine kinase activity in transgenic amyotrophic lateral sclerosis mice. *Free Radic Biol Med* 32 (9): 920–6.

Wijesekera, L. C., and P. N. Leigh. 2009. Amyotrophic lateral sclerosis. *Orphanet J Rare Dis* 4: 3.

Wilken, B., J. M. Ramirez, I. Probst, D. W. Richter, and F. Hanefeld. 1998. Creatine protects the central respiratory network of mammals under anoxic conditions. *Pediatr Res* 43 (1): 8–14.

Wyss, M., and R. Kaddurah-Daouk. 2000. Creatine and creatinine metabolism. *Physiol Rev* 80 (3): 1107–213.

Wyss, M., and T. Wallimann. 1994. Creatine metabolism and the consequences of creatine depletion in muscle. *Mol Cell Biochem* 133-134: 51–66.

12 Theanine, Mood, and Behavior

Jessica E. Smith and Peter J. Rogers

CONTENTS

Introduction 237
Background 238
Theanine Intake 238
Theanine Pharmacology 240
The Pharmacokinetics of Theanine 240
The Neuropharmacological Actions of Theanine 241
The Pharmacokinetics and Mechanisms of Action of Caffeine 245
Theanine and Relaxation 245
Theanine Alertness and Cognition 247
Electroencephalographic (EEG) Alpha Effects of Theanine 249
Theanine (Tea) Consumption and Risk of Cognitive Impairment and Decline 254
Blood Pressure Lowering Effects of Theanine 254
Neuro- and Memory-Protective Effects of Theanine 256
Theanine and Weight Control 258
Miscellany 259
A Short Note on Coffee, Cola, and Caffeine 260
General Conclusions and Implications for Public Health 260
Acknowledgments 262
References 262

INTRODUCTION

It is usual to begin an article on this subject by noting the worldwide popularity of tea, the only significant natural source of theanine in the human diet. Increasingly, though, theanine is being used as a dietary supplement as well as being consumed through intake of foods and drinks other than tea (as discussed in this chapter). Nevertheless, tea, after water, is the most widely consumed beverage in the world, with black tea representing the majority of tea consumption, and green tea being consumed mainly in Asia (Graham 1992), although it is becoming more popular in Western countries. Tea is reported to have a variety of beneficial effects on health and wellbeing (e.g., anxiety and stress relief/relaxation, cognitive health and neuroprotection, cardiovascular health, and cancer prevention), and theanine has been implicated in these effects. Tea and coffee are the major dietary sources of caffeine worldwide (Fredholm et al. 1999). Given the unique co-occurrence of caffeine and theanine in tea, and since theanine is most frequently consumed in this form, a review of this topic would not be complete without a discussion of the effects of caffeine and its interactions with theanine. The globally high intake of tea means small effects of theanine and caffeine on the health and well-being of individual consumers could sum to large effects at a population level. In this chapter, after providing some background information on theanine intake and its pharmacology, we discuss three interrelated aspects of the effects of theanine by drawing on differences between the effects of tea and other caffeine-containing drinks that unlike tea are devoid of theanine; first we discuss theanine and relaxation, then the acute effects of theanine on alertness and cognition, and third the possible longer term cognitive effects of tea consumption with emphasis on the role of theanine. Some other miscellaneous research findings on theanine are then discussed briefly.

BACKGROUND

Theanine Intake

Theanine, known also as gamma(γ)-glutamylethylamide or 5-*N*-ethylglutamine, is a nonproteinic amino acid that is found naturally in tea. The only other known natural source of theanine is the inedible mushroom *Xerocomus badius* (Casimir et al. 1960). In Asian, U.S., and European markets, theanine is now readily available in a pure form as a dietary supplement, and in a range of theanine-supplemented products including fruit juices, bottled water, sport drinks, sweets, and chocolate. Despite the recent proliferation of theanine supplements and theanine-supplemented food and drink products, tea, at present, remains the predominant source of theanine, particularly in the United Kingdom, where it is a very popular drink. Within tea leaves theanine is biosynthesized from glutamic acid (glutamate) and ethylamine (Ekborg et al. 1997), and it constitutes >50 percent of the amino acid content of tea and about 1.0–2.5 percent of its dry weight (Ekborg-Ott et al. 1997). The L-form of theanine (N-ethyl-L-glutamine) makes up the majority (around 98 percent) of the total theanine content of tea leaves (Ekborg-Ott et al. 1997). It is in this form that theanine has been administered in most of the studies that will be discussed in this chapter, and thus here after we will simply refer to it as theanine. But note that some commercially available theanine products contain a mixture of D- and L-theanine (Desai and Armstrong 2004).

Since theanine is a readily soluble amino acid, it is likely that most of the theanine in tea leaves dissolves in water when the tea beverage is prepared. From the above data, it is possible to estimate that a single cup of tea contains 20–50 mg of theanine (Rogers et al. 2008), and it is often assumed that the theanine content of black tea is less than that of other forms such as green tea. However, in a recent study that assessed the theanine content of various popular tea products in the United Kingdom, we found that the theanine content of teas was somewhat lower than would be predicted and, consistent with earlier studies (i.e., Eckborg-Ott et al. 1997; Neumann and Montag 1982), black teas tended to have higher levels of theanine compared to green teas (Keenan et al. 2010). Specifically, the average values per

200 ml cup of tea included the following: black tea, 24.2 mg; black specialty tea (e.g., Twinings), 10.9 mg; green tea, 7.9 mg; and white tea, 11.5 mg. We also found that the theanine content of the tea beverage was not affected by the temperature or pH level (3–9 pH) of the water, nor by the addition of small amounts of milk or sugar (Keenan et al. 2010).

In its pure form theanine has been found to be stable in solution under neutral (7 pH) and acidic (3–5 pH) conditions, and when stored in the dark at room temperature (25°C) it was reported to have a shelf-life of at least eighteen months (GRAS Monograph for Suntheanine®). With regard to the safety of theanine, two toxicological (acute and subacute) studies and a tumorigenicity study conducted in animals reported no adverse effects of (high doses) theanine (Borzelleca et al. 2006; Fujii and Inai 2008; Schmidt et al. 2005). As alluded to above, a variety of companies have begun to market theanine as a nutraceutical and/or food and drink additive. It was first approved for universal use (except in infant foods) in Japan by the Japanese Ministry of Health and Welfare in 1964, with no limits being placed on dietary intake of theanine by the Japanese Food Additive Association. It is now possible to buy over fifty different food and drink products that are supplemented with theanine in Japan. Suntheanine® (a pure L-theanine product) was introduced to the United States market as a dietary supplement ingredient in 1999 by Taiyo International. More recently, Taiyo International (in 2007) and Ethical Naturals (in 2010) have received confirmation from the U.S. Food and Drug Administration (FDA) regarding their self-affirmed GRAS (generally recognized as safe) status of the use of their branded pure L-theanine ingredients (Suntheanine® and AlphaWave™) in food and drink categories, levels up to 250 mg per serving were being granted, with no limitations placed on the maximum number of servings. Suntheanine® is also available in chewable tablet form for pet dogs and cats (Anxitane®). In Europe, food and food supplements are regulated by the European Commission (EC) and individual member states on the advice of the European Food Safety Authority (EFSA). In Germany, for example, theanine is classed as a novel food, and in 2003, the German Federal Institute for Risk Assessment objected to theanine added in isolated form to drinks. In the United Kingdom, theanine is permitted for use as a food supplement and comes under UK Food Laws. Its legal "status" is covered generally by the EC Food Supplements Directive 2002/46/EC. As yet, theanine is only available in the United Kingdom as a dietary supplement.

As tea provides the main vehicle for theanine intake, the amino acid is most often consumed in combination with caffeine. Data on average caffeine content of foods and drinks are readily available; the caffeine content of a single cup of black or green tea is about 40 mg per 200 ml cup (Heatherley et al. 2006a), although this varies considerably with brewing method, brand, and so on. These dietary data have been used to gain reasonable estimates of the amounts of caffeine consumed by individuals on a daily or weekly basis at the regional and national population level (Heatherley et al. 2006ab; Fredholm et al. 1999). In contrast, this has seldom been done for theanine because until our recent study (Keenan et al. 2010) there has been very little information on theanine content of teas in a form that can be easily applied to consumption. So far reports on dietary theanine intake come from the GRAS Monograph for Suntheanine® prepared for Taiyo International, which calculated that in the

United States an average of 152–382 mg of theanine is consumed daily through tea, with the top 10 percent of tea drinkers estimated to consume between 330 and 825 mg of theanine per day. However, these estimates seem unrealistically high, especially for the United States. Based on the highest average value of theanine content of tea (approximately 25 mg per 200 ml cup), our data would suggest that in the United Kingdom, where tea drinking is very popular, average daily theanine intake through tea is no more than 75 mg. Among tea drinkers the average per day is no more than 95 mg and, for the top 10 percent of tea drinkers, it is about 130 mg (unpublished data from the Dietary Caffeine and Heath Study; Heatherley et al. 2006b). Admittedly, these estimates do not include theanine consumed through intake of dietary supplements and theanine-supplemented products, although as previously mentioned, at present, this probably contributes relatively little to overall theanine intake, at least in the United Kingdom. It is also worth treating all these estimates of intake with some degree of caution at the moment given the lack of data available on theanine content of teas. Also, dietary data (including some of ours) tend not to adequately distinguish between tea brands and varieties, the volume of tea consumed and brewing method (e.g., teabag first, stirring, dunking, and squeezing) and time, which all can impact on the amount of theanine present (Keenan et al. 2010).

Theanine Pharmacology

Little is known about the pharmacokinetics of theanine and its neuropharmacological actions after consumption of tea or theanine-containing products. Available data are limited in the main to animal studies, in which theanine was administered in a pure form in a range of doses either orally/intragastrically (o.p./i.g.; 2–8,000 mg/kg) or intraperitoneally (i.p.; 1–2,000 mg/kg). It is worth pointing out that the majority of these studies administered theanine in amounts well above those found in several cups of tea, although some beneficial physiological and behavioral effects of theanine have been reported at doses relevant to tea consumption (see below). Theanine has also been administered directly into the brain (e.g., Yamada et al. 2009; Yokogoshi et al. 1998a) and, in vitro effects of theanine have been examined in both animal (e.g., Kakuda et al. 2002, 2008; Nagasawa et al. 2004) and human brain cells (e.g., Cho et al. 2008; Di et al. 2010).

The Pharmacokinetics of Theanine

Findings of studies in animals suggest that after oral administration theanine moves into the bloodstream via a common sodium-coupled transporter in the intestinal brush-border membrane (Kitaoka et al. 1996), from where it is distributed, fairly rapidly, to tissues, including the liver, kidney, and brain (Terashima et al. 1999; Unno et al. 1999). Concentration of theanine in the bloodstream has been found to peak within 30–60 minutes after oral/intragastric administration (Terishma et al. 1999; Unno et al. 1999). Transportation of theanine through the blood–brain barrier was reported to occur dose-dependently via the leucine-preferring transport system (Yokogoshi et al. 1998a), with a significant concentration of theanine (administered intragastrically) detected in the brain within one hour (or thirty minutes administered interperitoneally (Kimura and Murata, 1971), reaching its maximum concentration within five hours, and not diminishing substantially until at least eight hours later (Terishma et al. 1999). The metabolic pathway of theanine is unclear, although it would appear that it is metabolized into ethylamine and glutamate predominantly in the kidney by the enzyme phosphate-independent glutaminase (Terashima et al. 1999; Tsuge et al. 2003; Unno et al. 1999). The resarch suggests theanine and its metabolites are gradually eliminated from the body, mainly in urine, over the course of twenty-four hours (Terashima et al. 1999).

The Neuropharmacological Actions of Theanine

Thus far, animal studies have indicated a variety of effects of theanine on concentrations, receptors and/or functions of several neurotransmitter systems in the brain. These include actions on brain monoamines, the inhibitory neurotransmitters systems γ-aminobutyric acid (GABA) and glycine, and the major excitatory neurotransmitter system glutamate, which are discussed in the sections that follow.

Effects on Monoamines

The monoamine neurotransmitters, which include dopamine, noradrenaline, and serotonin, influence the function of widespread regions of the brain, increasing or decreasing activity of particular brain functions and modulating neuro-endocrine and vital organ function and a broad range of behaviors (Goldstein 1998), some of which are specified in the text that follows. Studies report effects of theanine on all three monoamine neurotransmitters. Its reported effects on noradrenaline include the modulation of concentration and activity of this neurotransmitter in the brain. Noradrenaline facilitates behaviorally appropriate arousal and attention as well as learning and memory in response to salient aspects of the environment (Ressler and Nemeroff 1999). In response to threat, noradrenaline triggers hyperactivity that contributes to anxiety and panic (Millan 2003). It also raises blood pressure by increasing vascular resistance (Barcroft and Konzett 1949). An initial study found that theanine reduced the stimulatory action of noradrenaline on cyclic adenosine monophosphate (cAMP) formation, which occurs due to the breakdown of the main energy source for living cells, adenosine triphosphate (ATP). This effect of theanine was unaffected by the presence of caffeine (Kimura and Murata 1980). In a later study, theanine was shown to reduce the concentration of noradrenaline in the brain, although it was found that this effect was reversed by caffeine (Kimura and Murata 1986). In a more recent study, theanine was found to induce a small but statistically significant rise in the concentration of noradrenaline in the striatum, but concentrations in other brain areas were found to be largely unaffected by theanine (Yokogoshi et al. 1998a).

Studies, including some of those discussed above, have examined theanine's effects on concentrations of serotonin in the brain, and the findings suggest these effects may also vary according to whether whole-brain concentration or concentrations on specific brain areas is being studied. Again these studies indicate that the presence of caffeine can influence the effects of theanine. Most attention on serotonin has focused on its role in emotion regulation, anxiety, and depression, although the precise nature of the mechanisms involved in these functions is complicated (Millan 2003; Hariri and Holmes 2006). Its role in arousal and attention involves the control of overall behavioral response output to relevant environmental stimuli (Robbins 2000). Consistent with this there is growing evidence in support of a role of serotonin in cognitive processes, including learning and memory (Meneses 1999). In addition, effects of serotonin in the brain have been implicated in regulating blood pressure, although again its role in controlling cardiovascular function is complex (Kuhn et al. 1980). Regarding theanine's effects, one study found that theanine did not affect concentrations of brain serotonin or 5-hydroxyindoleacetic acid (5-HIAA), the main metabolite of serotonin, but it did suppress the increase in these substances induced by caffeine (Kimura and Murata 1986). Other later studies reported that following theanine administration the concentrations of serotonin and 5-HIAA in the brain decreased (Yokogoshi et al. 1995, 1998b), but brain concentration of tryptophan, the precursor to serotonin, increased (Yokogoshi et al. 1998b). Another study, though, found that theanine increased serotonin concentrations in specific areas of the brain including the hippocampus, striatum, and hypothalamus, while concentrations of 5-HIAA in these and other brain regions were unaffected by theanine (Yokogoshi et al. 1999a). This last study also demonstrated that theanine increased the concentration of dopamine in the striatum alone (Yokogoshi et al. 1998a). Furthermore, when theanine was administered directly into this brain region in this and two other studies, striatal dopamine release into extracellular fluid was found to increase by up to 300 percent (Yokogoshi et al. 1998a; Yamada et al. 2005, 2009). Dopamine in the striatum has primarily been associated with the control of motor function (Salamone 1992).

Effects on the Glycine System

As well as increasing dopamine release in the striatum, theanine markedly increased striatal glycine release in two microdialysis studies (Yamada et al. 2005, 2009). Glycine is known as an inhibitory neurotransmitter and is involved in processing sensory and motor information that control movement, vision, and audition (Lopez-Corcuera et al. 2001). In animals, glycine administered directly into the striatum has been found to stimulate striatal dopamine release and extracellular concentrations of dopamine metabolites (Yadid et al. 1993). In one of the two studies, it was reported that the enhancing effect of theanine on dopamine release was found to be blocked by a glycine receptor antagonist (Yamada et al. 2009).

Effects on the GABA System

An early study reported that theanine increased the concentration of GABA in the brain (Kimura and Murata 1971). GABA, the major inhibitory neurotransmitter in the central nervous system (Millan 2003), or $GABA_A$ergic drugs, can produce anxiolytic or sedative effects via actions on $GABA_A$ receptors (Vinkers et al. 2009). Activation of $GABA_A$ receptors has also been associated with neuroprotection, including the prevention of cell damage or death during brain ischemia (e.g., Schwartz-Bloom and Sah 2001) and other neurological insults characterized by hyperactivation of glutamate neurotransmission such as toxicity of amyloid beta (Aβ) (Louzada et al. 2004). Consistent with this, a more recent study found that a protective effect of theanine on ischemic brain damage (see the "Neuro- and Memory-Protective Effects of Theanine" section of this chapter) was found to be prevented when it was administered in combination with a $GABA_A$ receptor antagonist (Egashira et al. 2007), implicating $GABA_A$ receptors in this beneficial effect of theanine.

Effects on the Glutamate System

A substantial amount of attention has focused on effects of theanine on glutamate, and this has been mostly in relation to theanine's neuroprotective effects. Glutamate is the principal excitatory neurotransmitter within the nervous system and can be found in all parts of the brain (Meldrum, 2000). Glutamate is largely because theanine is structurally very similar to glutamate, and because of the evidence demonstrating neuroprotective effects of theanine (referred to in the preceding section, later in the current section, and in more detail in the "Neuro- and Memory-protective Effects of Theanine" section). It is important for maintaining physiological balance (Danysz et al. 1995). It is also closely involved in learning and memory (McEntee and Crook 1993). However, under certain abnormal conditions, overactivation of glutamate neurotransmission is neurotoxic, and it follows that brain structures thought to underpin learning and memory, including the cortex and hippocampus, are particularly vulnerable to the effects of this (Danysz et al. 1995; McEntee and Crook 1993). Its excitotoxic effects are involved in cell damage caused by acute insults like ischemia, and more mild but chronic dysfunction of glutamate systems has been implicated in several neurodegenerative disorders, including Alzheimer's disease (AD) (Danysz et al. 1995), which is the most frequent cause of cognitive impairment in old age (Hynd et al. 2004). One factor that is thought might contribute to the involvement of glutamate in AD is the 39–43 amino acid peptide Aβ, which is the main constituent of senile plaques in AD patients (Danysz et al. 1995; Hynd et al. 2004). It has been found to directly induce neuronal death and, although the mechanism for this is unclear, it is thought to involve a neurotoxic cycle that includes the overstimulation of glutamate neurotransmission (Danysz et al. 1995; Hynd et al. 2004; Louzada et al. 2004). Hyperactivity of the glutamate system also contributes to psychiatric symptoms such as anxiety and panic (Bergink et al. 2004). Actions of glutamate are mediated via ionotropic glutamate receptors (iGluRs), namely, N-methyl-D-aspartate (NMDA), alpha(α)-amino-3-hydroxy-5-methylisoxazole-4-propionic acid (AMPA), and kainite, as well as via several types of metabotropic glutamate receptors (mGluRs group I, II, and III) (Meldrum 2000). Excessive activation of NMDA and/or AMPA receptors is implicated in the excitotoxic effects of glutamate (Danysz et al. 1995). Glutamine is a precursor of glutamate and its entry into neurons via glutamine transporters is crucial for restoring the neurotransmitter pool of glutamate in neurons ready for subsequent release upon stimulus (Dolinska et al. 2004). As theanine is similar in structure to glutamate and has been found to have neuroprotective effects, which include the prevention of brain damage caused by ischemic insult induced via occlusion of cerebral artery, acute exposure to glutamate, and direct injection of Aβ (see the "Neuro- and Memory-Protective Effects of Theanine" section), there has been some interest in the possibility that it acts on glutamate receptors as an antagonist blocking glutamate neurotransmission (e.g., Kakuda 2002). Indications of this effect of theanine were first reported in studies in crayfish (Maruyama and Takeda 1994; Shinozaki and Ishida 1978). A later study found that theanine bound to all ionotropic glutamate receptors and that the binding activity of theanine for the AMPA and kainite receptors was tenfold higher than that for the NMDA receptor (Kakuda et al. 2002). It is worth noting, though, that the binding capacity of theanine for these three glutamate receptor subtypes was 80- to 30,000-fold less than that of glutamate (Kakuda et al. 2002). These findings led the authors of the study to suggest that antagonism of the actions of glutamate by blockade of these receptor binding sites was possible but may only represent a very mild effect of theanine on the brain (Kakuda 2002). They do suggest that the binding capacity of theanine for the AMPA receptor may be useful in neuroprotection (Kakuda 2002; Kakuda et al. 2002). In view of the fact that theanine is more similar in structure to glutamine than to glutamate, Kakuda and colleagues went on to examine effects of theanine on glutamine transport as an alternative mechanism for theanine's neuroprotective effects (Kakuda et al. 2008). The findings of this study showed that sustained exposure (three days) to theanine reduced the concentration of extracellular glutamate released by neurons possibly via inhibiting intracellular transport of glutamine (Kakuda et al. 2008).

Some of the studies that have found neuroprotective effects of theanine (see "Neuro- and Memory-Protective Effects of Theanine") reported that theanine reduced elevated brain concentrations of amyloid beta (Aβ) (Di et al. 2010; Kim et al. 2009), and inhibited glutamate- and/or Aβ-induced neurotoxic effects in brain cells (Di et al. 2010; Kim et al. 2009; Nagasawa et al. 2004). The results of a study demonstrating that theanine and a non-competitive NMDA receptor antagonist had a similar protective effect on cell damage, indicated support for the notion that such effects of theanine might involve inhibition of the NMDA subtype of the glutamate receptors (Di et al. 2010). However, in another of these studies, the inhibition of glutamate-induced oxidative stress and neuronal cell death by theanine was found to be similar to that of group 1 mGluR agonists and abolished by group 1 mGluR antagonists (Nagasawa et al. 2004). Providing stronger evidence for another possible mechanism of action for theanine's neuroprotective effects, the findings of this last study indicate that theanine might act on some metabotropic glutamate receptor subtypes as an agonist leading to an increase in antioxidant activity and in turn a reduction in oxidative stress-induced cell death caused by glutamate (Nagasawa et al. 2004).

Furthermore, the findings of two other studies described earlier implicate the activation of ionotropic glutamate (NMDA and AMPA) receptors in the effects of theanine on dopamine and glycine release in the striatum (Yokogoshi et al. 1998a; Yamada et al. 2009), with the latter demonstrating that such actions may occur without causing changes in glutamate neurotransmission (Yamada et al. 2009). Instead, additional tests in this study showing inhibitory effects of a glycine and an AMPA receptor antagonist on the effects of theanine on striatal dopamine and glycine release suggested that by activating AMPA receptors theanine stimulated striatal glycine release which, via actions on glycine receptors of dopamine neurons in the striatum, increased striatal dopamine release (Yamada et al. 2009). Also worth mentioning is that studies on the chemotherapeutic effects of theanine (see the "Miscellany" section of this chapter) provide some support for an inhibitory effect of theanine on glutamate transporters, with reduced levels of glutamate uptake observed in tumor cells (e.g., see Sugiyama and Sadzuka 2003). However, Yamada et al. (2005) reported that the effects of theanine on neurotransmitter release in the striatum do not appear to be mediated via glutamate transporter blockade (Yamada et al. 2005).

Overall no clear picture of the neuropharmacological profile of theanine has emerged. The studies reviewed in the "Neuropharmacological Actions of Theanine" section have reported that a whole host of pharmacological effects of theanine on the brain. These findings do support the possibility that theanine could have significant physiological and behavioral effects. The best evidence for this would appear to be in relation to neuroprotection.

The Pharmacokinetics and Mechanisms of Action of Caffeine

In contrast to the little that is known about the pharmacology of theanine, the pharmacokinetics and mechanism of action of caffeine are better understood. After drinking tea, coffee, or other caffeine-containing drinks, caffeine is distributed rapidly throughout the body, reaching its highest concentration in the bloodstream and in the brain within 30–40 minutes (Rogers 2007). It passes easily from blood to brain via simple diffusion and saturable carrier-mediated transport (McCall et al. 1982). Together with its metabolites, including the psychoactive metabolite paraxanthine, it is then gradually eliminated from the body, mainly in the urine (Rogers 2007). In the amounts consumed in tea, coffee, cola, and so on, the physiological and behavioral effects of caffeine occur primarily via antagonism of the neuromodulator adenosine at adenosine A_1 and A_{2A} receptors which are distributed throughout the body, including the brain (Fredholm et al. 1999, 2005). There may also be important secondary effects of caffeine on neurotransmitter systems, because adenosine acting at adenosine A_1 and A_{2A} receptors modulates the release of virtually all neurotransmitters and the postsynaptic responsiveness and actions of other receptors systems (Cuhna 2001). By blocking the actions of endogenous adenosine caffeine has significant behavioral, cardiovascular, cerebrovascular, renal, gastrointestinal, and metabolic effects. Adenosine is also closely involved in the regulation of sleep and wakefulness (Basheer et al. 2004), and it plays an important role in preventing cell death during brain ischemia (Fredholm et al. 2005). Common targets for caffeine and theanine may include noradrenaline, dopamine, serotonin, glutamate, and GABA, as indicated by the findings of some of the studies discussed in this chapter. In response to regular caffeine consumption, however, there are changes in adenosine signaling that serve to counter the effects of caffeine and, at least in part, maintain normal functioning. Studies examining development of tolerance to the effects of theanine are not currently available.

THEANINE AND RELAXATION

Tea, the main source of theanine, has the reputation of being a relaxing drink, especially in comparison to coffee, which, like tea, contains caffeine, but, unlike tea, does not contain theanine. For example, in our cross-sectional survey of caffeine consumption habits (Bristol Dietary Caffeine and Health Study; Heatherley et al. 2006b), respondents who drank both tea and coffee endorsed the statement "relaxes me" more strongly for tea than for coffee (unpublished results). In a study that tested the relaxing properties of tea scientifically, it was found that tea consumed for six weeks reduced cortisol (a neuroendocrine stress response) and increased relaxation ratings associated with recovery from a stressful task (Steptoe et al. 2007).

Against this, caffeine can increase anxiety and jitteriness, perhaps especially in susceptible individuals (Goldstein et al. 1969; Rogers et al. 2010) and notably when administered in the form of coffee (Botella and Para, 2006). The stress-relieving effects of tea consumption reported by Steptoe et al. (2007) were based on comparisons with a placebo drink matched for caffeine content. What then can explain the reported relaxing effects of tea? Could theanine be the answer? Indeed, the ability of theanine to induce relaxation has received much public attention. The Google search engine yields a vast array of websites promoting theanine as a natural remedy for reducing anxiety and stress that can take the "edge" off caffeine. All theanine supplements and theanine-supplemented food and drink products now available both online and in drug and health food stores are marketed for this purpose (e.g., see Rao et al. 2007). While the extent of use of these so-called nutraceuticals is thought to be relatively minor in comparison to tea consumption (Shimbo et al. 2005), relatively few scientific studies have tested their claims empirically.

The reported neuropharmacological effects of theanine would appear to suggest that it might downregulate excitatory brain activity and function to some extent. Actually, support for a relaxing effect of theanine seems to have originated from electroencephalographic (EEG) studies investigating effects in humans of theanine on electrical brain activity. Specifically, increased alpha activity (8–13 Hz) observed across occipital and parietal brain regions after oral administration of theanine (200 mg) was interpreted as promoting a relaxed state without drowsiness (Ito et al. 1998, summarized in English in Juneja et al. 1999). More recently, this effect of theanine was observed when theanine was administered at a lower dose (50 mg) (Nobre et al. 2008), although it is not entirely clear whether the effect is limited to individuals classified as highly anxious. Alone though, this measure provides only an indirect and relatively crude indicator of relaxation (see the "Theanine Alertness and Cognition" section of this chapter for a more detailed discussion on brain alpha). As regards the mood and behavioral effects of theanine, so far four studies have been conducted using self-report measures of anxiety and stress in humans, and the findings of these studies are varied. One study, for example, found that theanine (200 mg) affected one of several self-report measures of anxiety (participants reported being more tranquil using the tranquil-troubled scale) under a relaxed condition; but not when anxiety was raised by the threat of an electric shock (Lu et al. 2004). Another study found self ratings of stress and anxiety made immediately after performing a twenty-minute arithmetic task, as well as two physiological measures of stress (i.e., heart rate and salivary immunoglobulin A), were found to be lowered by theanine (200 mg; Kimura et al. 2007). Despite the unique co-occurrence of caffeine and theanine in tea, and evidence that these compounds may have opposite effects on anxiety, our study and one other are the only studies to date that have investigated interactions between caffeine and theanine specifically in relation to anxiety. Adding to the mixed results, we found no main effects of theanine (200 mg) on self-report measures of anxiety and it had no effect on feelings of jitteriness induced by 250 mg of caffeine (Rogers et al. 2008). The other study that administered theanine (250 mg) alone and in combination with caffeine (150 mg) also found no effects of theanine on measures of mood (calm, relaxed, and jittery) (Haskell et al. 2008). Also relevant, there are claims that theanine alleviates premenstrual mood and physical symptoms. But again these claims do not appear to be adequately supported by science-based evidence; we can find only two reports (abstracts) on theanine and premenstrual syndrome (PMS) (Timmcke et al. 2008; Ueda et al. 2001, summarized in English in Rao et al. 2007), and the information provided by these reports makes it difficult to judge the effectiveness of theanine in mitigating effects of PMS.

Consistent with the introduction of theanine products (e.g., Anxitane®) to the pet market, there are several reports (abstracts and a journal article) of theanine being effective in reducing distress and anxiety-related emotional disorders in dogs (noise phobia and fear of humans; Araujo et al. 2010; Michelazzi et al. 2010), cats (physical and behavioral signs of anxiety; Dramard et al. 2007), and chicks (stress-induced analgesia and activity; Yang et al. 2007). In relation to a specific relaxing effect of theanine in humans, though, the relatively few studies that have put product claims regarding its efficacy as a "Unique Anxiety Reducer" to the test do not appear to provide strong evidence in support of this. It is worth pointing out that the recently introduced EC regulation 1924/2006 on nutrition and food claims now requires that a dossier of compelling scientific evidence be submitted in support of any commercially communicated claim for assessment by EFSA. From the literature reviewed in the current section it would seem unlikely that the published evidence to date on theanine and relaxation would be sufficient to warrant approval of the use of the claim that theanine 'promotes relaxation from anxiety'.

It also worth mentioning at this point that caffeine nonconsumers show an increase in anxiety and jitteriness in response to caffeine, while frequent consumers do not. Furthermore, these effects of caffeine in nonconsumers are observed at doses present in single cups of coffee (100–150 mg) but may not occur at doses found in single cups of tea. One possibility is that nonconsumers are predisposed to this effect, and this is why they largely avoid caffeine. Consistent with this, recent studies have discovered an association between caffeine-induced anxiety and variation (a single-nucleotide polymorphism, or SNP) in the gene coding for the adenosine A_{2A} receptor (*ADORA2A*). The significance of this vulnerability to caffeine-induced anxiety for caffeine intake in the wider population is, however, unclear. The main reason to doubt its relevance as a deterrent of dietary caffeine consumption is that substantial tolerance appears to develop this effect of caffeine, even at rather modest levels of daily caffeine intake and even in susceptible individuals. This is discussed in detail elsewhere (see Rogers and Smith in press). As for a role of theanine in understanding differences between effects of tea and coffee on mood, then, the relevance of an anxiety effect of theanine is diminished by the relatively low amounts of caffeine, develop substantial tolerance, found in single cups of tea, and the finding that regular tea and coffee drinkers become tolerant to the anxiogenic (anxiety inducing) effects of caffeine (Rogers et al. 2010).

THEANINE ALERTNESS AND COGNITION

Recently there has been growing interest in the ability of theanine to modify the alerting and cognitive performance-enhancing effects of caffeine (e.g., see Bryan

2008). There are many published studies measuring these psychostimulant effects of caffeine, and a clear result is that caffeine compared with placebo increases alertness, reduces fatigue and sleepiness, and improves performance, especially on vigilance and psychomotor tasks (see review by Rogers and Smith in press). Fewer studies have examined effects of theanine alone and in combination with caffeine on alertness and performance measures; but most have reported a positive effect. That is, alone theanine is found have little, or no, or even a detrimental effect on alertness and performance, but when it is combined with caffeine, the alerting/performance enhancing effect of the combination is reported to be greater than the sum total of the effect of each compound alone. In an extensive demonstration of this, Haskell et al. (2008) found that theanine (250 mg) impaired performance on a serial sevens subtraction task, considered to be largely an attention task, and increased headache ratings. As expected there were also several main effects of caffeine (150 mg). On further investigation of these effects (using paired *t*-tests to compare the "active" treatments to placebo), the authors reported that theanine interacted with caffeine on a number of measures. While theanine was reported to reduce the positive effects of caffeine on one reaction time task (a digit vigilance task), comparisons with placebo also revealed that the caffeine–theanine combination, but not caffeine alone, speeded reaction time on a simple reaction time task and two memory tasks (numeric working memory and delayed word recognition tasks), improved sentence verification accuracy, and increased alertness and decreased tiredness ratings (Haskell et al. 2008). However, in all but one instance (delayed word recognition task), it is not clear that the caffeine–theanine combination used in this study produced improvements in alertness and performance above those of caffeine alone. In order to convincingly demonstrate a synergistic or antagonistic effect of caffeine and theanine, it is necessary to obtain a caffeine × theanine interaction effect in an analysis of variance (ANOVA) analysis (e.g., see blood pressure results in Rogers et al. 2008). This is absent in several studies, as well as the study by Haskell et al. (2008), that claim or suggest synergistic effects of caffeine and theanine on, for example, alertness and attention (Einöther et al. 2010; Kelly et al. 2008; Parnell et al. 2006).

Returning specifically to tea and the perceived differences between its effects and those of coffee, respondents that took part in our Bristol Dietary Caffeine and Health Survey also endorsed the statements "makes me feel more alert" and "interferes with my sleep" more strongly for coffee than for tea, and they believed that coffee more than tea "improves mental performance" (Heatherley et al. 2006b, unpublished data), despite single cups of both tea and coffee (as drunk in the United Kingdom) containing amounts of caffeine well within the range of doses found to have alerting and psychomotor effects (Smit and Rogers 2000). The caffeine–theanine combinations that have been tested do not seem to produce reliable effects on alertness compared to placebo, which might be suggestive of theanine reducing alertness (Parnell et al. 2006; Einöther et al. 2010; Rogers et al. 2008), although, as noted above, this evidence is inconclusive. A main effect of theanine on alertness and performance would be more convincing, and has been found by several studies, including one that included caffeine. In an EEG study, theanine (250 mg) altered attention-related alpha brain activity and markedly slowed reaction time (by as much as sixty seconds) on the auditory, but not the visual, component of the attention task (Gomez-Ramirez et al. 2007). We too found that theanine (200 mg) slowed reaction time performance on an attention task and this was a visual task of attention (visual dot probe task) and irrespective of the presence of caffeine (Rogers et al. 2008). The finding that theanine slows performance on tasks of attention is consistent with the notion that theanine might reduce alertness, and these alertness and performance effects of theanine can occur irrespective of whether it is administered with or without caffeine. It is also possible that slowed performance after theanine reflects greater deliberation by participants when they are confronted with stimuli during cognitive tasks (Gomez-Ramirez et al. 2007), and perhaps this occurs as a consequence of feeling more relaxed during task performance? For example, we previously found evidence that relaxation induced by a low dose of alcohol led to improvements in cognitive performance (Lloyd and Rogers 1997). Measures of mood were not included in the study by Gomez-Ramirez et al. (2007), but despite finding that theanine slowed cognitive performance, as mentioned earlier, we found no effect of theanine on mood, including anxiety and relaxation (Rogers et al. 2008). Also if participants were taking more care when making their decisions about such stimuli, we might expect an improvement in task accuracy, and Gomez-Ramirez et al. (2007) did not find this. Moreover, a study that examined executive control of attention during a visual attention task following exposure to a psychological stressor (a highly arousing and negatively valenced video), it found that theanine impaired performance whereas caffeine improved performance (Mahoney et al. 2010). It is unclear, though, from the report (abstract) whether theanine had any real impact on anxiety and stress.

There are also several nonhuman animal studies showing that theanine can inhibit some of the stimulatory actions of caffeine. Results were that theanine suppressed the excitatory effects of caffeine on cortical neurons in vitro (Umerzara et al. 1995), and in vivo studies demonstrated inhibiting effects of theanine on convulsive actions (Kimura and Murata 1971) and spontaneous motor activity (Sagesaka et al. 1991) caused by caffeine. Interestingly, theanine was not found to suppress the convulsive actions of stimulatory agents other than caffeine (Kimura and Murata 1971). Both studies, however, administered caffeine and theanine at megadoses, that is, in amounts equivalent to more than 100-fold higher than would be consumed by humans through tea or coffee. At doses more relevant to tea consumption, Kakuda et al. (2000a) demonstrated in rats that theanine (0.87 mg/kg intravenous injection (i.v.) into the caudal [tail] vein) almost completely suppressed the substantial increase in brain activity in the beta band, indicative of increased alertness and excitation, caused by caffeine (0.97 mg/kg i.v.). Surprisingly, this study also found that a small dose of theanine (0.34 mg/kg i.v.) administered alone had a stimulatory EEG effect similar to that of caffeine (Kakuda et al. 2000a).

Electroencephalographic (EEG) Alpha Effects of Theanine

In humans, a number of EEG studies demonstrate effects of theanine on brain activity in the alpha band, although the nature of these effects of theanine varies from study to study. Specifically, alpha refers to oscillatory brain activity in the frequency of 8-14 Hz recorded over cortical, or more precisely, somatosensory (occipital and parietal), areas of the brain by the electroencephalogram (Klimesch et al. 2007). Of note the alpha-like variant observed over the motor cortex is referred to as mu (Klimesch et al., 2007). Effects of theanine have now been reported on two aspects of alpha band activity known as tonic and phasic. The latter term refers to short-term changes in activity that occur in response to specific moment-to-moment cognitive events or stimuli, while the former term refers to the baseline of activity that is not immediately related to any particular external event or stimulus and that changes more slowly over time (Dockree et al. 2007; Gomez-Ramirez et al. 2009). Thus the enhancing effect of theanine on alpha brain activity, described in the last section, refers to tonic activity and was found under conditions of rest (Ito et al. 2008; Nobre et al. 1998). Noteworthy is that activity in the alpha band is complex and the interpretation of it is still of some debate (Dockree et al. 2007; Klimesch et al. 2007). While this has generally been associated with a restful or relaxed state (Juneja et al. 1999), activity in the alpha band is increasingly being hypothesized to reflect attention-sensitive processes (Cooper et al. 2003, 2006; Dockree et al. 2007; Klimesch et al. 1007). This well-founded hypothesis argues that increases and decreases in alpha have been interpreted as inhibition and activation of processing in those brain regions, respectively (Klimesch et al., 2007). Tonic alpha is considered as cortical inhibition and activation preceding engagement in cognitive tasks (Klimesch et al. 2007). During task engagement, changes in alpha activity are hypothesized to reflect inhibition of task-irrelevant (increase in alpha) or activation of task-relevant (decrease in alpha) information processing. Phasic alpha has been further divided into induced and evoked activity, which refer to activity related to engagement in a particular task (induced) and activity time locked to a particular stimulus (evoked) (Cooper et al. 2003, 2006; Klimesch et al. 2007). One relevant (evoked) phasic alpha effect is that involved in the deployment of the selective attention during a cognitively demanding intersensory (i.e., audiovisual) attention task. Specifically, it has been found that cueing attention to the auditory features of an imminent compound audiovisual stimulus results in a phasic increase in alpha activity in brain regions involved in processing visual information (i.e., parieto-occipital region) in the anticipatory period preceding the onset of the stimulus compared to when attention is cued to the visual features of the stimulus (Foxe et al. 1998). Consistent with the hypothesis described above, this relative cue-related increase or decrease in alpha power reflects the disengagement or engagement of the visual attentional system in preparation for anticipated attentionally relevant information respectively (Foxe et al. 1998). Similar cue-related phasic alpha effects have been observed during a visuospatial selective attention task (Kelly et al. 2006).

Gomez-Ramirez and colleagues found that 250 mg of theanine enhanced the cued differential phasic alpha effect related to the audiovisual selective attention task described above (Gomez-Ramirez et al. 2007), but they did not find a similar effect of theanine (250 mg) for the visuospatial selective attention task (Gomez-Ramirez et al. 2009). Both of these studies reported that theanine decreased tonic alpha activity (Gomez-Ramirez et al. 2007, 2009). At first sight, this appears to contradict the enhancing effect of theanine reported in earlier studies (i.e., Juneja et al. 1999; Nobre et al. 2008). However, the apparent discrepancy in the results may be due to the misnaming of the effect as an effect on tonic activity as opposed induced phasic activity. That is, the effect reported by Gomez-Ramirez et al. was observed under conditions of task engagement while the effect reported in the earlier studies was observed under resting conditions (Ito et al. 1998; Nobre et al. 2008). While Gomez-Ramirez et al. note that the lowering effect of theanine on alpha occurred irrespective of which features of the stimulus were being attended to (Gomez-Ramirez et al. 2007, 2009), they reported that in one of their studies the effect was specific to task-relevant visual areas (Gomez-Ramirez et al. 2009). It was the EEG study reporting (evoked and induced) phasic effects of theanine on alpha during the audiovisual attention task that reported an accompanying behavioural effect of theanine, showing that theanine slowed reaction time on the auditory component of the task without affecting performance accuracy (Gomez-Ramirez et al. 2007). Yet despite the EEG study by Gomez-Ramirez et al. (2009) finding that theanine decreased alpha activity during the visuospatial attention task, theanine was not found to affect behaviour (Gomez-Ramirez et al. 2009). Using the same visuospatial task, another EEG study tested the effects of a lower dose of theanine (100 mg) alone and in combination with caffeine (50 mg) (Kelly et al. 2008). This study found neither EEG nor task performance effects of theanine alone. Instead, they reported that theanine in combination with caffeine lowered alpha activity during task engagement (induced phasic alpha) compared to placebo and theanine alone, but note that this effect was not statistically different to the effect of caffeine alone. Again, this effect was reported to occur irrespective of stimulus type, and the authors do not specify whether it was restricted to specific task-relevant brain areas. Also, the caffeine–theanine combination did not affect cued attention-related phasic alpha activity. Regarding behavioral effects, Kelly et al. (2008) reported that the caffeine–theanine combination improved task performance (compared to placebo), but the data indicated that while the effect of the combination was different to the effect of caffeine alone on one measure of performance (hit rate), it was not different on another (discriminability index).

The tentative conclusion drawn by the authors of these EEG studies is that theanine can facilitate processing in the brain's attentional systems. This seems at odds with the findings of behavioral studies, detailed above, showing no clear positive and even negative effects of theanine on alertness, attention, and cognitive performance. The EEG studies themselves provide inconsistent effects of theanine on task-related alpha activity, though these effects may be dependent on the dose administered, the task used, and/or the presence of caffeine. When they do find effects of theanine on alpha, these studies often fail to find accompanying improvements in task performance. Thus, at present the effects of theanine on alpha that have been reported are not sufficiently unambiguous to support robust conclusions regarding its possible impact on the brain attentional system. Even the authors of these EEG studies urge caution in drawing conclusions over whether these effects reflect improved brain function, and while warning against inferring any implications for behavior. Returning to the finding of increased alpha after theanine under rest conditions (tonic alpha), although one way of thinking about this is that increased alpha reflects increased internalization and concomitant reduction of reliance upon external sensory and cognitive information, in other words an "inattention" to external events or an "internalized attention" (Cooper et al. 2003, 2006; Hester et al. 2004; Klimesch et al. 2007). For example, some forms of meditation are mediated by a "switching-off" mechanism of external attention and are also associated with increased alpha activity on the EEG (Takahashi et al. 2005). However, this relationship has been shown to be task-specific. For example, it has been found that external (stimulus intake) tasks of attention such as perception performance is enhanced if the cortex is already activated (ie., low tonic alpha), whereas internal (stimulus rejection) tasks of attention such as memory or visualisation performance is enhanced if the cortex is deactivated (i.e., high tonic alpha) before the task is performed (Klimesch et al. 2007). Increases in alpha are also observed during performance of internally, but not externally, driven attention tasks (Cooper et al. 2003, 2006). This leads us to speculate that the enhancing effect of theanine found on alpha might indicate that theanine could enhance performance on internal attention tasks. The tasks employed in the previously described EEG studies and the most of the behavioral studies required externally directed attention. It would be of interest then to examine effects of theanine on alpha and performance during internally directed attention tasks. Perhaps this research would produce more favourable performance effects of theanine. The finding of increased alpha during rest has also been observed during marihuana-induced euphoria (Lukas et al. 1995). Thus, while the facilitatory effect of theanine on tonic alpha during rest appears to be reliable, perhaps it is indicative of lowered arousal or attention to external information. Paradoxically, this disengagement from irrelevant external input can serve to facilitate sustained and selective attention during task engagement and this is thought to underpin the positive relationship often found between tonic alpha levels and cognitive performance (Klimesch et al. 1999, 2007). However, this relationship has been shown to be task-specific. For example, it has been found that external (stimulus intake) tasks of attention such as perception performance is enhanced if the cortex is already activated (ie., low tonic alpha), whereas internal (stimulus rejection) tasks of attention such as memory or visualisation performance is enhanced if the cortex is deactivated (i.e., high tonic alpha) before the task is performed (Klimesch et al., 2007). Increases in alpha are also observed during performance of internally, but not externally, driven attention tasks (Cooper et al., 2003, 2006). This leads us to speculate that the enhancing effect of theanine found on alpha might indicate that theanine could enhance performance on internal attention tasks. The tasks employed in the previously described EEG studies and the most of the behavioral studies required externally directed attention. It would be of interest then to examine effects of theanine on alpha and performance during internally directed attention tasks. Perhaps this research would produce more favourable performance effects of theanine.

In the context of a role for theanine in accounting for differences between the properties of tea and coffee then, we speculate that at least part of the reason why tea is perceived to be a more relaxing drink and coffee a more stimulating drink is that theanine reduces (externally directed) arousal/alertness and slows behavioral responses. Perhaps theanine also takes the "edge" off the psychostimuluant (i.e., alert and performancing-enhancing) effects rather than the anxiety effects of caffeine. Equally, the lack of studies producing robust effects of theanine on alertness and performance could be taken to suggest that theanine plays little or no part in the explanation.

Clearly more research is needed to draw any firm conclusions. Indeed, relative to research on caffeine and alertness/performance, it is apparent that the studies on theanine alone and in combination with caffeine so far have not produced reliable effects. It is worth reiterating that careful thought regarding the design of studies and analysis of data is needed to gain a better picture of the actions of caffeine and theanine in combination. Other reasons for why tea has the reputation of being less stimulating than coffee may include the lower doses of caffeine present in tea. While acute psychostimulant effects of caffeine appear not to be strongly dose dependent (Smit & Rogers, 2000) making the peak effects of tea and coffee similar, the onset of increased alertness and improved performance may be faster and these effects may persist for longer after higher doses present in coffee. Another explanation may be that tea and coffee are used in somewhat different contexts, and in fact perceptions of tea and coffee may vary across cultures. Differences in the concentrations, or presence, or absence, of other substances apart from caffeine and theanine in tea and coffee may also prove to be important in explaining their different alerting, mood, and behavioral effects.

Regarding theanine's alerting and cognitive performance effects, and its effects on alpha brain activity, clearly more research is needed in order to draw any firm conclusions. Such research may also help us to discern its relevance, if any, in understanding the effects of tea. Relative to research on caffeine and alertness and performance, the studies on theanine and theanine in combination with caffeine so far have not produced reliable effects of either theanine with and without caffeine. It is apparent though that virtually no unequivocally positive effects of theanine have been demonstrated—theanine certainly does not appear to enhance alertness or speed up performance. If anything, it slows performance, and it is not clear this effect or its effects on alpha activity can be seen as beneficial. The nature of its interactions with caffeine on alertness, cognitive performance, and brain activity is also unclear. It is worth reiterating that careful thought regarding the design of studies and analysis of their data is needed to gain a better picture of the actions of caffeine and theanine in combination. But on balance the various findings would suggest that theanine either has no impact on the psychostimulant effects (i.e., increased alertness and cognitive performance) of caffeine or that it reduces them.

Lastly, it seems that the effects of caffeine on alertness and cognitive performance do not represent a net benefit for functioning but merely withdrawal reversal. Specifically, what is occurring is that in frequent caffeine consumers acute (e.g., overnight) caffeine withdrawal lowers alertness and degrades performance, and caffeine consumption restores functioning to, but not above, "normal" levels (Rogers and Richardson 1993). Counter-regulatory changes in adenosine receptors and/or increased levels of endogenous adenosine resulting from chronic exposure to caffeine (Fredholm et al. 1999) would be a plausible mechanism underlying such withdrawal effects. Evidence for withdrawal reversal comes from studies that have attempted to take into account the negative effects of acute caffeine withdrawal, and is reviewed by James and Rogers (2005) and discussed in detail elsewhere (Rogers and Smith in press). In sum, it does not look good for the regular tea and/or coffee drinker in the short term. The positive psychostimulant effects of tea and coffee *are* most likely due to caffeine, but these effects reflect merely the reversal of withdrawal effects after overnight caffeine abstinence; theanine appears to have little or no obvious beneficial effects on its own, and it may even reduce these effects of caffeine. On the other hand, as will be shown in the next section, over the longer term tea consumption appears to be beneficial for cognitive health.

THEANINE (TEA) CONSUMPTION AND RISK OF COGNITIVE IMPAIRMENT AND DECLINE

Observational (correlational) studies have found higher levels of black and green tea consumption to be associated with a lower prevalence of cognitive impairment and decline in older populations (> 55 years) (Kuriyama et al. 2006; Ng et al. 2008). Moreover, in one study cognitive function was also reassessed after 1–2 years, and again, levels of tea consumption showed an inverse relationship with cognitive decline (Ng et al. 2008). All these associations were significant after controlling for some potential important confounding factors, including level of education and overall health. These studies also provide another source of information on theanine intake and cognitive function. Despite having few positive effects on cognition acutely, it is still plausible that theanine is important in understanding the positive relationship found between tea consumption and cognition over the longer term. One possibility is that, rather than an acute effect, this is due to chronic protective effects of theanine and also other constituents of tea—theanine is just one of thousands of compounds present in tea.

Blood Pressure Lowering Effects of Theanine

At first sight, tea consumption might be expected to be harmful for cognitive health because caffeine increases blood pressure, and raised blood pressure in middle age increases risk of cognitive decline and dementia in later life (Stewart 1999). Indeed, the blood pressure effect of caffeine is potentially very important for health. For example, it has been argued that by increasing blood pressure (due to its vasoconstrictive effect), caffeine consumption may contribute substantially to the prevalence of cardiovascular disease and stroke (James 2004), and by implication to an increased risk of cognitive impairment in old age. One major reason for this is increased risk of transient ischemic episodes linked to underlying vascular disease (atherosclerosis) (O'Brien et al. 2004). Brain ischemia is the loss of glucose and oxygen supply to the brain, and can lead to cell death and in turn cognitive impairment (see "Neuro- and memory-protective effects of theanine" section). It is reassuring, however, that tea consumption has not generally been found to be associated with these adverse effects. Indeed, some studies even suggest that consumption of tea may reduce the risk of hypertension (Yang et al. 2004), cardiovascular and cerebrovascular disease (Arts et al. 2005). In view of this, tea consumption is now considered part of a healthy diet, for example, The American College of Cardiology Foundation Task Force recommends moderate tea intake as part of nutritional advice for risk reduction in cardiovascular disease (Vogel et al. 2005). These findings suggest that other components of tea must directly oppose the blood pressure raising effect of caffeine or outweigh the consequences of this effect. For example, polyphenols present in tea, including catechins, may reduce risk via vasorelaxant effects, and effects on blood cholesterol, blood coagulation, and inflammatory processes (Hodgson 2006). Furthermore, we found that theanine antagonized the blood pressure raising effect of caffeine (Rogers et al. 2008). Specifically, theanine acutely reduced systolic and diastolic blood pressure only when it was raised by caffeine within four to five minutes (Rogers et al. 2008). Two other studies have been published on the effects of theanine on blood pressure in animals (Yokogoshi et al. 1995; Yokogoshi and Kobayashi 1998). The reduction in blood pressure was found for spontaneously hypertensive rats, but not for normotensive rats and a closely related compound was found to have an even more potent effect (Yokogoshi and Kobayashi 1998). On the other hand, in a previous unpublished study we found that 300 mg of theanine reduced diastolic blood pressure by 3 mmHg from a normotensive baseline (69 mmHg) (Rogers et al. unpublished). In both our studies the oral doses of theanine (200 and 300 mg approximately equate to 3 and 5 mg/kg) administered were much lower than the dose found to reduce blood pressure in rats (2000 mg/kg i.p.). To put this into context, two hundred milligrams of theanine is about twice the theanine consumption of tea drinkers in the United Kingdom (see the "Background" section of this chapter). Also significant is a study that examined the effects of a decaffeinated tea composition containing 100 mg of theanine and 200 mg of catechin green tea extract consumed twice daily for 3 months on risk factors for cardiovascular disease (Nantz, Rowe, Bukowski, & Percival, 2009). This combination reportedly reduced systolic and diastolic blood pressure after 3 weeks by 5 and 4 mmHg, respectively, and systolic blood pressure remained lower by 3 mmHg after 3 months. Furthermore, the decreases observed in this study appear to have occurred from a pre-hypertensive (blood pressure that is higher than normal but not to the level considered to be hypertension) baseline (131/80 mmHg). As well as reducing blood pressure, it was reported that the combination affected three other independent risk factors, which included reducing total and low-density lipoprotein (LDL) cholesterol, lowering a biomarker of oxidative stress and lipid peroxidation (blood and urine malondialdehyde, or MDA), as well as decreasing a marker of chronic inflammation (blood amyloid-α). It is not possible to ascertain whether theanine was responsible to any extent for these benefits. As alluded to previously, tea has been associated with a variety of the health protection properties, including cholesterol-lowering, antioxidant, and anti-inflammatory effects, and these effects have been mainly attributed to catechins present in tea (e.g., see Dufresne and Farnworth 2001; Valcic et al. 1999; Sutherland et al. 2006), which were included in the tea composition administered in this last study. Although polyphenols present in tea, including catechins, may account for the majority of the beneficial effects of tea on these biomarkers of health, there is evidence that suggests that theanine might contribute to some extent as well (see "Neuro- and Memory-Protective Effects of Theanine" section and the "Miscellany" section).

The mechanism(s) of action relevant to the blood pressure effect of theanine is unclear. Yokogoshi et al. (1995) speculated that the blood pressure-lowering effects of theanine in rats reflected actions of theanine on pathways of the nervous system and peripheral blood vessels and not on brain serotonin concentration. It would seem unlikely that the effects are mediated via actions on glutamate systems as, unlike theanine, glutamate was not found to affect blood pressure in rats (Yokogoshi et al. 1995; Yokogoshi and Kobayashi 1998). In our study, raised blood pressure after caffeine was accompanied by increased jitteriness, but theanine only antagonized caffeine's effect on blood pressure (Rogers et al. 2008). In addition, in the unpublished study we did not administer caffeine or induce anxiety or other aspects of mood (Rogers et al. unpublished). Together, these findings may suggest a direct effect on vascular tone, heart rate and/or the baroreceptor reflex, rather than, for example, a lowering of blood pressure by reduction of anxiety. There is some evidence that theanine may affect heart rate. Also significant is a study that examined the effects of a decaffeinated tea composition containing 100 mg of theanine and 200 mg of catechin green tea extract consumed twice daily for 3 months on risk factors for cardiovascular disease (Nantz, Rowe, Bukowski, & Percival, 2009). This combination reportedly reduced systolic and diastolic blood pressure after 3 weeks by 5 and 4 mmHg, respectively, and systolic blood pressure remained lower by 3 mmHg after 3 months. Furthermore, the decreases observed in this study appear to have occurred from a pre-hypertensive (blood pressure that is higher than normal but not to the level considered to be hypertension) baseline (131/80 mmHg). This comes from a study in which increased heart rate following an arithmetic task was lowered by theanine (200 mg; Kimura et al. 2007). It is, however, puzzling that a near maximal effect occurred only 5 minutes or less after oral administration. An interesting parallel might be drawn between the effects of theanine and those of beta-blockers (e.g., propanol), which have been widely used in the treatment of hypertension and for cardioprotection (e.g., against heart attack) (Bangalore et al. 2007). There is some evidence that theanine may affect heart rate. This comes from a study in which increased heart rate following an arithmetic task was lowered by theanine (200 mg; Kimura et al., 2007). It is, however, puzzling that a near maximal effect occurred only 5 minutes or less after oral administration. However, they are often ineffective in treating anxiety disorders per se and are recommended for use alongside conventional anti-anxiety treatments (Hayes and Schulz 1987). All the above possibilities remain to be studied further in humans. At present, the clearest evidence is in relation to the blood pressure reducing effects of theanine, which, in turn, may potentially play an important role in explaining the relationship between tea consumption and cognitive health, with one mechanism being reduction in risk of vascular disease and associated brain injury.

Neuro- and Memory-Protective Effects of Theanine

Theanine might also be involved in neuroprotection during and/or after brain ischemia. During ischemia there is a large increase in extracellular glutamate, which, acting via glutamate receptor NMDA and/or AMPA subtypes, increases intracellular calcium leading to immediate and delayed (via activation of calcium related enzymes and nitric oxide and superoxide radicals) cell death in the brain (Choi and Rothman 1990; Danyz et al. 1995). Brain ischemia also disrupts $GABA_A$ neurotransmission which is thought to contribute to ongoing excitatory neuronal activity and possibly cell death (Schwartz-Bloom and Sah 2001). Several animal studies found that theanine reduced ischemic brain damage, including in the hippocampus, a brain region involved in memory, when it was administered acutely prior to and/or after the ischemic insult (Kakuda et al. 2000b; Egashira et al. 2004, 2007, 2008). As discussed in the "The Neuropharmacological Actions of Theanine" section of this chapter, the mechanism of action underlying this neuroprotective effect of theanine is uncertain, though there is some evidence to suggest that it may involve selective modulation of glutamate and/or GABA receptor function and/or effects on glutamine transport (see "The Neuropharmacological Actions of Theanine," and also see below). It is also worth noting that changes in cerebral blood flow and body temperature as an explanation for the protective effect of theanine against ischemic brain damage were considered unlikely in one study (Egashira et al. 2004). By reducing ischemic cell damage and death, theanine should reduce associated cognitive impairment, and indeed this is what has been found in animal studies assessing the effects of theanine on ischemic hippocampal cell death and related memory impairment (Egashira et al. 2008). Remarkably, theanine reduced ischemic brain damage and the physiological and/or cognitive protective effects of theanine occurred at doses relevant to the theanine consumed through tea intake in humans (0.3 and 1.0 mg/kg; Egashira et al. 2004, 2007, 2008). Also low doses of theanine (2 and 4 mg/kg) administered to animals chronically (five weeks) protected against induction of memory impairment by injection of Aβ (Kim et al. 2009). Improvements in learning and memory ability have also been reported in animals following chronic (three weeks to four months) administration of theanine (180–400 mg/day) (Juneja et al. 1999; Yamada et al. 2008b; Yokogoshi and Terashima 2000). These positive effects stand in contrast to the findings of studies of acute effects of theanine in humans detailed in the preceding section. One possibility is that the benefits reflect longer term neuroprotective effects of theanine, including neuroprotective effects described in the current section and/or a blood pressure lowering effect discussed int he previous section, as opposed to direct acute effects on cognition.

Additional but not mutually exclusive mechanisms by which theanine may contribute to (long-term) cognitive health via neuroprotection include its ability to (1) potentiate endogenous antioxidants in the brain (Kim et al. 2009; Nagasawa et al. 2004; Nishida et al. 2008); (2) reduce elevated brain concentrations of Aβ; (3) attenuate neurotoxic effects, primarily oxidative cell damage, caused by excessive glutamate and/or Aβ concentrations (Di et al. 2010; Kim et al. 2009; Nagasawa et al. 2004); and (4) inhibit inflammatory responses (Kim et al. 2008). These effects may involve actions on some types of glutamate and/or GABA receptors (see "The Neuropharmacological Actions of Theanine," this chapter). Furthermore, some of these possibilities (i.e., antioxidant and anti-inflammatory effects) have also been suggested as mechanisms of action for the neuroprotective effects of catechins during brain ischemia and brain atrophy (Kakuda 2002; Sutherland et al. 2006; Unno et al. 2007). Of note, two studies have reported neuroprotective effects of theanine in cultured human cells, including one that used glutamate-induced excitotoxicity in the human amyloid precursor protein (APP) transgenic SH-SY5Y cell, as an in vitro model of Alzheimer's disease (Di et al. 2010), and another that used the human dopamine SH-SYSY cells treated with Parkinson's disease-related neurotoxins (Cho et al. 2008).

Theanine has also been reported to have antioxidant effects in body tissues other than the brain, including the liver and heart (Sugiyama and Sadzuka 2003, 2004). Its potential for reducing oxidative (side) effects of chemotherapeutic drugs in healthy body tissues such as these has been suggested as important for modulating the antitumor efficacy of these drugs (Sugiyama and Sadzuka 2003, 2004), which is addressed in the miscellany section. Lastly, lipid anti-oxidation effects of theanine, including in relation to LDL, have been reported in several studies (Yokowaza and Dong 1997; Kim et al. 2009; Nantz et al. 2009). Though these effects may be weaker than those of tea polyphenols (Yokowaza and Dong 1997; Nantz et al. 2009), they might be important in protecting against hypercholesterolemia and vascular disease (e.g., atherosclerosis).

Theanine and Weight Control

It is the case that atherosclerosis, hypertension, and stroke are, among other factors, all associated with obesity. Avoidance of obesity might then be expected to reduce risk of cognitive decline in later life. Recently there have been reports that suggest a role of tea in controlling body weight and preventing obesity, mainly through studying the effects of components of tea such as catechins and caffeine (see reviews by Bolling et al. 2009; Cooper et al. 2005a). Anti-obesity effects of theanine have also been investigated in animal studies, and these studies possibly provide partial support for a role of theanine (Zheng et al. 2004, 2005; Yamada et al. 2008a). For example, Zheng et al. (2005) found that body weight increases, fat accumulation (measured as interperitoneal adipose tissue weight), and lipid concentrations in the blood and liver of mice were suppressed by a diet containing theanine at a concentration of 0.04 percent administered for sixteen weeks. These effects occurred in the absence of changes in food intake measured at four-week intervals. It is curious that these effects of theanine were only observed at a concentration of 0.04 percent and not at higher (0.08 or 0.16 percent) or lower (0.01 or 0.02 percent) theanine concentrations (Zheng et al. 2005). Yamada et al. (2008a) found that 4000 mg/kg of theanine administered orally suppressed food intake in rats four hours after administration, while blood insulin concentration and body weight gain were found to be reduced within one hour of its administration. In the same study, it was reported theanine administered thirty minutes prior to the administration of glucose (3000 mg/kg) reduced the rise in blood insulin concentration and increased the concentrations of corticosterone and free fatty acids but did not affect the concentration of blood glucose (Yamada et al. 2008a). In another study conducted in rats, hyperlipidemia as a result of a hepatic tumor growth was reduced by theanine administered in the rats' diet at a concentration of 0.1 percent for two weeks (Zhang et al. 2002). However, this blood lipid-lowering effect may have occurred due to its ability to suppress the tumor growth (Zhang et al. 2002). Much more research, especially long-term studies conducted in humans, is needed to draw any conclusions on theanine's role in weight control. In the context of tea, it is more likely that effects of catechins and caffeine underpin any possible weight control benefits of the drink (Zheng et al. 2004; Dulloo et al. 1999).

In sum, the health effects of tea consumption appear to be mainly longer term ones, and, reassuringly, these are generally positive. Theanine may be important in understanding these effects of tea. While it may provide little benefit acutely, theanine may help to preserve cognition over the longer term, especially for people who are at increased risk of hypertension, ischemic brain and heart disease, and/or neurodegenerative disease such as dementia.

MISCELLANY

Effects of theanine, other than those discussed in the preceding sections, have drawn some interest. The one receiving the most attention is its reputed ability to modulate the chemotherapeutic effects of anticancer drugs (Sugiyama and Sadzuka 2003). The association between tea consumption and reduced risk of cancer has received a great deal of interest (see, for example, reviews by Bolling et al. 2009; Cooper et al. 2005b; Dufresne and Farnworth 2001). Again most studies implicate tea catechins as mediators of this relationship, but there is growing evidence that theanine might be another contributing factor. Thus far, the studies on theanine's role in this area are limited to those conducted in animals, of which there are several, mostly conducted by the same research group. They suggest theanine can both enhance the efficacy and reduce the toxicity of cancer treatments. For a review of the findings of this research, see Sugiyama and Sadzuka (2003) and their 2004 paper and also Eschenauer and Sweet (2006). Two other studies report therapeutic effects of theanine on the development cancers, including a study in rats mentioned in this chapter (Zhang et al. 2002), and also a study that used human lung cancer and leukemia cells (Liu et al. 2009).

A related area of interest is theanine's potential immune function benefits. Daily consumption of tea (5–6 cups) for 2–4 weeks (Kamath et al. 2003), and daily supplementation with a capsule containing theanine and EGCG (doses were not specified) for three months (Rowe et al. 2007), or theanine (280 mg) and l-cystine (700 mg) for two weeks (Miyagawa et al. 2008), has been associated with enhanced immune responses. With two of the studies reporting improved function of a specific subset of T cells, known as gamma delta (γδ) T cells, in healthy participants (Kamath et al. 2003; Rowe et al. 2007). These T cells are considered to be the first line of defense against infection, some inflammatory responses, and tumor cells (see Burowski and Percival 2008; Nantz et al. 2009; Percival et al. 2008). It is also thought that increased immunity might lead to a reduced risk of cancer (cf. Percival et al. 2008). Though health consequences were not examined following the tea consumption intervention (Kamath et al. 2003), Rowe et al. (2007) reported that the theanine–EGCG intervention somewhat reduced cold and flu symptoms. There was also some evidence of an enhanced response to the influenza virus in a subgroup of elderly participants following the theanine–cystine intervention (Miyagawa et al. 2008). The rationale for immune function effects of theanine is derived from the idea that theanine is a precursor of the antigen ethylamine, which has been shown to prime γδ T cells to react to bacteria. This is discussed in a review by Percival et al. (2008). However, it is not possible to discern this potential role of theanine based on the studies published so far.

Lastly, there are some reports (abstracts) that theanine (200–300 mg) taken before bed for a around a week can improve feelings of sleep quality without affecting total sleep time (Ozeki et al. 2003; Shirakawa et al. 2006). However, the information provided by these reports is limited, and as the results are based primarily on self-report measures of sleep, and within a small sample, they may not actually reflect actual sleep patterns. Again more information and research is required to adequately assess the merits of the reported effects and draw conclusions of the effectiveness of theanine in this area.

A SHORT NOTE ON COFFEE, COLA, AND CAFFEINE

It is worth noting that coffee consumption, with its higher level of caffeine intake, has not been associated with cognitive impairment or decline (Ng et al. 2008), and in several studies it has even been associated with a decreased risk (e.g., Jarvis 1993; Johnson-Kozlow et al. 2002; Ritchie et al. 2007). Consistent with this, consumption of coffee, despite its lack of theanine, does not appear to have the increased risk of hypertension or cardiovascular disease that we might have expected from its caffeine content (Cornelis and El-Sohemy 2007; Winkelmayer et al. 2005). On the other hand, cola intake, whether sugar containing or low calorie, and after controlling for Body Mass Index (BMI), has been found to be associated with greater risk of hypertension (Winkelmayer et al. 2005). While the caffeine content of coffee may be over twice that of cola, coffee, unlike cola, contains thousands of other compounds, including antioxidants such as flavonoids and phenols, which may be protective against the blood pressure raising effects of caffeine and compounds other than caffeine present in coffee that might increase risk (Vinson 2006). Furthermore, coffee intake, whether caffeinated or decaffeinated, has been associated with reduced risk of type 2 diabetes (van Dam and Hu 2005), which itself is a risk factor for cognitive decline (Stewart and Liolista 1999).

Another interesting possibility is that, despite its blood pressure raising effects, caffeine may in part explain the positive association between tea and coffee consumption and cognitive health in the long term, due to it affects on adenosine. For example, during brain ischemia there is a large increase in extracellular adenosine which, acting via adenosine A_1 and A_{2A} receptors, helps to counter some of the key pathophysiological processes, including excitatory neurotransmitter release, that lead to ischemic neuronal cell death (Freholm et al. 2005). Because caffeine blocks adenosine A_1 and A_{2A} receptors, it should exacerbate ischemic damage, and indeed this is what has been found in animal studies when caffeine is administered acutely prior to the ischemic insult (Jacobson et al. 1996). However, chronic pretreatment with caffeine reduces ischemic brain damage (Jacobson et al. 1996), suggesting that a protective effect is gained through upregulation of adenosine receptors or other related adaptive change. In other words, frequent exposure to caffeine may modify the adenosine system to increase its neuroprotective function.

GENERAL CONCLUSIONS AND IMPLICATIONS FOR PUBLIC HEALTH

Overall, the findings of the studies reviewed in this chapter indicate that theanine, a compound found naturally in tea, can affect brain activity, physiology, and behavior.

Though there is still much to learn, theanine does not appear to have any strong unwanted or harmful (side) effects. Furthermore, theanine's ability to reach the brain fairly rapidly, with acute effects observed soon after oral administration, and its high solubility and lack of noticeable taste or flavor effects make it a convenient added ingredient. At present, then, its use as a dietary supplement is perhaps of little concern to public health. Moreover, the extent to which these and theanine supplemented food and drink products contribute to overall theanine consumption is likely to be small in comparison to tea, the world's most frequently consumed drink apart from water.

Naturally research on the potential effects of theanine has tended to be discussed with reference to the putative effects of tea, which include various mood and health benefits, one being anxiety and stress relief or relaxation. However, a review of the small number of studies investigating theanine's specific relaxing properties reveals a lack of strong support. This should provoke a reconsideration of the strength of the message on which theanine, in the form of dietary supplements and supplemented products, is specifically marketed. By contrast, there is perhaps greater public and commercial demand for products that offer the possibility of increased alertness and enhanced mental performance. Caffeine's reputation as a relatively harmless psychostimulant makes it the most popular, and currently perhaps the only, functionally significant ingredient for such products. Growing interest in theanine and its possible interactions with caffeine in relation to alertness and cognitive performance has not clearly established an acute cognitive benefit to be gained from adding theanine to these products.

Where theanine may be more effective is in preserving cognition in later life through its chronic protective effects. Important in this respect are theanine's potential effects in relation to vascular health and neuroprotection (including possibly even antioxidant effects), with their implications for healthy aging. There is ample evidence that diet affects cognitive health, a major mechanism being avoidance of vascular disease. We have discussed this ("a healthy body and healthy mind"), together with the longer term dietary influences on mood, in more detail elsewhere (Rogers 2001). The potential of food ingredients to be used as natural alternatives for treatment of hypertension and corresponding reduction of the risk of cardiovascular disease and stroke has received considerable interest of late. Pre-hypertensive patients (individuals whose blood pressure is marginally high but not high enough for use of prescription of blood pressure lowering medications) are expected to benefit from this the most (e.g., Chen et al. 2009). Theanine's (and/or tea's) usefulness as a "functional" ingredient may then lie in its ability to reduce blood pressure. It is worth noting that while with frequent intake tolerance develops to the modest anxiety effects of caffeine, there is at most partial tolerance to its blood pressure raising effects (James 2004). Though other constituents present in tea and coffee may also outweigh these effects of caffeine, in caffeine-containing products where these are not present, such as in cola and energy drinks, the addition of theanine may be particularly useful.

Another equally pressing concern is the epidemic of increasing obesity throughout the world and the problems it brings for general heath and wellbeing. Sufficient evidence for a role of theanine in weight management remains to be seen, but plausibly it may to some degree offset some of the complications associated with obesity, including, as summarized above, hypertension and vascular disease and with that cognitive decline. In view of the contribution of cardiovascular and cerebrovascular disease as well as mood disorders to the global burden of disease (Murray and Lopez 1996), and a recent report suggesting that dementia poses the most significant health and social crisis of the twenty-first century (The World Alzheimer's Report 2010; Wimo and Prince 2010), these areas should be given high priority in future nutrition and behavior research.

Finally, it is important not to overstate the effects of theanine. It is just one of the many compounds in tea, and its beneficial effects may be relatively weak in comparison, for example, to tea polyphenols, such as catechins. The absence of theanine in coffee, which despite its higher caffeine content appears if anything to reduce risk of vascular disease and cognitive decline, affirms this. Furthermore, potential tolerance and withdrawal effects associated with frequent theanine intake have yet to be examined. Tea is commercially important, and through its widespread consumption, together with the varied physiological and psychoactive effects of its constituent compounds, including theanine, it potentially has a significant impact on population health and wellbeing. Nevertheless, when taking either an individual or population perspective it is important to place such effects in the context of a broad range of modifiable risk factors. Thus, the typical consumption of tea and increasingly theanine-supplemented products probably sums to a very small impact on health compared with, for example, quitting smoking, reducing overweight, or increasing level of physical activity.

ACKNOWLEDGMENTS

Sue Heatherley and Emma Mullings contributed to the previously unpublished research described in this chapter. We are grateful to Kit Pleydell-Pearce for helpful comments on a previous version of this manuscript.

REFERENCES

Araujo, J. A., de Rivera, C., Ethier, J. L., Landsberg, G. M., Denenberg, S., Arnold, S., et al. (2010). Anxitane® tablets reduce fear of human beings in a laboratory model of anxiety-related behavior. *J Vet Beha*, *5*, 268–75.

Arts, I. C. W., and Hollman, P. C. H. (2005). Polyphenols and disease risk in epidemiologic studies. *Am J Clin Nutr, 81 (Suppl 1)*: 317S–325S.

Bangalore, S., Messerli, F. H., Kostis, J. B., and Pepine, C. J. (2007). Cardiovascular protection using beta-blockers: a critical review of the evidence. *J Am Coll Card, 50*, 563–572.

Barcroft, H., and Konzett, H. (1949). On the actions of noradrenaline, adrenaline and isopropyl noradrenaline on the arterial blood pressure, heart rate and muscle blood flow in man. *J Physiol, 110*, 194–204.

Basheer, R., Strecker, R. E., Thakkar, M. M., and McCarley, R. W. (2004). Adenosine and sleep-wake regulation. *Prog Neurobiol, 73*, 379–396.

Bergink, V., van Megen, H., and Westenberg, H. G. M. (2004). Glutamate and anxiety. *Eur Neuropsychopharmacol, 14*, 175–183.

Bolling, B. W., Chen, C. Y. O., and Blumberg, J. B. (2009). Tea and health: preventive and therapeutic usefulness in the elderly? *Curr Opin Clin Nutr Metab Care, 12*, 42–48.

Borzelleca, J. F., Peters, D., and Hall, W. (2006). A 13-week dietary toxicity and toxicokinetic study with l-theanine in rats. *Food Chem Toxicol, 44*, 1158–1166.

Botella, P., and Parra, A. (2003). Coffee increases state anxiety in males but not in females. *Hum Psychopharmacol, 18*, 141–143.

Bryan, J. (2008). Psychological effects of dietary components of tea: caffeine and L-theanine. *Nutr Rev, 66*, 82–90.

Bukowski, J. F., and Percival, S. S. (2008). L-theanine intervention enhances human gamma delta T lymphocyte function. *Nutr Rev, 66*, 96–102.

Casimir, J., Jadot, J., and Renard, M. (1960). Separation and characterisation of N-ethyl-gamma-glutamine in Xerocomus badius. *Biochim Biophys Acta, 39*, 462–468.

Chen, Z. Y., Peng, C., Jiao, R., Wong, Y. M., Yang, N., and Huang, Y. (2009). Anti-hypertensive nutraceuticals and functional Foods. *J Agric Food Chem, 57*, 4485–4499.

Cho, H. S., Kim, S., Lee, S. Y., Park, J. A., Kim, S. J., and Chun, H. S. (2008). Protective effect of the green tea component, L-theanine on environmental toxins-induced neuronal cell death. *Neurotoxicology, 29*, 656–662.

Cooper, N. R., Burgess, A. P., Croft, R. J., and Gruzelier, J. H. (2006). Investigating evoked and induced electroencephalogram activity in task-related alpha power increases during an internally directed attention task. *Neuroreport, 17*, 205–208.

Cooper, N. R., Croft, R. J., Dominey, S. J. J., Burgess, A. P., and Gruzelier, J. H. (2003). Paradox lost? Exploring the role of alpha oscillations during externally versus internally directed attention and the implications for idling and inhibition hypotheses. *Int J Psychophysiol, 47*, 65–74.

Cooper, R., Morre, D. J., and Morre, D. M. (2005a). Medicinal benefits of green tea: Part I. Review of noncancer health benefits. *J Altern Complement Med, 11*, 521–528.

Cooper, R., Morre, J., and Morre, D. M. (2005b). Medicinal benefits of green tea: Part II. Review of anticancer properties. *J Altern Complement Med, 11*, 639–652.

Cornelis, M. C., and El-Sohemy, A. 2007. Coffee, caffeine, and coronary heart disease. *Curr Opin Clin Nutr Metab Care, 10*, 745–751.

Cunha, R. A. (2001). Adenosine as a neuromodulator and as a homeostatic regulator in the nervous system: different roles, different sources and different receptors. *Neurochem Int, 38*, 107–125.

Danysz, W., Parsons, C. G., Bresink, I., and Quack, G. (1995). Glutamate in CNS disorders. *Drug News and Perspect, 8*, 261–277.

Desai, M. J., and Armstrong, D. W. (2004). Analysis of derivatized and underivatized theanine enantiomers by high-performance liquid chromatography/atmospheric pressure ionization-mass spectrometry. *Rapid Commun Mass Spectrom, 18*, 251–256.

Di, X., Yan, J., Zhao, Y., Zhang, J., Shi, Z., Chang, Y., et al. (2010). L-Theanine Protects the APP (Swedish mutation) transgenic SH-SY5Y cell against glutamate-induced excitotoxicity via inhibition of the NMDA receptor pathway. *Neuroscience, 168*, 778–786.

Dockree, P. M., Kelly, S. P., Foxe, J. J., Reilly, R. B., and Robertson, I. H. (2007). Optimal sustained attention is linked to the spectral content of background EEG activity: greater ongoing tonic alpha (similar to 10 Hz) power supports successful phasic goal activation. *Eur J Neurosci, 25*, 900–907.

Dolinska, M., Zablocka, B., Sonnewald, U., and Albrecht, J. (2004). Glutamine uptake and expression of mRNA's of glutamine transporting proteins in mouse cerebellar and cerebral cortical astrocytes and neurons. *Neurochem Int, 44*, 75–81.

Dramard, V., Kern, L., Hofmans, J., Halsberghe, C., and Rème, C. A. (2007). Clinical efficacy of L-theanine tablets to reduce anxiety-related emotional disorders in cats: A pilot open-label clinical trial. *J Vet Beha, 2,* 85–6.

Dufresne, C. J., and Farnworth, E. R. 2001. A review of latest research findings on the health promotion properties of tea. *Journal of Nutritional Biochemistry* 12:404–421.

Dulloo, A. G., Duret, C., Rohrer, D., Girardier, L., Mensi, N., Fathi, M., et al. (1999). Efficacy of a green tea extract rich in catechin polyphenols and caffeine in increasing 24-h energy expenditure and fat oxidation in humans. *Am J Clin Nutr, 70*, 1040–1045.

Egashira, N., Hayakawa, K., Mishima, K., Kimura, H., Iwasaki, K., and Fuji/wara, M. (2004). Neuroprotective effect of gamma-glutamylethylamide (theanine) on cerebral infarction in mice. *Neurosci Lett, 363*, 58–61.

Egashira, N., Hayakawa, K., Osajima, M., Mishima, K., Iwasaki, K., Oishi, R., et al. (2007). Involvement of $GABA_A$ receptors in the neuroprotective effect of theanine on focal cerebral ischemia in mice. *J Pharmacol Sci, 105*, 211–214.

Egashira, N., Ishigami, N., Pu, F., Mishima, K., Iwasaki, K., Orito, K., et al. (2008). Theanine prevents memory impairment induced by repeated cerebral ischemia in rats. *Phytother Res, 22*, 65–68.

Einöther, S. J. L., Martens, V. E. G., Rycroft, J. A., and De Bruin, E. A. (2010). L-Theanine and caffeine improve task switching but not intersensory attention or subjective alertness. *Appetite, 54*, 406–409.

Ekborg-Ott, K. H., Taylor, A., and Armstrong, D. W. (1997). Varietal differences in the total and enantiomeric composition of theanine in tea. *J Agric Food Chem, 45*, 353–363.

Eschenauer, G., and Sweet, B. V. (2006). Pharmacology and therapeutic uses of theanine. *Am J Health Syst Pharm, 63*, 26–30.

Foxe, J. J., Simpson, G. V., and Ahlfors, S. P. (1998). Parieto-occipital similar to 10 Hz activity reflects anticipatory state of visual attention mechanisms. *Neuroreport, 9*, 3929–3933.

Fredholm, B. B., Battig, K., Holmen, J., Nehlig, A., and Zvartau, E. E. (1999). Actions of caffeine in the brain with special reference to factors that contribute to its widespread use. *Pharmacol Rev, 51*, 83–133.

Fredholm, B. B., Chen, J-F., Cunha, R. A., Svenningsson, P., and Vaugeois, J-M. (2005). Adenosine and brain function. *Int Rev Neurobio, 63*, 191–270.

Fujii, S., and Inai, K. (2008). Tumorigenicity study of l-theanine administrated orally to mice. *Food Chem, 110*, 643–646.

Goldstein, D. S., Eisenhofer, G. and McCarty, R. (1998). *Catecholamines: Bridging Basic Science with Clinical Medicine*. San Diego, CA: Academic Press.

Goldstein, A., Kaizer, S., and Whitby, O. (1969). Psychotropic effects of caffeine in man. IV. Quantitative and qualitative differences associated with habituation to coffee. *Clin Pharmacol Ther,* 10, 489–497.

Gomez-Ramirez, M., Higgins, B. A., Rycroft, J. A., Owen, G. N., Mahoney, J., Shpaner, M., et al. (2007). The deployment of intersensory selective attention: A high-density electrical mapping study of the effects of theanine. *Clin Neuropharmacol, 30*, 25–38.

Gomez-Ramirez, M., Kelly, S. P., Montesi, J. L., and Fox, E. (2009). The effects of L-theanine on alpha-band oscillatory brain activity during a visuo-spatial attention task. *Brain Topogr, 22*, 44–51.

Graham, H. N. (1992). Green tea composition, consumption, and polyphenol chemistry. *Prev Med, 21*, 334–350.

Hariri, A. R., and Holmes, A. (2006). Genetics of emotional regulation: the role of the serotonin transporter in neural function. *Trends Cogn Sci, 10*, 182–191.

Haskell, C., Kennedy, D., Milne, A., Wesnes, K., and Scholey, A. B. (2008). Caffeine at levels found in decaffeinated beverages is behaviourally active. *Appetite, 50*, 559.

Hayes, P. E., and Schulz, S. C. (1987). Beta-blockers in anxiety disorders. *Journal of Affective Disorders, 13*, 119–130.

Heatherley, S. V., Mullings, E. L., Tidbury, M. A., and Rogers, P. J. (2006a). Caffeine consumption among a sample of adults. *Appetite, 47*, 266.

Heatherley, S. V., Mullings, E. L., Tidbury, M. A., and Rogers, P. J. (2006b). The Dietary Caffeine and Health Survey: Administration of a large postal survey in Bristol. *Appetite, 47*, 266.

Heimbach J. T. (2005). GRAS Monograph for Suntheanine®. www.accessdata.fda.gov/scripts/fcn/gras_notices/615880A.pdf

Hester, R. L., Murphy, K., Foxe, J. J., Foxe, D. M., Javitt, D. C., and Garavan, H. (2004). Predicting success: Patterns of cortical activation and deactivation prior to response inhibition. *J Cogn Neurosci, 16*, 776–785.

Hodgson, J. M. (2006). Effects of tea and tea flavonoids on endothelial function and blood pressure: a brief review. *Clin Exp Pharmacol Physiol, 33*, 838–841.

Hynd, M. R., Scott, H. L., and Dodd, P. R. (2004). Glutamate-mediated excitotoxicity and neurodegeneration in Alzheimer's disease. *Neurochem Int, 45*, 583–595.

Ito, K., Nagato, Y., Aoi, N., Juneja, L. R., Kim, M., Yamamoto, T., et al. (1998). Effects of L-theanine on the release of alpha-brain waves in human volunteers. *Nippon Nogeikagaku Kaishi* (*In Japanese, see Juneja et al. 1999 for English summary*), *72*, 153–157.

Jacobson, K. A., vonLubitz, D., Daly, J. W., and Fredholm, B. B. (1996). Adenosine receptor ligands: Differences with acute versus chronic treatment. *Trend Pharmacol Sci, 17*, 108–113.

James, J. E. (2004). Critical review of dietary caffeine and blood pressure: A relationship that should be taken more seriously. *Psychosom Med, 66*, 63–71.

James, J. E., and Rogers, P. J. (2005). Effects of caffeine on performance and mood: withdrawal reversal is the most plausible explanation. *Psychopharmacology, 182*, 1–8.

Jarvis, M. J. (1993). Does caffeine intake enhance absolute levels of cognitive performance. *Psychopharmacology, 110*, 45–52.

Johnson-Kozlow, M., Kritz-Silverstein, D., Barrett-Connor, E., and Morton, D. (2002). Coffee consumption and cognitive function among older adults. *Am J Epidemiol, 156*, 842–850.

Kakuda, T. (2002). Neuroprotective effects of the green tea components theanine and catechins. *Bio Pharm Bull, 25*, 1513–1518.

Kakuda, T., Hinoi, E., Abe, A., Nozawa, A., Ogura, M., and Yoneda, Y. (2008). Theanine, an ingredient of green tea, inhibits [H-3]glutamine transport in neurons and astroglia in rat brain. *J Neurosci Res, 86*, 1846–1856.

Kakuda, T., Nozawa, A., Sugimoto, A., and Niino, H. (2002). Inhibition by theanine of binding of [3H]AMPA, [3H]kainate, and [3H]MDL 105,519 to glutamate receptors. *Biosci Biotechnol Biochem,* 66, 2683–2686.

Kakuda, T., Nozawa, A., Unno, T., Okamura, N., and Okai, O. (2000a). Inhibiting effects of theanine on caffeine stimulation evaluated by EEG in the rat. *Biosci Biotechnol Biochem, 64*, 287–293.

Kakuda, T., Yanase, H., Utsunomiya, K., Nozawa, A., Unno, T., and Kataoka, K. (2000b). Protective effect of γ-glutamylethylamide (theanine) on ischemic delayed neuronal death in gerbils. *Neurosci Lett, 289*, 189–192.

Kamath, A. B., Wang, L. S., Das, H., Li, L., Reinhold, V. N., and Bukowski, J. F. (2003). Antigens in tea-beverage prime human Vγ2Vδ2 T cells in vitro and in vivo for memory and nonmemory antibacterial cytokine responses. *Proc Natl Acad Sci U S A, 100*, 6009–6014.

Keenan, E. K., Finnie, M. D. A., Jones, P. S., Rogers, P. J., and Priestley, C. M. (2010). How much theanine in a cup of tea? Effects of tea type and method of preparation. *Food Chem,* 125, 588–594.

Kelly, S. P., Gomez-Ramirez, M., Montesi, J. L., and Foxe, J. J. (2008). L-theanine and caffeine in combination affect human cognition as evidenced by oscillatory alpha-band activity and attention task performance. *J Nutr, 138*, 1572S–1577S.

Kelly, S. P., Lalor, E. C., Reilly, R. B., and Foxe, J. J. (2006). Increases in alpha oscillatory power reflect an active retinotopic mechanism for distracter suppression during sustained visuospatial attention. *Journal of Neurophysiology,* 95, 3844–3851.

Kim, T. I., Lee, Y. K., Park, S. G., Choi, I. S., Ban, J. O., Park, H. K., et al. (2009). L-Theanine, an amino acid in green tea, attenuates β-amyloid-induced cognitive dysfunction and neurotoxicity: Reduction in oxidative damage and inactivation of ERK/p38 kinase and NF-κB pathways. *Free Radic Biol Med, 47*, 1601–1610.

Kimura, R., and Murata, T. (1971). Influence of alkylamides of glutamic acid and related compounds on the central nervous system. I. Central depressant effect of theanine. *Chem Pharm Bull, 19*, 1257–1261.

Kimura, R., and Murata, T. (1980). Influence of alkylamides of glutamic acid and related compounds on the central nervous system .IV. Effect of theanine on adenosine-3′,5′-monophosphate formation in rat cerebral cortex. *Chem Pharm Bull*, 28, 664–666.

Kimura, R., and Murata, T. (1986). Effect of theanine on norepinephrine and serotonin levels in rat brain. *Chem Pharm Bull,* 34, 3053–3057.

Klimesch, W. (1999). EEG alpha and theta oscillations reflect cognitive and memory performance: a review and analysis. *Brain Res Rev,* 29, 169–195.

Klimesch, W., Sauseng, P., and Hanslmayr, S. (2007). EEG alpha oscillations: The inhibition-timing hypothesis. *Brain Res Rev, 53*, 63–88.

Kuhn, D. M., Wolf, W. A., and Lovenberg, W. 1980. Review of the role of the central serotonergic neuronal system in blood pressure regulation. *Hypertension* 2:243–255.

Kuriyama, S., Hozawa, A., Ohmori, K., Shimazu, T., Matsui, T., Ebihara, S., et al. 2006. Green tea consumption and cognitive function: a cross-sectional study from the Tsurugaya Project. *American Journal of Clinical Nutrition* 83:355–361.

Liu, Q., Duan, H., Luan, J., Yagasaki, K., and Zhang, G. 2009. Effects of theanine on growth of human lung cancer and leukemia cells as well as migration and invasion of human lung cancer cells. *Cytotechnology* 59:211–217.

Lloyd, H. M., and Rogers, P. J. (1997). Mood and cognitive performance improved by a small amount of alcohol given with a lunchtime meal. *Behavioural Pharmacology, 8* (2–3), 188–195.

Lopez-Corcuera, B., Geerlings, A., and Aragon, C. (2001). Glycine neurotransmitter transporters: an update. *Mol Membr Biol, 18,* 13–20.

Louzada, P. R., Lima, A. C. P., Mendonca-Silva, D. L., Noel, F., De Mello, F. G., and Ferreira, S. T. (2004). Taurine prevents the neurotoxicity of beta-amyloid and glutamate receptor agonists: activation of GABA receptors and possible implications for Alzheimer's disease and other neurological disorders. *FASEB J, 18,* 511–518.

Lu, K., Gray, M. A., Oliver, C., Liley, D. T., Harrison, B. J., Bartholomeusz, C. F., et al. (2004). The acute effects of L-theanine in comparison with alprazolam on anticipatory anxiety in humans. *Hum Psychopharmacology, 19*, 457–465.

Lukas, S. E., Mendelson, J. H., and Benedikt, R. (1995). Electroencephalographic correlates of marijuana-induced euphoria. *Drug Alcohol Depend, 37*, 131–140.

Mahoney, C. R., Brunyé, T. T., Giles, G., and Kanarek, R. (2010). Theanine reverses caffeine-induced enhancement control following exposure to stressor. In 40th annual meeting of the Society for Neuroscience, San Diego, CA, November 13–17.

Maruyama, M., and Takeda, K. (1994). Electrophysiologically potent non-competitive glutamate antagonists at crayfish neuromuscular junctions are also potent inhibitors of [3H] MK801 binding to synaptic membranes from rat central nervous system. *Comp Biochem Physiol C Pharmacol Toxicol Endocrinol, 107*, 105–110.

McCall, A. L., Millington, W. R., and Wurtman, R. J. (1982). Blood-brain barrier transport of caffeine: Dose-related restriction of adenine transport. *Life Sci, 31*, 2709–2715.

McEntee, W., and Crook, T. (1993). Glutamate: its role in learning, memory, and the aging brain. *Psychopharmacology, 111*, 391–401.

Meldrum, B. S. (2000). Glutamate as a neurotransmitter in the brain: Review of physiology and pathology. *J Nutr, 130*, 1007S–1015S.

Meneses, A. (1999). 5-HT system and cognition. *Neurosci Biobehav Rev, 23*, 1111–1125.

Michelazzi, M., Berteselli, G., Minero, M., and Cavallone, E. (2010). Effectiveness of L-theanine and behavioral therapy in the treatment of noise phobias in dogs. *J Vet Beha, 5*, 34–35.

Millan, M. J. (2003). The neurobiology and control of anxious states. *Progress in Neurobiology* 70:83–244.

Miyagawa, K., Hayashi, Y., Kurihara, S., and Maeda, A. (2008). Co-administration of L-cystine and L-theanine enhances efficacy of influenza vaccination in elderly persons: Nutritional status-dependent immunogenicity. *Geriatr Gerontol Int, 8*, 243–250.

Murray, C. J. L., and Lopez, A. D. (1997). Global mortality, disability, and the contribution of risk factors: Global Burden of Disease Study. *Lancet, 349*, 1436–1442.

Nagasawa, K., Aoki, H., Yasuda, E., Nagai, K., Shimohama, S., and Fujimoto, S. (2004). Possible involvement of group I mGluRs in neuroprotective effect of theanine. *Biochem Biophys Res Commun, 320*, 116–122.

Nantz, M. P., Rowe, C. A., Bukowski, J. F., and Percival, S. S. (2009). Standardized capsule of *Camellia sinensis* lowers cardiovascular risk factors in a randomized, double-blind, placebo-controlled study. *Nutrition, 25*, 147–154.

Neumann, K., and Montag, A. (1982). Quantitative determination of theanine in tea extracts. *Dtsch Lebensmitt Rundsch, 78*, 160–164.

Ng, T. P., Feng, L., Niti, M., Kua, E. H., and Yap, K. B. (2008). Tea consumption and cognitive impairment and decline in older Chinese adults. *Am J Clin Nutr, 88*, 224–231.

Nishida, K., Yasuda, E., Nagasawa, K., and Fujimoto, S. (2008). Altered levels of oxidation and phospholipase C isozyme expression in the brains of theanine-administered rats. *Biol Pharm Bull, 31*, 857–860.

Nobre, A. C., Rao, A., and Owen, G. N. (2008). L-theanine, a natural constituent in tea, and its effect on mental state. *Asia Pac J Clin Nutr, 17,* 167–168.

O'Brien, J., Ames, D., Gustafson, L., Folstein, D., and Chiu, E. (2004). *Cerbrovascular Disease, Cognitive impairment and Dementia*. London: Martin Dunitz.

Ozeki, M., Juneja, L. R., and Shirakawa, S. 2003. The effects of theanine on sleep with physiological evaluation using actigraph. *J Physiol Anthropol Appl Human Sci, 23*, 58. (*Meeting abstract*).

Parnell, H., Owen, G. N., and Rycroft, J. A. (2006). Combined effects of L-theanine and caffeine on cognition and mood. *Appetite, 47*, 273.

Percival, S. S., Bukowski, J. E., and Milners, J. (2008). Bioactive food components that enhance γδ T cell function may play a role in cancer prevention. *J Nutr, 138,* 1–4.

Rao, A., Ozeki, M., and Juneja, L. R. (2007). Suntheanine: A pure and safe L-theanine dietary supplement for relaxation and stress relief. *NutraCos Articles—B5srl,* January/February, 26–30.

Ressler, K. J., and Nemeroff, C. B. (1999). Role of norepinephrine in the pathophysiology and treatment of mood disorders. *Biol Psychiatry, 46*, 1219–1233.

Ritchie, K., Carriere, I., de Mendonca, A., Portet, F., Dartigues, J. F., Rouaud, O., et al. (2007). The neuroprotective effects of caffeine: a prospective population study (the Three City Study). *Neurology, 69*, 536–545.

Robbins, T. W. (2000). Neural systems underlying arousal and attention. *Eur J Neurosci, 12*, 443–443.

Rogers, P. J. (2001). A healthy body, a healthy mind: long-term impact of diet on mood and cognitive function. *Proc Nutr Soc, 60,* 135–143.

Rogers, P. J. (2007). Caffeine, mood and mental performance in everyday life. *Nutr Bull, 32 Suppl. 1*, 84–89.

Rogers, P. J., Hohoff, C., Heatherley, S. V., Mullings, E. L., Maxfield, P. J., Evershed, R. P., et al. (2010). Association of the anxiogenic and alerting effects of caffeine with ADORA2A and ADORA1 polymorphisms and habitual level of caffeine consumption. *Neuropsychopharmacology, 35*, 1973–1983.

Rogers, P. J., and Richardson, N. J. (1993). Why do we like drinks that contain caffeine? *Trends Food Sci Technol, 4*, 108–111.

Rogers, P. J., Smith, J. E., Heatherley, S. V., and Pleydell-Pearce, C. W. (2008). Time for tea: mood, blood pressure and cognitive performance effects of caffeine and theanine administered alone and together. *Psychopharmacology, 195*, 569–577.

Rogers, P. J., and Smith, J. E. (in press). Caffeine, mood and cognition. In D. Benton (Ed.), *Lifetime nutritional influences on cognition, behaviour and psychiatric illness*. Woodhead.

Sagesaka Y., Kakuda, K., and Kawamura, K. (1991). Pharmacological effect of theanine. Proceedings of the International Symposium on Tea Science, Shizuoka, August 1991, 362–365.

Salamone, J. D. (1992). Complex motor and sensorimotor functions of striatal and accumbens dopamine: involvement in instrumental behavior processes. *Psychopharmacology, 107*, 160–174.

Schmidt, M., Schmitz, H. J., Baumgart, A., Guédon, D., Netsch, M. I., Kreuter, M. H., et al. (2005). Toxicity of green tea extracts and their constituents in rat hepatocytes in primary culture. *Food Chem Toxicol, 43*, 307–314.

Schwartz-Bloom, R. D., and Sah, R. (2001). γ-Aminobutyric $acid_A$ neurotransmission and cerebral ischemia. *Journal of Neurochemistry, 77*, 353–371.

Shimbo, M., Nakamura, K., Jing Shi, H., Kizuki, M., Seino, K., Inose, T., et al. (2005). Green tea consumption in everyday life and mental health. *Public Health Nutr, 8*, 1300–1306.

Shinozaki, H., and Ishida, M. (1978). Theanine as a glutamate antagonist at a crayfish neuromuscular junction. *Brain Res, 151*, 215–219.

Shirakawa, S., Nagata, Y., and Yasuoka, H. (2006). The improvement effects of theanine on sleep of women with job. *J Physiol Anthropol, 25*, 388. (*Meeting abstract*).

Smit, H. J., and Rogers, P. J. (2000). Effects of low doses of caffeine on cognitive performance, mood and thirst in low and higher caffeine consumers. *Psychopharmacology, 152*, 167–173.

Steptoe, A., Gibson, E. L., Vuononvirta, R., Williams, E. D., Hamer, M., Rycroft, J. A., et al. (2007). The effects of tea on psychophysiological stress responsivity and post-stress recovery: a randomised double-blind trial. *Psychopharmacology, 190*, 81–89.

Stewart, R. (1999). Hypertension and cognitive decline. *Br J Psychiatry, 174*, 286–287.

Stewart, R., and Liolitsa, D. (1999). Type 2 diabetes mellitus, cognitive impairment and dementia. *Diabet Med, 16*, 93–112.

Sugiyama, T., and Sadzuka, Y. (2003). Theanine and glutamate transporter inhibitors enhance the antitumor efficacy of chemotherapeutic agents. *Biochim Biophys Acta, 1653*, 47–59.

Sugiyama, T., and Sadzuka, Y. (2004). Theanine, a specific glutamate derivative in green tea, reduces the adverse reactions of doxorubicin by changing the glutathione level. *Cancer Lett, 212*, 177–184.

Sutherland, B. A., Rahman, R. M. A., and Appleton, I. (2006). Mechanisms of action of green tea catechins, with a focus on ischemia-induced neurodegeneration. *J Nutr Biochem, 17*, 291–306.

Takahashi, T., Murata, T., Hamada, T., Omori, M., Kosaka, H., Kikuchi, M., et al. (2005). Changes in EEG and autonomic nervous activity during meditation and their association with personality traits. *Int J Psychophysiol, 55*, 199–207.

Terashima, T., Takido, J., and Yokogoshi, H. (1999). Time-dependent changes of amino acids in the serum, liver, brain and urine of rats administered with theanine. *Biosci Biotechnol Biochem, 63*, 615–618.

Timmcke, J. Q., Juneja, L. R., and Kapoor, M. P. (2008). Efficacy and short-term safety of L-theanine in randomized, double-blind, parallel-group study. *FASEB J, 22,* 760. (*Meeting Abstract*).

Tsuge, H., Sano, S., Hayakawa, T., Kakuda, T.,and Unno, T. (2003). Theanine, γ-glutamylethylamide, is metabolized by renal phosphate-independent glutaminase. *Biochim Biophys Acta, 1620*, 47–53.

Tyrer, P. (1988). Current status of beta-blocking drugs in the treatment of anxiety disorders. *Drugs, 36*, 773–783.

Ueda, T., Ozeki, M., Okubo, T., Chu, D., Juneja, L. R., Yokogoshi, H., et al. (2001). Improving effects of L-theanine on premenstrual syndrome. *JSPOG 6*, 234–239. (In Japanese; see Rao et al. 2007 for English translation.)

Umezara, K., Kobayashi, k., Muramoto, K., Kawahara, M., Mitzutani, A., Kakuda, T., et al. (1995). Theanine, a glutamate analog stimulates NMDA-receptors but suppresses excitatory effect of caffeine on cortical neurons. In 25th annual meeting of the Society for Neuroscience, San Diego, CA, November 11–16, 21:835.

Unno, K., Takabayashi, F., Yoshida, H., Choba, D., Fukutomi, R., Kikunaga, N., et al. (2007). Daily consumption of green tea catechin delays memory regression in aged mice. *Biogerontology, 8*, 89–95.

Unno, T., Suzuki, Y., Kakuda, T., Hayakawa, T., and Tsuge, H. (1999). Metabolism of theanine, γ-glutamylethylamide, in rats. *J Agric Food Chem, 47*, 1593–1596.

Valcic, S., Muders, A., Jacobsen, N. E., Liebler, D. C., and Timmermann, B. N. (1999). Antioxidant chemistry of green tea catechins. Identification of products of the reaction of (–) -epigallocatechin gallate with peroxyl radicals. *Chem Res Toxicol, 12*, 382–386.

van Dam, R. M., and Hu, F. B. (2005). Coffee consumption and risk of type 2 diabetes: a systematic review. *JAMA, 294*, 97–104.

Vinkers, C. H., Klanker, M., Groenink, L., Korte, S. M., Cook, J. M., Van Linn, M. L., et al. (2009). Dissociating anxiolytic and sedative effects of $GABA_A$ergic drugs using temperature and locomotor responses to acute stress. *Psychopharmacology, 204*, 299–311.

Vinson, J. A. (2006). Caffeine and incident hypertension in women. JAMA, *295*, 2135; author reply.

Vogel, J. H. K., Bolling, S. F., Costello, R. B., et al. (2005). Integrating complementary medicine into cardiovascular medicine: A report of the American College of Cardiology Foundation Task Force on Clinical Expert Consensus Documents (Writing Committee to Develop an Expert Consensus Document on Complementary and Integrative Medicine). *J Am Coll Cardiol, 46*, 184–221.

Wimo, A., and Prince, M. (2010). *The World Alzheimer's Report 2010. The Global Economic Impact of Dementia.* Alzheimer's Disease International. www.alz.co.uk/research/files/WorldAlzheimerReport2010.pdf

Winkelmayer, W. C., Stampfer, M. J., Willett, W. C., and Curhan, G. C. (2005). Habitual caffeine intake and the risk of hypertension in women. *JAMA, 294*, 2330–2335.

Yadid, G., Pacak, K., Golomb, E., Harveywhite, J. D., Lieberman, D. M., Kopin, I. J., et al. (1993). Glycine stimulates striatal dopamine release in conscious rats. *Br J Pharmacol, 110*, 50–53.

Yamada, T., Nishimura, Y., Sakurai, T., Terashima, T., Okubo, T., Juneja, L. R., et al. (2008a). Administration of theanine, a unique amino acid in tea leaves, changed feeding-relating components in serum and feeding behavior in rats. *Biosci Biotech Biochem, 72*, 1352–1355.

Yamada, T., Terashima, T., Honma, H., Nagata, S., Okubo, T., Juneja, L. R., et al. (2008b). Effects of theanine, a unique amino acid in tea leaves, on memory in a rat behavioral test. *Biosci Biotech Biochem, 72*, 1356–1359.

Yamada, T., Terashima, T., Kawano, S., Furuno, R., Okubo, T., Juneja, L., et al. (2009). Theanine, γ-glutamylethylamide, a unique amino acid in tea leaves, modulates neurotransmitter concentrations in the brain striatum interstitium in conscious rats. *Amino Acids, 36*, 21–27.

Yamada, T., Terashima, T., Okubo, T., Juneja, L. R., and Yokogoshi, H. (2005). Effects of theanine, r-glutamylethylamide, on neurotransmitter release and its relationship with glutamic acid neurotransmission. *Nutr Neurosci, 8*, 219–226.

Yang, Y-C., Lu, F-H., Wu, J-S., Wu, C-H., and Chang, C-J. (2004). The protective effect of habitual tea consumption on hypertension. *Arch Intern Med, 164*, 1534–1540.

Yang, Q.-S., Xu, P.-X., Li, Y.-H., Jiang, S., Zhang, X., and Xue, M. (2007). Effects of theanine and houpu extract in 7-day chick social separation-stress procedure. *Zhongguo Zhong Yao Za Zhi, 32*, 2040–2043. *(Abstract available in English.)*

Yokogoshi, H., Kato, Y., Sagesaka, Y. M., Takiharamatsuura, T., Kakuda, T., and Takeuchi, N. (1995). Reduction effect of theanine on blood pressure and brain 5-hydroxyindoles in spontaneously hypertensive rats. *Biosci Biotechnol Biochem, 59*, 615–618.

Yokogoshi, H., and Kobayashi, M. (1998). Hypotensive effect of γ-glutamylmethylamide in spontaneously hypertensive rats. *Life Sci, 62*, 1065–1068.

Yokogoshi, H., Kobayashi, M., Mochizuki, M., and Terashima, T. (1998a). Effect of theanine, r-glutamylethylamide, on brain monoamines and striatal dopamine release in conscious rats. *Neurochem Res, 23,* 667–673.

Yokogoshi, H., Mochizuki, M., and Saitoh, K. (1998b). Theanine-induced reduction of brain serotonin concentration in rats. *Biosci Biotechnol Biochem, 62*, 816–817.

Yokogoshi, H., and Terashima, T. (2000). Effect of theanine, r-glutamylethylamide, on brain monoamines, striatal dopamine release and some kinds of behavior in rats. *Nutrition, 16*, 776–777.

Yokozawa, T., and Dong, E. B. (1997). Influence of green tea and its three major components upon low-density lipoprotein oxidation. *Exp Toxicol Pathol, 49*, 329–335.

Zhang, G. Y., Miura, Y., and Yagasaki, K. (2002). Effects of dietary powdered green tea and theanine on tumor growth and endogenous hyperlipidemia in hepatoma-bearing rats. *Biosci Biotechnol Biochem, 66*, 711–716.

Zheng, G., Bamba, K., Okubo, T., Juneja, L. R., Oguni, I., and Sayama, K. (2005). Effect of theanine, γ-glutamylethylamide, on bodyweight and fat accumulation in mice. *Anim Sci J, 76*, 153–157.

Zheng, G. D., Sayama, K., Okubo, T., Juneja, L. R., and Oguni, I. (2004). Anti-obesity effects of three major components of green tea, catechins, caffeine and theanine, in mice. *In Vivo 18*, 55–62.

13 Caffeine
Practical Implications

Andrew P. Smith

CONTENTS

Introduction 272
What Is Caffeine? 272
Where Does It Come From? 272
How Does It Act on the Body? 272
How Much Can We Have? 273
What External Factors Affect Its Metabolism? 273
Behavioral Effects of Acute Caffeine Ingestion 273
A Profile of the Effects of Caffeine on Mood and Cognitive Performance 273
Caffeine and Physical Performance 274
Mechanisms Underlying the Behavioral Effects of Caffeine 275
Reversal of Impairments Due to Low Levels of Arousal 277
Sleep Deprivation 277
Maintenance of Wakefulness Test (MWT) and Multiple Sleep Latency Test (MSLT) 277
Performance and Subjective Alertness 277
A Review of Randomized Trials of Caffeine and Cognitive Performance 279
Caffeine and Circadian Troughs 279
Night Work 279
Postlunch Dip 279
Early Morning 279
Evening 279
Caffeine and Sustained Work 280
Sustained Military Operations 280
Caffeine and the Common Cold 281
Beneficial Effects of Habitual Consumption Of Caffeine 281
Caffeine at Work 282
Caffeine Outside of Work 282
Caffeine and Social Behavior 282
Caffeine and Driving 283
Prevention of Cognitive Decline in the Elderly 284
Mental Health 284

Caffeine and Children and Adolescents 285
Health Effects of Caffeine Consumption 286
Conclusions 286
References 286

INTRODUCTION

There have been many articles written about caffeine, and the present chapter has the aim of discussing some of the practical implications of caffeine ingestion. This first section gives some brief information about caffeine and then describes the structure of the chapter.

What Is Caffeine?

Caffeine (1,3,7-trimethylxanthine) is a member of a class of naturally occurring substances termed methylxanthines (two other methylxanthines found in food are theobromine, which is present in cocoa, and theophylline, which is found in small amounts in coffee and tea).

Where Does It Come From?

Caffeine occurs naturally in a number of foods, is employed as a food additive, and is added to medications. It occurs naturally in coffee, tea, and cocoa, and the exact amount present will depend on growing conditions and preparation. Some rough approximations are shown below:

- Filter coffee: 100–150 mg (5 oz cup)
- Instant coffee: 50–60 mg
- Tea: 35–45 mg
- Chocolate: milk, up to 15 mg; dark, up to 35 mg

It is also added to soft drinks (e.g., cola has 40 mg in a 12 oz serving). Higher amounts are added to energy drinks (70 mg+). Caffeine is also added to over-the-counter (OTC) medications (usually in the region of 30–50 mg per tablet). Caffeine tablets can also be purchased, and the recommended dose to increase alertness is 100–200 mg.

How Does It Act on the Body?

Caffeine acts by blocking the effects of the naturally occurring neuromodulator adenosine. This produces an increase in central nervous system (CNS) activity which is associated with changes in many neurotransmitter systems (e.g., an increased turnover of central noradrenaline). Caffeine has other effects (e.g., influencing blood flow to the brain) but many of these (e.g., calcium mobilization and prostaglandin antagonism) are unlikely to occur with the amounts consumed by humans.

How Much Can We Have?

Average intake varies as a function of country and demographic factors. Several estimates suggest that average intake is about 200 mg per day. Problems usually only occur when excessive amounts are consumed (500 mg a day plus), and these problems usually take the form of an increase in anxiety. Some individuals are very sensitive to effects of caffeine, and even small amounts can cause adverse reactions. Most individuals control their consumption of caffeine. For example, consumption usually occurs when alertness is reduced (e.g., early in the morning, after prolonged work, and after lunch) and is reduced at times when high alertness is undesirable (e.g., before going to sleep).

What External Factors Affect Its Metabolism?

Plasma levels of caffeine peak 15–45 minutes after ingestion, and the half-life is between five and six hours. A variety of factors influence the metabolism. For example, in pregnant women the half-life can increase to eighteen hours. Oral contraceptive use increases the half-life to eleven hours. In contrast, in cigarette smokers the half-life is only about three hours, which may account for the high level of consumption found in this group.

BEHAVIORAL EFFECTS OF ACUTE CAFFEINE INGESTION

There has been extensive research on the effects of caffeine on human behavior (for reviews, see Lieberman 1992; Smith 2002, 2005). In general, there is good agreement about the behavioral changes that occur following caffeine ingestion. Like most topics that have been studied extensively, there is variation in the reported effects. This often reflects methodological features of the studies (e.g., sample size, design and analysis, procedure, and choice of tasks) and there is a need to examine the literature so that findings can be weighted according to their methodological rigor. Once this has been achieved, secondary analyses based on large data sets can determine which effects are going to have a significant impact on real-life behavior. Similarly, meta-analyses, based on these robust findings, can address this issue in another way. What is presented below is a less formal approach to the topic based on years of reviewing the topic and conducting studies in the laboratory.

A Profile of the Effects of Caffeine on Mood and Cognitive Performance

Substantial research has shown that ingestion of moderate doses of caffeine (typically between 100 and 300 mg) increases alertness and the ability to respond to signals occurring at unexpected times or in unknown places. Smith et al. (2003) distinguish two different types of effect: the first occurs in low-alertness situations (e.g., when circadian alertness is low, after prolonged work, when the person is sleep deprived, or when the person has a cold), and caffeine improves performance on tasks known to be impaired by reduced alertness (e.g., variable foreperiod simple reaction time tasks, vigilance tasks, and self-paced serial response tasks). A second type of effect can be seen even in alert individuals and involves caffeine increasing the speed of encoding of new information. This faster processing can then lead to benefits in choice reaction time tasks or tasks involving detection of stimuli which are degraded (Smith et al. 1999) or presented rapidly.

Effects of caffeine on other functions, such as episodic memory, are rarely observed. However, caffeine may improve working memory (e.g., performance of a logical reasoning task; Smith et al. 1992, 1994) and retrieval from semantic memory (Smith et al. 1992, 1994, 1999). These effect sizes are usually smaller than those found using selective and sustained attention tasks. Similarly, it is unclear whether the effects are really due to changes in memory or just to a change in the speed of encoding the information.

While the main effect of caffeine is to increase alertness (especially at the end of a test battery when volunteers are fatigued), some studies have also shown that caffeine increases hedonic tone (feeling happier, feeling more sociable). Caffeine can also increase anxiety when given either in high doses (over 300 mg; Lieberman 1992) or to sensitive individuals. It has been suggested that this may lead to impairments in fine motor performance, but other research (e.g., Lieberman et al. 1987) has failed to confirm this. Given that caffeine increases alertness, it is not surprising that large amounts consumed late in the evening can interfere with sleep (prevent individuals from going to sleep and reduce sleep duration). Indeed, most individuals control their consumption to avoid this. While it is quite easy to demonstrate effects of late-night caffeine on sleep, it is more difficult to find evidence that high consumption per se will affect sleep. Indeed, Sanchez-Ortuno et al. (2005) conducted a survey of 1,498 French workers and found that habitual use of up to seven cups of coffee (600 mg caffeine) per day was not associated with a decreased duration of sleep

A number of studies have shown that beneficial effects of caffeine can be observed using low doses typically found in commercial products (e.g., Lieberman et al. 1987; Durlach 1998; Smith et al. 1999). A linear dose–response curve has also been shown in a number of studies (Amendola et al. 1998; Brice and Smith 2001), although others suggest that the dose–response curve is flat at the low-dose end (Smit and Rogers 2000).

The effects described above appear to be robust in that they can be observed when caffeine is given to nonconsumers, withdrawn consumers, or those who have recently consumed caffeine. Most studies of caffeine have involved administration of a single large dose which is often not the way we consume it. Brice and Smith (2002) found that the enhanced mood and performance seen after a single dose of 200 mg caffeine were also obtained if volunteers consumed four separate doses of 65 mg over the day (which resulted in an identical final caffeine level to the 200 mg dose).

Caffeine and Physical Performance

This issue has often been considered in relation to sports performance (its relation to sustained operations will be discussed in this chapter). A recent position paper on this topic from the International Society of Sports Nutrition (Goldstein et al. 2010) can be summarized as follows:

1. Caffeine is effective for enhancing sport performance in trained athletes when consumed in low-to-moderate dosages (~3–6 mg/kg) and overall does not result in further enhancement in performance when consumed in higher dosages (≥ 9 mg/kg).
2. Caffeine exerts a greater ergogenic effect when consumed in an anhydrous state as compared to coffee.
3. It has been shown that caffeine can enhance vigilance during bouts of extended exhaustive exercise, as well as periods of sustained sleep deprivation.
4. Caffeine is ergogenic for sustained maximal endurance exercise, and has been shown to be highly effective for time-trial performance.
5. Caffeine supplementation is beneficial for high-intensity exercise, including team sports such as soccer and rugby, both of which are categorized by intermittent activity within a period of prolonged duration.
6. The literature is equivocal when considering the effects of caffeine supplementation on strength–power performance, and additional research in this area is warranted.
7. The scientific literature does not support caffeine-induced diuresis during exercise, or any harmful change in fluid balance that would negatively affect performance.

The next section considers whether there are plausible mechanisms for the observed effects of caffeine.

MECHANISMS UNDERLYING THE BEHAVIORAL EFFECTS OF CAFFEINE

Mechanisms underlying the behavioral effects of caffeine can be considered at several different levels (see Fredholm et al. 1999). If one starts with effects on the CNS, one finds that most of the data suggest that caffeine, in the doses that are commonly consumed, acts primarily by blocking adenosine A_1 and A_{2a} receptors. Even though the primary action of caffeine may be to block adenosine receptors, this leads to very important secondary effects on many classes of neurotransmitters, including noradrenaline, acetylcholine, dopamine, serotonin, glutamate, and GABA. Such effects show that caffeine has the ability to increase alertness, a possible reason underlying why people consume caffeine-containing beverages. There are other effects of caffeine on the CNS (e.g., direct release of intracellular calcium, and effects on alkaline phosphatase), but many of these only occur at doses well above the range of human consumption.

Studies from humans show that mechanisms underlying the beneficial effects seen in mood and performance may reflect changes in a number of neurotransmitter systems. The effects seen in low-alertness situations have been mimicked using drugs that reduce the turnover of central noradrenaline (Smith et al. 2003) or increase sedation through changes in the GABA/benzodiazepine system (File et al. 1982). The faster encoding of new information produced by caffeine may reflect cholinergic changes, and caffeine has been shown to reduce the impairments produced by scopolamine (Riedel et al. 1995).

Recent studies have examined effects of caffeine on blood flow to the brain. It is likely that caffeine exerts a complex pattern of effects, but the net effect is reduced cerebral blood flow. This has been demonstrated using positron emission tomography (PET), magnetic resonance imaging (MRI), and trans-cranial Doppler methodologies. The relevance of these changes for the behavioral effects requires further investigation.

Another approach has tried to elucidate the stages of processing influenced by caffeine. For example, Lorist and Snel (1997) have shown that target detection and response preparation are enhanced by caffeine, and Ruijter et al. (1999) have demonstrated that the quantity of information processed is greater after caffeine. Smith et al. (1999) have shown that caffeine increases the speed of processing new stimuli, confirming results reported by Streufert et al. (1997). In contrast to its effects on encoding and sustained attention, caffeine has not been shown to reduce resistance to distraction (Kenemans and Verbaten 1998). Similarly, caffeine appears to have little effect on output processes (e.g., movement time; Lorist 1998), although occasional reports of caffeine-induced impairments on hand steadiness can be found (e.g., Bovim et al. 1995). One interpretation of effects of caffeine in low-alertness situations is that caffeine reduces the occasional long responses which may be the result of error correction or stimulus uncertainty. This means that the tail of the reaction time distribution will be much more sensitive to effects of caffeine as will responses to stimuli where the fore-period is long (or, in the case of vigilance tasks, targets occurring at irregular rather than regular intervals). Effects of caffeine seen when individuals are alert have been interpreted in terms of faster encoding of new information. This can be examined by considering responses which are the same as the one made on the previous trial (little new encoding needed, little effect of caffeine) or different responses (new stimuli and new response, large improvement with caffeine). Similarly, degraded stimuli which are hard to encode show greater benefits after caffeine than those which are easy to encode (Smith et al. 1999).

Another type of explanation has been to suggest that caffeine has no direct effects but reverses the negative effects of withdrawal (James 1994; James and Rogers 2005). Smith (2005) has questioned this view and argued that it is unlikely to be correct for the following reasons. First, it cannot account for the behavioral effects seen in animals or nonconsumers (Addicott and Laurienti 2009; Childs and de Wit 2006; Haskell et al. 2005; Hewlett and Smith 2006; Smith et al. 2006), where withdrawal cannot occur. Second, caffeine withdrawal cannot account for behavioral changes following caffeine consumption after a short period of abstinence (Warburton 1995; Smith et al. 1994) or the greater effects of caffeine when arousal is low. Finally, claims about the negative effects of caffeine withdrawal require closer examination as they can often be interpreted in ways other than caffeine dependence (e.g., expectancy; see Smith 1996; Rubin and Smith 1999). Indeed, in most of the studies that have demonstrated increases in negative affect following caffeine withdrawal, the volunteers have not been blind but have been told or even instructed to abstain from caffeine. This is clearly very different from the double-blind methodology typically used to study effects of caffeine challenge.

Other studies (e.g., Comer et al. 1997) suggest that effects of withdrawal are restricted to mood and that performance is unaltered. Like many areas of caffeine research, some of the effects that have been attributed to withdrawal are open to other interpretations. For example, Lane (1997), Phillips-Bute and Lane (1997), and Lane and Phillips-Bute (1998) compared days when midmorning coffee was either caffeinated or decaffeinated and found that caffeine consumption was associated with better performance and mood. The authors interpret this as a negative effect of caffeine withdrawal, whereas one could interpret it as a positive effect of caffeine. Other studies of caffeine withdrawal effects have methodological problems such as the lack of predrink baselines (e.g., James 1998; Robelin and Rogers 1998) or failure to consider possible asymmetric transfer when using within-subject designs (e.g., James 1998).

In summary, there is good agreement about the behavioral effects of caffeine. Plausible mechanisms have been put forward to account for these results, and it is likely that caffeine induces changes through more than one pathway. Given robust effects and plausible underlying mechanisms, it is now appropriate to consider the practical implications of ingesting caffeine. The next section examines the extent to which caffeine can remove or alleviate low alertness due to factors such as sleep deprivation, working at night, prolonged work, circadian dips (early morning and after lunch), and having a minor illness like the common cold.

REVERSAL OF IMPAIRMENTS DUE TO LOW LEVELS OF AROUSAL

Sleep Deprivation

It is well established that caffeine increases alertness and that this can interfere with sleep (for a review, see Bonnet and Arand 1994). Given this effect, it appears plausible that caffeine can reduce the effects of lack of sleep, an issue that has relevance to shift work and sustained operations. Sleep deprivation is both common and critically relevant in society. Sleep loss is common in a broad range of occupations and occurs in normal healthy individuals. Sleepiness poses an increased risk when driving and when carrying out other safety-critical activities. For those for whom sleep loss is inevitable (e.g., emergency services, and military personnel in prolonged operations), judicious use of stimulants such as caffeine is warranted. The American Academy of Sleep Medicine (Bonnet et al. 2005) has examined the efficacy and safety of caffeine use during sleep loss, and their main conclusions are summarized below.

Maintenance of Wakefulness Test (MWT) and Multiple Sleep Latency Test (MSLT)

The most commonly reported measure used to study this topic is the ability to stay awake (as measured by the MWT) or fall asleep (as measured by the MSLT). Fourteen out of fifteen studies reviewed have shown increased wakefulness measured by sleep latency tests following ingestion of caffeine by sleep-deprived volunteers.

Performance and Subjective Alertness

Choice reaction time performance of sleep-deprived individuals has been improved by caffeine in eight studies. Similarly, working memory performance of sleep-deprived individuals (e.g., digit symbol substitution or grammatical reasoning) has been shown to be improved by caffeine in over ten studies. The effects of caffeine during sleep loss have been examined over a dose range from 75 to 1200 mg per twenty-four hours. Recommended doses are usually in the range of 200–300 mg, mainly because side effects are more prevalent with higher doses (e.g., 600 mg may lead to tachycardia, particularly in women; Lagarde et al. 2000). Caffeine administration typically improves performance during sleep loss as compared with placebo, but performance and alertness often continue to decline even when caffeine is given due to further sleep loss, circadian variations, and caffeine half-life.

Self-reported alertness decreases with sleep loss, and ratings of fatigue increase. Studies that have monitored mood typically show that caffeine ameliorates these changes with a similar time course to that seen for performance variables. However, some studies have not obtained these results, and it has been suggested that some "subjective tolerance" to caffeine may develop with prolonged testing.

Studies have generally shown that doses of 200–300 mg caffeine produce few side effects, whereas higher doses (600 mg+) may increase mild symptoms (e.g., gastrointestinal upset, nervousness, and muscle twitching). Based on these findings, the review concluded that caffeine can increase alertness and improve performance at doses of 75–150 mg after acute restriction of sleep and at doses of 200–600 mg after a night or more of total sleep loss. Caffeine is unlikely to have major disruptive effects on sleep that follows eight hours or longer after administration. Prolonged administration is not recommended due to the increasing likelihood of side effects with high doses.

This topic has continued to be studied since the above review. Results show that similar effects can be observed in rats (Alhaider et al. 2010) and monkeys (van Vliet et al. 2008), which suggests that benefits do not reflect removal of effects of caffeine withdrawal (Keane and James 2008). Recent studies have shown that caffeine does not remove all of the negative effects of sleep loss (e.g., caffeine may not prevent detrimental effects of sleep on more complex cognitive functions [Gottsleig et al. 2006] or the integration of emotion and decision making [Killgore et al. 2007]). Research has also suggested individual differences in the effects of caffeine on sleep-deprived volunteers (e.g., females given caffeine were worse than males; Killgore et al. 2008). Others (e.g., Anderson and Horne 2008) have shown that similar effects to caffeine may be produced by telling a person given a placebo that he or she took caffeine. Other research has focused on individuals with pathological sleepiness and shown that caffeine can reduce performance impairments seen in sleep apnea (Norman et al. 2008).

Another recent development has been examination of how caffeine influences a model of fatigue, sleep deprivation, and circadian rhythms (Benitez et al. 2009). This model highlights patterns in data suggesting that there is a performance inhibitor that increases and saturates over a period of continuous wakefulness. Caffeine produces competitive inhibition of this inhibitor, whereas there is a multiplicative relationship between circadian rhythm and the performance inhibitor. Finally, another study (Jay et al. 2006) has extended our knowledge of caffeine and sleep loss by examining functional energy drinks. The results showed that energy drinks may be effective in reducing sleepiness associated with a single night shift.

The next section considers applied situations and the first addresses whether caffeine can prevent injuries and errors in shift workers

A Review of Randomized Trials of Caffeine and Cognitive Performance

Ker et al. (2010) have reviewed the effects of caffeine for preventing injuries, errors, and cognitive problems caused by impaired alertness in persons with jet lag or doing shift work. Thirteen trials were included, but none measured injury, two measured error, and the remaining trials assessed cognitive performance. The trials assessing the impact on errors found that caffeine significantly reduced the number of errors compared to placebo. Caffeine improved concept formation and reasoning, memory, orientation, attention, and perception. No benefit was found for language skills or verbal functioning. The results were largely from studies involving young participants under simulated conditions, and further research is needed on older workers and real-world shift work. Ker et al. conclude "Based on the current evidence, there is no reason for healthy individuals who already use caffeine within recommended levels to improve their alertness to stop doing so" (2010: p. 1).

Caffeine and Circadian Troughs

Night Work

Sleep deprivation has been the most widely studied low-arousal state in caffeine studies. Other research (e.g., Smith et al. 1993) has examined beneficial effects of caffeine over the course of the night. The results again show that caffeine can reduce the decline in alertness and psychomotor performance, although a similar circadian pattern is observed in caffeine and no-caffeine groups.

Postlunch Dip

Another time of day associated with reduced alertness is the early afternoon. Again, studies involving ingestion of caffeine after lunch have shown that caffeine can reduce the postlunch dip in alertness and sustained attention (Smith et al. 1991).

Early Morning

Many laboratory studies of the behavioral effects of caffeine have been carried out in the early morning. This is a time of day when alertness is below optimum, and this may be partially responsible for some of the observed benefits.

Evening

Circadian alertness starts to decline in the evening, and studies have shown that caffeine may improve performance and alertness at this time (Smith et al. 1993). However, fatigue in the evening is likely to be due to a combination of endogenous rhythms and fatigue due to the activity over the day. Indeed, research that has examined effects of prolonged work in the evening has been able to demonstrate both effects of caffeine seen in alert individuals and those effects that are only observed when alertness is reduced (Smith et al. 2005).

The next section reviews the effects of caffeine in counteracting the effects of prolonged work.

Caffeine and Sustained Work

One source of reduced alertness is prolonged work. Indeed, even in short laboratory studies, alertness usually decreases over the course of the task battery. This shows that even when we are studying reasonably alert individuals, we are often studying effects of caffeine on reduced alertness. Caffeine has a much bigger effect when alertness is reduced by work, and this can be seen by comparing effects on rating of alertness taken at the start of a test battery and those at the end (caffeine has a much bigger effect on the posttask ratings). It also explains why studies with pre-caffeine baselines can be more sensitive as the volunteer may be fatigued by completing the baseline test battery.

Laboratory studies of prolonged work have shown that caffeine can reduce fatigue. For example, Smith et al. (1994) examined effects of caffeine given after eight hours of performing laboratory tasks on performance over the subsequent four hours. The results showed that caffeine reduced the fatigue seen in the placebo group (after twelve hours, the caffeine group was at the same level of performance as it had been six hours earlier). Smith et al. (2005) examined effects of caffeine after a day of normal consumption on prolonged performance in the evening. The results showed initial benefits of caffeine consumption that became more widespread as the session continued.

Other research has examined effects of caffeine in simulations of extreme situations (usually military operations), and these are described in the next section.

Sustained Military Operations

Lieberman et al. (2002) investigated whether caffeine would reduce the adverse effects of sleep deprivation and exposure to severe environmental and operational stress. They studied U.S. Navy Sea–Air–Land trainees and found that even in the most adverse circumstances, moderate doses of caffeine improved vigilance, learning, memory, and mood state. A dose of 200 mg appeared to be optimal under such conditions. Lieberman et al. conclude, “When cognitive performance is critical and must be maintained during exposure to severe stress, administration of caffeine may provide a significant advantage” (2002: p. 250).

Other research has examined the beneficial effects of caffeinated tube food on pilot performance during a nine-hour simulated U-2 mission (Doan et al. 2006). The results showed that the caffeinated tube food (200 mg caffeine consumed every four hours) maintained cognitive performance at baseline levels over a nine-hour overnight period. Research has considered both cognitive and physical performance measures in sustained operations. McClellan et al. (2005) investigated performance during twenty-seven hours of sustained wakefulness in Special Forces personnel. They found that caffeine (200 mg caffeinated gum administered on three occasions) maintained performance of a reconnaissance vigilance task and also improved running times compared to placebo. However, caffeine had no effect on marksmanship. McClellan et al. (2007) conducted a similar study over a period of four days and three nights of sustained operations. The results showed caffeine maintained both vigilance and physical performance during sustained operations that require periods of overnight wakefulness and restricted opportunities for daytime sleep.

Alertness can be reduced in many different ways, and one factor that is frequent and widespread is having a minor illness such as the common cold. The next section summarizes the effects of the caffeine in removing the malaise associated with the common cold.

Caffeine and the Common Cold

There is considerable evidence that minor illnesses, such as the common cold, are associated with reduced alertness and impaired performance (see Smith [in press] for a review). Smith et al. (1997) examined whether caffeine (1.5 mg/kg) would remove this behavioral malaise. Measures taken prior to the ingestion of caffeine showed that those with a cold reported reduced alertness and had slower reaction times on a simple reaction time task and a five-choice serial response task. The volunteers were then retested following ingestion of caffeine or placebo. Volunteers with colds given caffeine reported a similar level of performance to healthy volunteers, whereas those with colds in the placebo condition continued to show reduced alertness and impaired psychomotor performance.

One must now ask whether beneficial effects of habitual caffeine consumption can be seen in other aspects of everyday life.

BENEFICIAL EFFECTS OF HABITUAL CONSUMPTION OF CAFFEINE

There has been far less research on the effects of regular caffeine consumption than on acute effects. However, a number of papers suggest that high consumers of caffeine demonstrate better performance (e.g., Loke 1988, 1989). The strongest evidence for beneficial effects of regular caffeine consumption comes from a study by Jarvis (1993). He examined the relationship between habitual coffee and tea consumption and cognitive performance using data from a cross-sectional survey of a representative sample of over 9,000 British adults. Participants completed tests of simple reaction time, choice reaction time, incidental verbal memory, and visuo-spatial reasoning, in addition to providing self-reports of usual coffee and tea intake. After controlling extensively for potential confounding variables, a dose–response trend for improved performance with higher levels of coffee consumption (best performance associated with about 400 mg caffeine per day) was found for all tests. Estimated overall caffeine consumption showed a dose–response relationship to improved cognitive performance that was strongest in those who had consumed high levels for the longest time period (the fifty-five years plus age group). Studies by Hogervorst et al. (1998) and Rogers and Dernoncourt (1998) have failed to replicate these effects using acute caffeine challenges, suggesting that the above effects reflect regular consumption patterns rather than recent intake of caffeine.

Other research has demonstrated that effects of caffeine on artificial laboratory tasks extend to simulations of real-life activities (e.g., driving; Horne and Reyner

1996; Brice and Smith 2001). A study of simulated assembly line work (Muehlbach and Walsh 1995) also demonstrated significant improvements after caffeine on five consecutive nights and showed no decrements when caffeine was withdrawn.

Caffeine at Work

Smith (2005) examined the impact of habitual caffeine consumption on performance and safety at work. In the first study, volunteers, all of whom were regular caffeine consumers, rated their alertness and carried out a simple reaction time task before and after work on a Monday and Friday. Caffeine consumption during the day was recorded, and volunteers were subdivided into low and high consumers based on a median split (220 mg/day). The results showed that those who consumed higher levels of caffeine reported significantly greater increases in alertness over the working day and a significantly smaller slowing of reaction time.

The second study involved secondary analyses of a database formed by combining the Bristol Stress and Health at Work and Cardiff Health and Safety at Work studies. In the first analyses, associations between caffeine consumption and frequency of cognitive failures were examined in a sample of 1,253 white-collar workers. The second set of analyses examined associations between caffeine consumption and accidents at work in a sample of 1,555 workers who were especially at risk of having an accident. The results from the second study demonstrated significant associations between greater caffeine consumption and fewer cognitive failures and accidents at work. After controlling for possible confounding factors, it was found that higher caffeine consumption was associated with about half the risk of frequent or very frequent cognitive failures and a similar reduction in risk for accidents at work. Overall, the results from the three analyses confirmed that caffeine consumption may have benefits for performance and safety at work.

Caffeine Outside of Work

Smith (2009) conducted secondary analyses of a large epidemiological database to examine associations between caffeine consumption and cognitive failures (errors of memory, attention, and action) in a nonworking sample. Associations between caffeine consumption and physical and mental health problems were also examined. After controlling for possible confounding factors, significant associations between caffeine consumption and fewer cognitive failures were observed. Overall, the results show that caffeine consumption may benefit cognitive functioning in a nonworking population. This confirms earlier findings from working samples. This beneficial effect of caffeine was not associated with negative health consequences.

Caffeine and Social Behavior

Most of the studies of caffeine have examined behavioral changes of an individual. Tse et al. (2009) examined the effects of caffeinated coffee on cooperative behavior. The results showed that caffeine improved social support and reduced negative affect. Animal studies show that caffeine can increase aggression in the rat (Wilson et al. 2000), whereas the only laboratory study of acute caffeine and aggression has shown a decrease in aggressive responding after caffeine (Cherek et al. 1984). Similarly, there is a lack of information on caffeine and risk taking. Recent studies have suggested a link between high consumption of energy drinks and risky or antisocial behaviors (Miller 2008; O'Brien et al. 2008; Jones and Lejuez 2005). These cross-sectional studies do not rule out the influence of other possible confounders or reverse causality (e.g., individuals who habitually take risks choose to consume energy beverages). Further research is needed on this topic.

Caffeine and Driving

It has been frequently shown that sleepiness is a major cause of road traffic accidents. The previous section showed that caffeine can reduce some of the negative effects of low alertness, and this has been examined using simulated driving (e.g., De Valck et al. 2003). For example, Horne and Reyner conducted a number of studies looking at (1) the efficacy of 200 mg caffeine with sleep-restricted and completely sleep-deprived drivers (Reyner and Horne 2000), (2) an energy drink containing caffeine with sleep-restricted drives (Horne and Reyner 2001), and (3) "a functional energy drink" and sleep-restricted drivers (Reyner and Horne 2002). The results from these studies showed that caffeine generally reduced the impaired driving performance seen in sleepy drivers given placebo.

Philip et al. (2006) extended these results by examining the effects of sleepiness and caffeine on real-life driving. Extended driving and sleepiness resulted in an increase in lane crossing, which was reduced by 200 mg of caffeine. As shown in previous sections of this chapter, fatigue can be induced in a number of ways. Brice and Smith (2001) conducted a study which involved one hour of simulated driving before and after either caffeine or placebo. Volunteers were also given a battery of tasks measuring subjective alertness and sustained attention. Caffeine reduced steering variability, which in real-life driving may lead to lane crossing, and increased subjective alertness and improved cognitive vigilance. This suggests that results found after caffeine with artificial laboratory tasks may be applicable to real-life activities involving similar functions.

Driving performance can be impaired by a number of factors, the most widely studied being alcohol. Liquori and Robinson (2001) examined whether caffeine would reduce an alcohol-induced impairment of simulated driving. The results suggested that caffeine may increase alertness and improve reaction time after alcohol use but will not completely counteract the alcohol impairments seen in driving.

James and Keane (2007) argue that many of the effects of caffeine seen in studies of driving can be interpreted in terms of reversal of the effects of caffeine withdrawal. One method of distinguishing a benefit of caffeine from a reversal of caffeine withdrawal is to compare caffeine consumers with nonconsumers. This could be studied using an epidemiological approach examining associations between caffeine consumption and road traffic accidents. For example, Smith (submitted) examined a community sample from South Wales (N = 6,648). The respondents provided information on involvement in road traffic accidents; 3.6 percent of nonconsumers of caffeine were involved in a road accident requiring medical attention, whereas only 2.2 percent of caffeine consumers were. Logistic regression analyses, including demographic, lifestyle, and psychosocial characteristics, showed that consumption of caffeine nearly halved the risk of being in a road accident (OR = 0.58 CI 0.35, 0.98). This result confirms previous research showing that caffeine reduces the risk of accidents (at work) and supports the existing literature and information campaigns about the positive benefits of caffeine for road safety.

Prevention of Cognitive Decline in the Elderly

Animal studies suggest that habitual caffeine consumption may prevent memory decline (Cunha and Agostinho 2010). Several epidemiological studies have examined associations between consumption of caffeine and dementia. A recent systematic review and meta-analysis (Santos et al. 2010) considered nine cohort and two case control studies. The outcomes examined were Alzheimer's disease (four studies), dementia or cognitive impairment (two studies), and cognitive decline (three studies). The summary relative risk for the association between caffeine intake and the different cognitive measures was 0.84 [95 percent CI: 0.72–0.99]. This suggests a trend toward a protective effect of caffeine, but the large methodological heterogeneity across a small number of studies precludes more definitive conclusions. Further research is clearly needed in this area.

Mental Health

Anecdotal evidence suggests that when individuals have consumed an excessive amount of caffeine, they may become anxious. Similarly, some psychiatric patients attribute their problems to consumption of caffeine, which has led to a diagnosis of "caffeinism." Other patients, especially those with anxiety disorders, report that caffeine may exacerbate their problems. The validity of these statements will now be assessed by consideration of the literature on these topics.

Lieberman stated that

> it appears that caffeine can increase anxiety when administered in single bolus doses of 300 mg or higher, which is many times greater than the amount present in a single serving of a typical caffeine-containing beverage. However, in lower doses it appears to have little effect on this mood-state or, under certain circumstances, it may even reduce anxiety levels. It has also been observed that caffeine reduces self-rated depression when administered in moderate doses (Lieberman, 1988). (1992: p. 63)

The literature supports Lieberman's view since only a small proportion of the studies reviewed show increases in anxiety following administration of caffeine. Overall, these results suggest that increases in anxiety following caffeine are often only found following consumption of amounts that would rarely be ingested by the majority of people.

It is important to assess whether caffeine leads to mood problems when the person ingesting it already has a high level of anxiety. It has been claimed that some people abstain from caffeinated drinks because of the accompanying jitteriness and nervousness (Goldstein et al. 1969). Other authors have even gone as far as to suggest that caffeine acts as a "fairly convincing model of generalised anxiety" (Lader and Bruce 1986). Caffeinism refers to a constellation of symptoms associated with very high caffeine intake that are virtually indistinguishable from severe chronic anxiety (Greden 1974). Caffeinism is usually associated with daily intakes of between 1000 and 1500 mg. However, it appears to be a rather specific condition, and there is little evidence for correlations between caffeine intake and anxiety in either nonclinical volunteers (Lynn 1973; Hire 1978) or psychiatric outpatients (Eaton and Mcleod 1984). Other research has investigated whether caffeine is capable of increasing the anxiety induced by other stressors. Shanahan and Hughes (1986) found that 400 mg of caffeine increased anxiety when paired with a stressful task. However, other research (e.g., Hasenfratz and Battig 1992; Smith et al. 1997b) has not been able to provide any evidence of interactive effects of caffeine and stress.

Recent research has shown an association between ADORA2A and DRD2 polymorphisms and caffeine-induced anxiety (Childs et al. 2008). Adenosine receptors functionally interact with dopamine receptors in the brain. Functional polymorphisms in the genes for either adenosine or dopamine receptors may, therefore, affect responses to caffeine. Childs et al. (2008) found that 50 mg caffeine didn't increase anxiety in any individuals, whereas 450 mg caffeine increased it in the majority of the volunteers. With a dose of 150 mg caffeine, anxiety was associated with ADORA2A and DRD2 polymorphisms.

In contrast, moderate caffeine intake has been associated with fewer depressive symptoms and a lower risk of suicide (see Lara 2010 for a review). This effect of caffeine on depression may have other knock-on effects with regard to health. Smith (in preparation) conducted secondary analyses of a large epidemiological database (N = 2,750) to examine associations between caffeine and both chronic and acute health outcomes. Many of the initial associations between caffeine and health were no longer significant when potential confounders were examined. However, caffeine consumption was still significantly associated with reduced depression in the final regressions. Caffeine consumption was also associated in a dose–response fashion with fewer upper respiratory tract symptoms. This suggests that caffeine may influence the immune system, either directly or by reducing depression (a well-established risk factor for immunosuppression).

CAFFEINE AND CHILDREN AND ADOLESCENTS

Ingestion of caffeine from naturally occurring sources has been largely restricted to adults, but it is now added, sometimes in large quantities, to drinks that are consumed by children. Our knowledge of the effects of caffeine on the behavior of children needs to be extended by further research. The current position on this topic can be briefly summarized as follows (for a more detailed account, see Temple 2009).

Older studies of the behavioral effects of caffeine on children have shown similar effects to those observed in adults (Bernstein et al. 1994; Elkins et al. 1981; Rapoport et al. 1981). Effects in children are often smaller than those observed in adults, which may reflect the smaller doses consumed.

It is generally agreed that caffeine intake by pregnant women should be kept at a low level (below 200 mg) (CARE Study Group 2008). However, there is no evidence showing that caffeine consumption during pregnancy or childhood influences brain development.

HEALTH EFFECTS OF CAFFEINE CONSUMPTION

It is important to conduct a cost–benefit analysis when considering the effects of caffeine. Benefits with respect to caffeine usually refer to behavioral outcomes, and costs reflect possible long-term health effects. Caffeine has been linked with a range of possible health problems, but most of these associations are not significant when confounding factors are adjusted for (Nawrot et al. 2003). Indeed, in recent years there have been suggestions that caffeine consumption may have health benefits. Daly (2007) argues that studies of caffeine have played a key part in defining the role of adenosine receptors, phosphodiesterases, and calcium release channels in physiological processes. Caffeine and various analogs, the latter designed to enhance potency and selectivity toward specific biological targets, are potential therapeutic agents for intervention in Alzheimer's disease, asthma, cancer, diabetes, and Parkinson's disease.

CONCLUSIONS

Research on the behavioral effects of caffeine has provided an established profile of the changes in mood and performance that occur after ingestion. A systems neuroscience approach provides plausible mechanisms for these effects. For example, the alerting effects seen in fatigued individuals can be accounted for by blockade of adenosine receptors. In addition, explanations based on changes in cognition can explain behavioral effects of caffeine in terms of faster encoding of new information and changes in the tail of the reaction time distribution. A systems neuroscience approach should also look at effects at the level of real-life behavior, both of the individual and in society. The research reviewed here has shown that impairments in real-life activities due to fatigue can be reduced by caffeine. Such effects may be especially important for safety-critical activities such as driving.

There is also evidence that caffeine may prevent or reduce the impact of diseases associated with behavioral problems (e.g., depression and Alzheimer's disease). Further research is required to determine effects of caffeine on well-being rather than removal of functional deficits. Similarly, more research is needed on the effects of caffeine on children and on a wider range of social behavior.

REFERENCES

Addicott, M. A., and P. J. Laurienti. 2009. A comparison of the effects of caffeine following abstinence and normal caffeine use. *Psychopharmacology* 207 (3):423–31.

Alhaider, I. A., A. M. Aleisa, T. T. Tran, et al. 2010. Chronic caffeine treatment prevents sleep deprivation-inducted impairment of cognitive function and synaptic plasticity. *Sleep* 33 (4):437–44.

Amendola, C. A., J. D. E. Gabrieli, and H. R. Lieberman. 1998. Caffeine's effects on performance and mood are independent of age and gender. *Nutritional Neuroscience* 1:269–80.

Anderson, C., and J. A. Horne. 2008. Placebo response to caffeine improves reaction time performance in sleepy people. *Human Psychopharmacology* 23 (4):333–6.

Benitez, P. L., G. H. Kamimore, T. J. Balkin, et al. 2009. Modeling fatigue over sleep deprivation, circadian rhythm and caffeine with a minimal performance inhibitor model. *Methods Enzymol* 454:405–21.

Bernstein, G. A., M. E. Carroll, R. D. Crosby, A. R. Perwein, F. S. Go, and N. L. Benowitz. 1994. Caffeine effects on learning, performance, and anxiety in normal school-aged children. *Journal of the American Academy of Child and Adolescent Psychiatry* 35:407–19.

Bonnet, M. H., and D. L. Arand. 1994. Impact of naps and caffeine on extended nocturnal performance. *Physiology and Behavior* 56:103–9.

Bonnet, M. H., T. J. Balkin, D. F. Dinges, et al. 2005. The use of stimulants to modify performance during sleep loss: a review by the sleep deprivation and stimulant task force of the American Academy of Sleep. *Sleep* 28 (9):1163–87.

Bovim, G., P. Naess, J. Helle, and T. Sand. 1995. Caffeine influence on the motor steadiness battery in neuropsychological tests. *Journal of Clinical and Experimental Neuropsychology* 17:472–6.

Brice, C. F., and A. P. Smith. 2001a. The effects of caffeine on simulated driving, subjective alertness and sustained attention. *Human Psychopharmacology* 16:523–31.

Brice, C. F., and A. P. Smith. 2001b. Caffeine levels in saliva: associations with psychosocial factors and behavioral effects. *Human Psychopharmacology* 16:507–21.

Brice, C. F., and A. P. Smith. 2002. Effects of caffeine on mood and performance: A study of realistic consumption. *Psychopharmacology* 164:188–92.

Bruce, M., N. Scott, P. Shine, and M. Lader. 1991. Caffeine withdrawal: a contrast of withdrawal symptoms in normal subjects who have abstained from caffeine for 24 hours and for 7 days. *Journal of Psychopharmacology* 5:129–34.

CARE Study Group. 2008. Maternal caffeine intake during pregnancy and risk of fetal growth restriction: a large prospective observational study. *British Medical Journal* 337:a2332.

Cherek, D. R., J. L. Steinberg, and R. V. Vines. 1984. Low doses of alcohol affect human aggressive responses. *Biological Psychiatry* 19 (2):263–7.

Childs, E., and H. de Wit. 2006. Subjective, behavioral, and physiological effects of acute caffeine in light, nondependent caffeine users. *Psychopharmacology* 185:514–23.

Childs, E., C. Hohoff, J. Deckert, et al. 2008. Association between ADORA2A and DRD2 polymorphisms and caffeine-induced anxiety. *Neuropsychopharmacology* 33 (12): 2791–2800.

Christopher, G., D. Sutherland, and A. Smith. (2005). Effects of caffeine in non-withdrawn volunteers. *Human Psychopharmacology Clinical and Experimental* 20:47–53.

Comer, S. D., M. Haney, R. W. Foltin, and M. W. Fischman. 1997. Effects of caffeine withdrawal in humans living in a residential laboratory. *Experimental and Clinical Psychopharmacology* 5:399–403.

Cunha, R. A., and P. M. Agostinho. 2010. Chronic caffeine consumption prevents memory disturbance in different animal models of memory decline. *Journal of Alzheimers Disease* 20 (suppl. 1):S95–116.

Daly, J. W. 2007. Caffeine analogs: biomedical impact. *Cellular and Molecular Life Sciences* 64 (16):153–69.

De Valck, E., E. De Groot, and R. Cluydts. 2003. Effects of slow-release caffeine and a nap on driving simulator performance after partical sleep deprivation. *Perceptual and Motor Skills* 96 (1):67–78.

Dews, P. B., C. P. O'Brien, and J. Bergman. 2002. Caffeine: behavioral effects of withdrawal and related issues. *Food and Chemical Toxicology* 40:1257–61.

Doan, B. K., P. A. Hickey, H. R. Lieberman, et al. 2006. Caffeinated tube food effect on pilot performance during a 9-hour, simulated nighttime U-2 mission. *Aviation, Space, and Environmental Medicine* 77 (10):1034–40.

Durlach, P. J. 1998. The effects of a low dose of caffeine on cognitive performance. *Psychopharmacology* 140:116–19.

Eaton, W. W., and J. McLeod. 1984. Consumption of coffee or tea and symptoms of anxiety. *American Journal of Public Health* 74:66–8.

Elkins, R. N., J. L. Rapoport, T. P. Zahn, M. S. Buchsbaum, H. Weingartner, I. J. Kopin, D. Langer, and C. Johnson. 1981. Acute effects of caffeine in normal prepubertal boys. *American Journal of Psychiatry* 138:178–83.

File, S. A., A. J. Bond, and R. G. Lister. 1982. Interaction between effects of caffeine and lorazepam in performance tests and self-ratings. *Journal of Clinical Psychopharmacology* 2:102–6.

Fredholm, B. B., K. Battig, J. Holmen, A. Nehlig, and E. E. Zvartau. 1999. Actions of caffeine in the brain with special reference to factors that contribute to its widespread use. *Pharmacological Reviews* 91:83–133.

Goldstein, A., S. Kaizer, and O. Whitby. 1969. Psychotropic effects of caffeine in man: iv. Quantitative and qualitative differences associated with habituation to coffee. *Clinical Pharmacology and Therapeutics* 10:489–97.

Goldstein, E. R., T. Ziegenfuss, D. Kalman, et al. 2010. International society of sport nutrition position stand: caffeine and performance. *Journal of International Society of Sports Nutrition* 7 (1):5.

Gottsleig, J. M., M. Adam, J. V. Retey, et al. 2006. Random number generation during sleep deprivation: effects of caffeine on response maintenance and stereotypy. *Journal of Sleep Research* 15 (1):31–40.

Greden, J. F. 1974. Anxiety or caffeinism: a diagnostic dilemma. *American Journal of Psychiatry* 131:1089–92.

Hasenfratz, M., and K. Battig. 1992. No psychophysiological interactions between caffeine and stress? *Psychopharmacology* 109:283–90.

Hasenfratz, M., and K. Battig. 1994. Acute dose-effect relationships of caffeine and mental performance, EEG, cardiovascular and subjective parameters. *Psychopharmacology* 114:281–87.

Haskell, C. F., D. O. Kennedy, K. A. Wesnes, and A. B. Scholey. 2005. Cognitive and mood improvements of caffeine in habitual consumers and habitual non-consumers of caffeine. *Psychopharmacology* 179:813–85.

Hewlett, P. J., and A. P. Smith. 2006. Acute effects of caffeine in volunteers with different patterns of regular consumption. *Human Psychopharmacology Clinical and Experimental* 21:167–80.

Hire, J. N. 1978. Anxiety and caffeine. *Psychological Reports* 42:833–4.

Hogervorst, E., W. J. Riedel, J. A. J. Scmitt, and J. Jolles. 1998. Caffeine improves memory performance during distraction in middle-aged but not in young or old subjects. *Human Psychopharmacology* 13:277–84.

Horne, J. A., and L. A. Reyner. 1996. Counteracting driver sleepiness: Effects of napping, caffeine, and placebo. *Psychophysiology* 33:306–9.

Horne, J. A., and L. A. Reyner. 2001. Beneficial effects of an "energy drink" given to sleepy drivers. *Amino Acids* 20:83–9.

James, J. E. 1994. Does caffeine enhance or merely restore degraded psychomotor performance? *Neuropsychobiology* 30:124–5.

James, J. E. 1998. Acute and chronic effects of caffeine on performance, mood, headache and sleep. *Neuropsychobiology* 38:32–41.

James, J. E., and M. A. Keane. 2007. Caffeine, sleep and wakefulness: implications of new understanding about withdrawal reversal. *Human Psychopharmacology* 22:549–58.

Jarvis, M. J. 1993. Does caffeine intake enhance absolute levels of cognitive performance? *Psychopharmacology* 110:45–52.

Jay, S. M., R. M. Petrilli, S. A. Ferguson, et al. 2006. The suitability of a caffeinated energy drink for night-shift workers. *Physiological Behavior* 87 (5):925–31.

Jones, H. A., and C. W. Leguez. 2005. Personality correlates of caffeine dependence: the roles of sensation seeking, impulsivity and risk-taking. *Experimental and Clinical Psychopharmacology* 13:259–66.

Juliano, L. M., and R. R. Griffiths. 2004. A critical review of caffeine withdrawal: empirical validation of symptoms and signs, incidence, severity and associated features. *Psychopharmacology* 126:1–29.

Keane, M., and J. J. James. 2008. Effects of dietary caffeine on EEG, performance and mood when rested and sleep restricted. *Human Psychopharmacology: Clinical and Experimental* 23:669–80.

Kenemans, J. L., and M. Verbaten. 1998. Caffeine and visuo-spatial attention. *Psychopharmacology* 135:353–60.

Ker, K., P. J. Edwards, L. M. Felix, K. Blackhall, and I. Roberts. 2010. Caffeine for the prevention of injuries and errors in shift workers (Review). In *The Cochrane Collaboration*. Chichester: Wiley.

Killgore, W. D., E. L. Kipizzi, G. H. Kamimore, et al. 2007. Caffeine effects on risky decision making after 75 hours of sleep deprivation. *Aviation, Space, and Environmental Medicine* 78 (10):957–62.

Killgore, W. D., A. E. Muckle, N. L. Grugle, et al. 2008. Sex differences in cognitive estimation during sleep deprivation: effects of stimulant countermeasures. *Int. J. Neuroscience* 118 (11):1547–57.

Lader, M., and M. Bruce. 1986. States of anxiety and their induction by drugs. *British Journal of Clinical Pharmacology* 22:252–61.

Lagarde, D., D. Batejat, B. Sicard, S, Trocherie, D. Chassard, M. Enslen, and F. Chauffard. 2000. Slow release caffeine: a new response to the effects of a limited sleep deprivation. *Sleep* 23:651–61.

Lane, J. D. 1997. Effects of brief caffeinated-beverage deprivation on mood, symptoms and psychomotor performance. *Pharmacology, Biochemistry and Behavior* 58:203–8.

Lane, J. D., and B. G. Phillips-Bute. 1998. Caffeine deprivation affects vigilance performance and mood. *Physiology and Behavior* 65:171–75.

Lara, D. R. 2010. Caffeine, mental health, and psychiatric disorders. *J. Alzheimers Dis* 20 (suppl. 1):S239–48.

Leathwood, P. D., and P. Pollet. 1982. Diet-induced mood changes in normal populations. *Journal of Psychiatric Research* 17:147–54.

Lieberman, H. R. 1992. Caffeine. In *Handbook of Human Performance*, ed. A. P. Smith and D. M. Jones. vol. 2, 49–72. London: Academic Press.

Lieberman, H. R., R. J. Wurtman, G. G. Emde, C. Roberts, and I. L. G. Covielle. 1987. The effects of low doses of caffeine on human performance and mood. *Psychopharmacology* 92:308–12.

Lieberman, H. R., W. J. Tharion, B. Shukitt-Hale, K. L. Speckman, and R. Tulley. 2002. Effects of caffeine, sleep loss, and stress on cognitive performance and mood during U.S. Navy SEAL training. *Psychopharmacology* 164:250–61.

Liquori, A., and J. H. Robinson. 2001. Caffeine antagonism of alcohol-induced driving impairment. *Drug and Alcohol Dependence* 63 (2):123–9.

Loke, W. H. 1988. Effects of caffeine on mood and memory. *Physiology and Behavior* 44:367–72.

Loke, W. H. 1989. Effects of caffeine on task difficulty. *Psychologica-Belgica* 29:51–62.

Loke, W. H. 1990. Effects of repeated caffeine administration on cognition and mood. *Human Psychopharmacology Clinical and Experimental* 5:339–48.

Loke, W. H., J. V. Hinrichs, and M. M. Ghoneim. 1985. Caffeine and diazepam: separate and combined effects on mood, memory, and psychomotor performance. *Psychopharmacology* 87:344–50.

Loke, W. H., and C. J. Meliska. 1984. Effects of caffeine use and indigestion on a protracted visual vigilance task. *Psychopharmacology* 84:54–7.

Lorist, M. M. 1998. Caffeine and information processing in man. In: *Nicotine, caffeine and social drinking: Behavior and brain function*, ed. J. Snel and M. M. Lorist, pp. 185–200. Amsterdam: Academic Press.

Lorist, M. M., and J. Snel. 1997. Caffeine effects on perceptual and motor processes. *Electroencephalography and Clinical Neurophysiology* 102:401–13.

Lorist, M. M., J. Snel, and A. Kok. 1994. Influence of caffeine on information processing stages in well-rested and fatigued subjects. *Psychopharmacology* 113:411–21.

Lorist, M. M., J. Snel, A. Kok, and G. Mulder. 1994. Influence of caffeine on selective attention in well-rested and fatigued subjects. *Psychophysiology* 31:525–34.

Lynn, R. 1973. National differences in anxiety and the consumption of coffee. *British Journal of Social Clinical Psychology* 12:92–3.

McLellan, Y. M., G. H. Kamimore, D. M. Voss, et al. 2005. Caffeine maintains vigilance and improves run times during night operations for Special Forces. *Aviation, Space, and Environmental Medicine* 76 (7):647–54.

McLellan, Y. M., G. H. Kamimore, D. M. Voss, et al. 2007. Caffeine effects on physical and cognitive performance during sustained operations. *Aviation, Space, and Environmental Medicine* 78 (9):871–7.

Miller, K. E. 2008. Wired: energy drinks, jock identity, masculine norms, and risk taking. *Journal of the American College of Health* 56:481–89.

Muehlbach, M. J., and J. K. Walsh. 1995. The effects of caffeine on simulated night-shift work and subsequent daytime sleep. *Sleep* 18:22–9.

Nawrot, P., S. Jordan, J. Eastwood, et al. 2003. Effects of caffeine on human health. *Food Additives and Contaminants* 20 (1):1–30.

Norman, D., Bardwell, W. A., Loredo, J. S. et al. 2008. Caffeine intake is independently associated with neuropsychological performance in patients with obstructive sleep apnea. *Sleep Breath* 12 (3):199–205.

O'Brien, M. C., T. McCoy, S. D. Rhodes, A. Wagoner, and M. Wolfson. 2008. Caffeinated cocktails: get wired, get drunk, get injured. *Academy of Emergency Medicine* 15:453–60.

Philip, P., J. Taillard, N. Moore, et al. 2006. The effects of coffee and napping on nighttime highway driving; a randomized trial. *Annals of Internal Medicine* 144 (11):785–91.

Phillips-Bute, B. G., and J. D. Lane. 1997. Caffeine withdrawal symptoms following brief caffeine deprivation. *Physiology and Behavior* 63:35–9.

Rapoport, J. L., R. Elkins, A. Neims, T. Zahn, and C. J. Berg. 1981. Behavioural and autonomic effects of caffeine in normal boys. *Developmental Pharmacology and Therapy* 3:74–82.

Regina, E. G., G. M. Smith, C. G. Keiper, and R. K. McKelvey. 1974. Effects of caffeine on alertness in simulated automobile driving. *Journal of Applied Psychology* 59:483–9.

Reyner, L. A., and J. A. Horne. 1997. Suppression of sleepiness in drivers: Combination of caffeine with a short nap. *Psychophysiology* 34:721–5.

Reyner, L. A., and J. A. Horne. 2000. Early morning driver sleepiness: Effectiveness of 200 mg caffeine. *Psychophysiology* 37:251–6.

Reyner, L. A., and J. A. Horne. 2002. Efficacy of a "functional energy drink" in counteracting driver sleepiness. *Physiology and Behavior* 75:331–5.

Riedel, W., E. Hogervorst, R. Leboux, and F. Verhey. 1995. Caffeine attenuates scopolamine-induced memory impairment in humans. *Psychopharmacology* 122:158–68.

Robelin, M., and P. J. Rogers. 1998. Mood and psychomotor performance effects of the first but not subsequent, cup-of-coffee equivalent doses of caffeine consumed after overnight caffeine abstinence. *Behavioral Pharmacology* 9:611–18.

Rogers, P. J., and C. Dernoncourt. 1998. Regular caffeine consumption: A balance of adverse and beneficial effects for mood and psychomotor performance. *Pharmacology, Biochemistry and Behavior* 59:1039–45.

Rogers, P. J., N. J. Richardson, and C. Dernoncourt. 1995a. Caffeine use: Is there a net benefit for mood and psychomotor performance? *Neuropsychobiology* 31:195–9.

Rubin, G. J., and A. P. Smith. 1999. Caffeine withdrawal and headaches. *Nutritional Neuroscience* 2:123–6.

Sanchez-Ortuno, M., N. Moore, J. Taillard, et al. 2005. Sleep duration and caffeine consumption in a French middle-aged working population. *Sleep Medicine* 6 (3):247–51.

Santos, C., J. Costa, J. Santos, et al. 2010. Caffeine intake and dementia: systematic review and meta-analysis. *Journal of Alzheimers Disease* 20 (suppl. 1):s187–204.

Santos, C., N. Lunet, A. Azevedo, et al. 2010. Caffeine intake is associated with a lower risk of cognitive decline: a cohort study from Portugal. *Journal of Alzheimers Disease* 20 (suppl. 1):s175–85.

Shanahan, M. P., and R. N. Hughes. 1986. Potentiation of performance-induced anxiety by caffeine in coffee. *Psychological Reports* 59:83–6.

Smit, H. J., and P. J. Rogers. 2000. Effects of low doses of caffeine on cognitive performance, mood and thirst in low and high caffeine consumers. *Psychopharmacology* 152:167–73.

Smith, A. P. 1995. Caffeine, caffeine withdrawal and psychomotor performance: a reply to James. *Neuropsychobiology* 31:200–1.

Smith, A. P. 1996. Caffeine dependence: An alternative view. *Nature Medicine* 2:494.

Smith, A. P. 2002. Effects of caffeine on human behaviour. *Food and Chemical Toxicology* 40:1243–55.

Smith, A. P. 2005. Caffeine at work. *Human Psychopharmacology,* 20:441–5.

Smith, A. P. 2005. Caffeine. In *Nutritional Neuroscience*, ed. H. Lieberman, R. Kanarek and C. Prasad, 335–9. New York: Taylor and Francis.

Smith, A. P. 2009. Caffeine, cognitive failures and health in a non-working community sample. *Human Psychopharmacology: Clinical and Experimental* 24:29–34. doi: 10.1002/hup.991.

Smith, A. P., C. F. Brice, J. Nash, N. Rich, and D. J. Nutt. 2003. Caffeine and central noradrenaline: effects on mood and cognitive performance. *Journal of Psychopharmacology* 17:283–92.

Smith, A. P., B. Brockman, R. Flynn, A. Maben, and M. Thomas. 1993. Investigation of the effects of coffee on alertness and performance during the day and night. *Neuropsychobiology* 27:217–23.

Smith, A., C. Christopher, and D. Sutherland. 2006. Effects of caffeine in overnight-withdrawn consumers and non-consumers. *Nutritional Neuroscience* 9:63–71.

Smith, A. P, R. Clark, and J. Gallagher. 1999. Breakfast cereal and caffeinated coffee: Effects on working memory, attention, mood and cardiovascular function. *Physiology and Behavior* 67:9–17.

Smith, A. P., A. M. Kendrick, and A. L. Maben. 1992. Effects of breakfast and caffeine on performance and mood in the late morning and after lunch. *Neuropsychobiology* 26:198–204.

Smith, A. P., A. M. Kendrick, A. L. Maben, and J. Salmon. 1994. Effects of breakfast and caffeine on cognitive performance, mood and cardiovascular functioning. *Appetite* 22:39–55.

Smith, A. P., A. L. Maben, and P. Brockman. 1994. Effects of evening meals and caffeine on cognitive performance, mood and cardiovascular functioning. *Appetite* 22:57–65.

Smith, A. P., and W. Phillips. 1993. Effects of low doses of caffeine in coffee on human performance and mood. In *15th International Scientific Colloquium on Coffee* Vol 2. *Association Scientifique Internationale de Cafe*, Paris, 461–9.

Smith, A. P., J. M. Rusted, P. Eaton-Williams, M. Savory, and P. Leathwood. 1991. Effects of caffeine given before and after lunch on sustained attention. *Neuropsychobiology* 23:160–3.

Smith, A. P., W. Sturgess, and J. Gallagher. 1999. Effects of a low dose of caffeine given in different drinks on mood and performance. *Human Psychopharmacology* 14:473–82.

Smith, A., D. Sutherland, and C. Christopher. 2005. Effects of repeated doses of caffeine on mood and performance of alert and fatigued volunteers. *Journal of Psychopharmacology* 19:620–6.

Smith, A. P., M. Thomas, K. Perry, and H. Whitney. 1997a. Caffeine and the common cold. *Journal of Psychopharmacology* 11:319–24.

Smith, A. P., H. Whitney, M. Thomas, K. Perry, and P. Brockman. 1997b. Effects of caffeine and noise on mood, performance and cardiovascular functioning. *Human Psychopharmacology* 12:27–33

Streufert, S., U. Satish, R. Pogash, D. Gingrich, R. Landis, J. Roache, and W. Severs. 1997. Excess coffee consumption in simulated complex work settings: Detriment or facilitation of performance? *Journal of Applied Psychology* 82:774–82.

Tse, W. S., C. C. Chan, S. Y. Shiu, et al. 2009. Caffeinated coffee enhances co-operative behavior in the Mixed Motive Game in healthy volunteers. *Nutritional Neuroscience* 12 (1):21–7.

Van Vliet, S. A., M. J. Jongsma, R. A. Vanwersch, et al. 2008. Efficacy of caffeine and modafinil in counteracting sleep deprivation in the marmoset monkey. *Psychopharmacology* 197 (1):59–66.

Warburton, D. M. 1995. Effects of caffeine on cognition and mood without caffeine abstinence. *Psychopharmacology* 119:66–70.

Warburton, D. M., E. Bersellini, and E. Sweeney. 2001. An evaluation of a caffeinated taurine drink on mood, memory and information processing in healthy volunteers without caffeine abstinence. *Psychopharmacology* 158:322–8.

Wilson, J. F., N. R. Nugent, J. E. Baltes, et al. 2000. Effects of low doses of caffeine on aggressive behavior of male rats. *Psychological Reports* 86 (3):941–6.

Yeomans, M. R., T. Ripley, L. H. Davies, J. M. Rusted, and P. J. Rogers. 2002. Effects of caffeine on performance and mood depend on the level of caffeine abstinence. *Psychopharmacology* 164:241–9.

14 Caffeine Effects on Aggression and Risky Decision Making

Caroline R. Mahoney, Tad T. Brunyé, and Grace E. Giles

CONTENTS

Introduction....293
Background....294
Caffeine and Aggressive Behavior....295
Caffeine and Risky Behavior....297
 Inhibitory Control and Risk Taking....298
 Active versus Reactive Measures....299
Caffeine Effects on Physiology and Implications for Aggressive and Risky Behavior....301
 Caffeine, Serotonin, and Aggression....301
 Effects of Caffeine on Serotonin....301
 Serotonin–Aggression Relationship....302
 Caffeine Withdrawal and Measures of Serotonin and Aggression....302
Caffeine, Dopamine, and Behavioral Control....303
Caffeine and Cortisol....305
Conclusion and Future Directions....306
References....307

INTRODUCTION

It is widely accepted that caffeine, in moderate doses, can exert positive effects on behavior and mood, such as enhancing vigilance, decreasing response time, and increasing alertness (Lieberman et al. 1987, 2002; Lieberman 2001; Nehlig et al. 1992; Smith 2002; Wesensten et al. 2005). Perhaps due to these positive effects, caffeine is the most widely consumed stimulant in the world, with most adults in the United States consuming it daily (Gilbert 1984). The extent of daily caffeine consumption appears to be on the rise as well, potentially due to increasing popularity of local and national coffee chains as well as increased caffeine content in commonly available retail beverages. In addition, consumption of heavily marketed energy drinks such as Red Bull, Amp, Monster, Full Throttle, Spike Shooter, and

Wired X344, which can contain up to 505 mg of caffeine per 12 oz serving, is rising (Reissig et al. 2009). The popular press has taken interest in the growing rate of caffeine consumption, especially in the young adult (college-aged) population, as indicated by numerous news stories claiming that high caffeine consumption is associated with aggressive and risky behavior. The veracity of these claims, however, is unclear, as they are not strongly supported by empirical work. This chapter discusses the existing literature in the areas of caffeine and aggression, and caffeine and risk-taking behavior, as well as related literature in the areas of caffeine effects on inhibitory control. In addition, it includes a brief discussion of caffeine effects on physiology and the implications for aggressive and risky behavior. The chapter begins with a discussion of three recent publications that have initiated much of the interest in the popular press and then continues with a review of the empirical literature in each of the related areas.

BACKGROUND

Increased interest in the relationship between caffeine consumption and aggressive and risky behaviors has been largely inspired by a wave of popular press reports (e.g., *New York Times*, MSNBC, *Science Daily*, *APA Monitor*, NaturalNews.com, etc.) on three recent studies. In the first study, Miller (2008) examined sex differences in risk-taking behavior, sport-related identity, masculine norms, and energy drink consumption. Undergraduate students at a large public university ($N = 795$) were surveyed about their typical energy drink consumption over the past month, sport-related identities, and conformity to masculine norms (using portions of the Mahalik's Conformity to Masculine Norms Inventory. which included risk-taking and violence subscales; Mahalik et al. 2003). The authors conducted correlational tests and linear regression analyses. The results indicated that frequency of energy drink consumption over a one-month period was associated with risk-taking behavior and that consumption of energy drinks predicted a greater tendency to engage in risky behaviors, such as unprotected sex.

A second study, conducted by O'Brien and colleagues (2008), examined the relationship between mixing alcohol with energy drinks and high-risk drinking behaviors (e.g., number of heavy episodic drinking days) and drinking-related consequences (e.g., choosing to ride with a driver who was under the influence or taking advantage of someone sexually). A total of 4,271 students across ten universities (eight public and two private) in North Carolina completed a private online survey. A multivariable linear mixed-effects regression model was used that adjusted for gender, age, race, fraternity or sorority status, and within-campus clustering. They report that the consumption of mixed alcohol and energy drinks was more strongly associated with both high-risk drinking behaviors and prevalence of alcohol-related consequences (e.g., perpetrating sexual assault) than consumption of alcohol without caffeinated energy drinks. This study suggested that caffeine consumption increases the tendency to engage in risky behaviors beyond those of consuming alcohol alone. However, based on this work, it is impossible to isolate the independent effects of caffeine from potential interactive effects of mixing caffeine and alcohol.

A third study, conducted by Jones and Lejuez (2005), examined several personality correlates of caffeine consumption and dependence. Specifically, the researchers identified sixty university students who were either "caffeine dependent—high consuming" or "caffeine nondependent—low consuming" and administered self-report or behavioral measures of sensation seeking, impulsivity, and risk taking. Correlational data showed that sensation seeking and impulsivity were related to consumer type, such that high caffeine consumers scored higher on the sensation-seeking scale and the measure of impulsivity. There was no relationship between consumer type and the behavioral measure of risk taking. When these measures were considered in a regression analysis, only sensation seeking had a significant relationship with consumer type, such that high sensation-seeking scores were associated with high consumption profiles (Jones and Lejuez 2005). As with the Miller study (2008), the direct relationship between caffeine and risky behavior is difficult to determine from these findings, and the lack of a pharmacological manipulation in these studies renders it impossible to ascertain a causal relationship.

CAFFEINE AND AGGRESSIVE BEHAVIOR

Elevated affective arousal is associated with aggression (Anderson et al. 1995). Since there is a large literature showing that caffeine clearly contributes to both emotional arousal and physiological arousal (for a review, see Nehlig et al. 1992; Smith et al. 1999) and that changes in mood associated with caffeine administration may follow a dose–response curve (Hasenfratz and Battig 1994), it is conceivable that moderate doses of caffeine may also contribute to aggressive behavior. Animal work has shown that caffeine can increase aggressive behavior (Cappell and Latane 1969; Eichelman et al. 1978; Emley and Hutchinson 1983) and this effect may follow an inverted-U dose–response curve (Wilson et al. 2000). However, few studies have addressed this issue in human volunteers. and the results are equivocal. In addition, much of the work has been done with special populations, and it is unclear how these data translate to normal populations. For example, high caffeine consumption has been associated with degraded mood, including increased anger, in a prisoner population (Hughs and Boland 1992). In this study, the authors surveyed 144 inmates from a maximum security penitentiary. Participants were asked to provide information about their daily caffeine and nicotine consumption as well as complete ratings about their mood and specific feelings of anger, anxiety, frustration, and irritability; level of concentration; appetite; and sleep quality. The authors then used factor analysis to create two variables, mood state and somatic state (sleep, concentration, and appetite). An analysis of variance was then conducted. They found an interaction between mood and habitual caffeine consumption (moderate versus high), such that high caffeine consumers (who were nonsmokers) had poorer mood ratings than the other groups.

It should be noted, however, that this study did not experimentally manipulate caffeine consumption and thus could not directly address the role of caffeine consumption in producing changes in anger and overall mood ratings. Other limited evidence in humans suggests that removal of caffeine from the diet of chronic psychiatric patients results in a reduction of nurse's reports of the inability to inhibit contextually inappropriate behaviors, such as hostility, anxiety, and irritability (De Freitas and Schwartz 1979). In this study, investigators switched patients' normal caffeinated coffee with decaffeinated coffee for a three-week period in a double-blind fashion, such that neither the patients nor the nursing staff providing the evaluations knew about the switch. All of the improvements reported with the decaffeinated coffee disappeared when the coffee was switched back to caffeinated. Removing caffeine from the diet has also been associated with a decrease in the number of aggressive outbursts in female mentally retarded patients (Podboy and Malloy 1977). Conversely, some evidence shows that caffeine reduces reports of impulsivity and aggression in children with attention deficit hyperactivity disorder (ADHD) (Leon 2000), improves mood in psychiatric patients (Furlong 1975), and decreases irritability when a very high dose (1000 mg) is administered in clinical populations (Stephenson 1977). These results, of course, are difficult to generalize to a healthy population.

The two laboratory studies, which have examined the relationship between acute caffeine consumption and aggressive behavior in normal adults, suggest a decrease in aggressive responding following caffeine consumption (Cherek et al. 1983, 1984). In the first study, eight male volunteers were administered 1 mg/kg, 2 mg/kg, and 4 mg/kg caffeine in capsule form and thirty minutes later completed a computerized task designed to measure aggressive behavior (Cherek et al. 1983). Treatment conditions were double-blind, and each volunteer completed four sessions for each condition (sixteen sessions total) on successive days. When participants consumed the highest dose of caffeine, aggressive responding was significantly reduced compared to placebo. A second study by the same authors (Cherek et al. 1984) was designed to examine the effects of caffeinated and decaffeinated coffee on aggressive behavior using the same methodology as the previous study. Coffee contains other ingredients as well as caffeine, and thus it is possible that it may exert different effects on aggressive behavior than caffeine alone. Eight male volunteers were administered one of three treatments dissolved in 500 ml of hot water: (1) four teaspoons of decaffeinated coffee, (2) two teaspoons decaffeinated and two teaspoons caffeinated (approximately 140 mg caffeine), and (3) four teaspoons caffeinated coffee (approximately 260 mg caffeine) and thirty minutes later completed a computerized task designed to measure aggressive behavior (Cherek et al. 1981, 1983). Treatment conditions were double-blind, and each volunteer completed three sessions for each condition (nine sessions total) on successive days. Consistent with the results from the previous study using caffeine capsules, when participants consumed the highest dose of caffeine, aggressive responding was significantly reduced compared to placebo. It should be noted that participants in both of the studies were regular caffeine consumers. In the first study, seven of the eight subjects consumed 100–250 mg caffeine daily and the eighth reported consuming 500–1000 mg caffeine a day. In the second study, half reported consuming more than 200 mg of caffeine a day, and the other half reported consuming less than 100 mg caffeine per day. Thus, data from both studies support the notion that caffeine, consumed alone or in coffee, leads to a decrease in aggressive behavior in individuals who normally consume caffeine and are in aggregate with recent work showing that in habitual consumers, higher doses may be necessary to elicit the same effects seen at lower doses in nonconsumers (Brunyé et al. 2010b). In addition, data from these studies may be specific to only regular caffeine consumers in two ways. First, regular caffeine consumers are more likely to report decreased irritability and positive changes in mood following caffeine consumption (Goldstein et al. 1969), and these positive effects may be responsible, in part, for the reduction in aggressive responses. Second, these data could reflect not a decrease in aggressive responding in the caffeine conditions, but rather an increase in aggressive responding in the placebo condition due to caffeine withdrawal. Finally, the notion that high consumers may respond differently to caffeine administration is supported by data from the second study which indicate that normal consumption habits appeared to influence the extent to which coffee suppressed aggressive responding. Participants who average over 200 mg caffeine/day showed a 58 percent reduction in aggressive responding when they had caffeine compared to placebo, while those averaging less than 100 mg/day showed only a 35 percent reduction in aggressive responses with caffeine compared to placebo. Thus, it is possible that these effects may not be generalizable to individuals who do not normally consume caffeine. The influence of caffeine on aggressive behavior may vary as a function of several factors, such as regular consumption profiles. In addition, more recent work, although it did not examine aggressive behavior, has shown that caffeine differentially affects self-reported tension and anger as a function of age. Ten younger men (19–26 years) and ten older men (65–80 years) who normally consume caffeine (Y = 126 ± 30 mg/d; O = 160 44 mg/d; mean ±−SEM) received caffeine and placebo on separate days following fasting, and then measures of mood, heart rate, and blood pressure were assessed. Older men report reduced feelings of tension and anger following caffeine consumption (5 mg/kg fat-free mass), and younger men report increased anger (Arciero et al., 1998).

In summary, based on the available literature, it is unclear what effects caffeine will have on aggressive behavior in healthy adults of varied ages and consumption profiles who participate in a well-controlled experimental paradigm that examines a wide range of doses. Studies report either relationships between typical caffeine consumption and self-reported or observed aggression or the effects of removing caffeine from the diet on subjective reports. In addition, these tend to be in special populations, and it is not clear how these data generalize to normal adult populations. Only two studies have examined empirically the influence of acute caffeine consumption on behavioral measures of aggression in normal adults. However, both of these studies examined only a small number of male subjects ($N = 8$) who were regular caffeine consumers and did not appear to control for regular caffeine consumption throughout the study. Future work is necessary to determine how a range of caffeine doses affects both male and female participants within a range of ages and varied consumption profiles. In addition, future research should attempt to determine if any observed changes in aggressive behavior are due to the beneficial effects of caffeine consumption or to the negative effects of caffeine withdrawal in moderate consumers.

CAFFEINE AND RISKY BEHAVIOR

Empirical work examining the relationship between caffeine consumption and risk taking does not support the notion that caffeine consumption leads to increased risk-taking behavior. Two recent studies specifically looking at "risk taking" in a laboratory setting following manipulation of caffeine consumption have examined the effects of sleep deprivation on risk-taking behavior and the ability of caffeine to attenuate these effects. In the first study, Killgore and colleagues (2007) examined the efficacy of caffeine to counteract increases in risk-taking behavior following a period of seventy-five hours of wakefulness. Participants consisted of twenty-six healthy adults (twenty-one men and five women). A between-subjects double-blind design was used. During each of the two overnight periods of the testing session, participants received either caffeinated chewing gum or a matched placebo gum every two hours (four times a night totaling 800 mg caffeine a night). At three hours (baseline), fifty-one hours, and seventy-five hours, a measure of risk taking was completed (i.e., the Iowa Gambling Task). Results showed that at baseline, volunteers easily learned to choose more frequently from advantageous low-risk card decks and avoid disadvantageous high-risk card decks. However, when participants were tested after fifty-one and seventy-five hours of sleep deprivation, they chose more frequently from the disadvantageous high-risk card decks, particularly during the latter half of the game. Caffeine had no significant effects on performance during sleep deprivation.

In a second study conducted by Killgore and colleagues (2008), fifty-four healthy adults (twenty-nine men and twenty-five women) were tested during a period of forty-six hours of sleep deprivation using a between-subjects double-blind design. Participants completed three daily measures of risk-taking propensity, the Brief Sensation Seeking Scale, the Evaluation of Risks Scale, and the Balloon Analog Risk Task. After forty-four hours of wakefulness, participants consumed caffeine (600 mg), modafinil (400 mg), dextroamphetamine (20 mg), or placebo and completed the risk measures two hours later. The placebo group showed a decline in risk-taking measures compared to baseline performance, and there were no significant effects of caffeine consumption. While it appears from this work that caffeine does not influence measures of risk taking, it is unclear how this work translates to normal consumption in well-rested individuals. No studies to date have specifically examined how manipulations in caffeine consumption influence measures of risk-taking behavior in well-rested volunteers with varying consumption profiles.

Inhibitory Control and Risk Taking

Additional insight into the relationship between risk taking and caffeine consumption comes from the literature on inhibitory control. Inhibitory control is a form of executive function and involves the ability to inhibit inappropriate impulses and actions in order to achieve an appropriate behavioral response. A diminished ability to control one's impulses and actions may contribute to risk-taking behavior. However, results from the limited work examining the influence of caffeine on measures of inhibitory control are equivocal. For example, some work suggests that caffeine can enhance inhibitory control (Hasenfratz and Battig 1992; Kenemans et al. 1999; Lorist et al. 1994, 1996), while other studies report contradictory findings (Foreman et al. 1989). However, other work has failed to find significant effects of caffeine on measures of inhibitory control (Kenemans and Verbaten 1998; Tieges et al. 2009). Thus, at initial examination, there is no clear agreement regarding caffeine's effects on inhibitory control. However, recent work indicates that when a distinction is made between active and reactive inhibitory control, a clearer picture emerges.

Active versus Reactive Measures

The distinction between active versus reactive inhibition control may account for some of the inconsistent findings in the literature to date. When inhibition is a deliberate, conscious action, it is active inhibition, whereas reactive inhibition results from interference due to some other competing behavioral response (Fillmore and Rush 2002). There is some evidence emerging in the literature that caffeine may improve reactive behavioral inhibition (Brunyé et al. 2010a; Hasenfratz and Battig 1992; Kenemans et al. 1999; Lorist et al. 1994, 1996) but not active inhibition (Kenemans and Verbaten 1998; Tieges et al. 2009). For example, Hasenfraz and colleagues (1992) examined the effects of caffeine and nicotine, alone and in combination, on the Stroop task, which assesses performance differences between processing conflicting and nonconflicting stimuli (reactive inhibition). This study used a within-subjects, pre-post cross-over design in which twenty female regular caffeine and nicotine users (mean coffee consumption, 5.4 cups/day; mean cigarettes smoked, 23.4 a day) received 250 mg caffeine alone, two cigarettes alone, both, or none on four separate test sessions. Two versions of the Stroop task were used, differing only in that in the "easy" version, the participant's response initiated a new trial after a one-second delay, and in the more "difficult" version, a new trial began immediately following the participant's response. Caffeine, but not nicotine or placebo, improved Stroop performance only in the more difficult version of the task. This study supports the notion that caffeine consumption may enhance reactive inhibition. In contrast, Barry and colleagues (2007) used a go/no go paradigm (active inhibition) to examine the effects of caffeine on active behavioral inhibition. A single oral dose of caffeine (250 mg) was administered using a within-subjects, double-blind, placebo-controlled design. Participants consisted of male and female university students (N = 24; sixteen female and eight male) who were nonsmokers and moderate caffeine users (2–4 cups of coffee or equivalent daily). Subjects consumed their normal amounts of caffeine in the morning and then were asked to abstain from further caffeine use four hours prior to testing. The two test sessions took place one week apart in the afternoon. When participants arrived, they were instrumented, consumed a capsule, and thirty minutes after consumption completed a go/no go task lasting approximately 2.5 minutes. Results from the go/no go task showed that caffeine reduced response time, but had no effect on commission errors, omission errors, or total errors, indicating that caffeine does not affect active inhibitory control.

A recent study by Tieges and colleagues (2009) specifically examined if caffeine differentially affects active versus reactive inhibition. In each of three experiments, a repeated-measures design was used with caffeine (3 mg/kg body weight [a very high dose] versus 0 mg caffeine) as the within-subjects factor. The first experiment employed a cued version of the continuous performance task (AX-CPT), which is similar to a cued go/no-go task. Both of these tasks require both execution and inhibition of motor response. Results indicated that while caffeine did not affect reaction time or error rate on the cued trials, it did decrease response time only on nontarget trials, providing some evidence that caffeine enhances inhibitory control. The second experiment used the stop signal task, which calculates participants' ability to inhibit a prepotent or current motor response. The stop signal task assesses active inhibition. Results showed that caffeine decreased choice reaction time as well as go-reaction times relative to placebo. However, caffeine only showed a trend toward producing faster stop signal reaction time (SSRT), which measures the reaction time in trials in which there is a delay before the go-cue. The authors interpret the results to indicate that although caffeine increases processing speed, it does not affect active inhibitory control. The third experiment utilized a flanker task, which assesses reactive inhibition by requiring participants to selectively inhibit incorrect responses. In this study, although caffeine again enhanced response time, it did not affect accuracy or the difference in response times or accuracy between congruent and incongruent trials. Thus, while the results from this work are consistent with previous work indicating that caffeine can exert positive effects on behavior and mood, such as enhancing vigilance and response times, as well as increased alertness (Lieberman et al. 1987, 2002; Lieberman 2001; Nehlig et al. 1992; Smith 2002; Wesensten et al. 2005), they also suggest that caffeine, at least 3 mg/kg body weight, does not influence inhibitory control processes. However, it should be noted that the participants in these studies were all regular caffeine consumers, averaging 2–4 cups of coffee a day. It is possible, therefore, that a dose of approximately 200 mg may be sufficient to elicit performance effects in participants who normally consume caffeine.

Caffeine's primary mechanism of action in the brain is via competitive inhibition of adenosine receptors. Chronic consumption of caffeine increases the number of adenosine receptors in the brain, which suggests that in high consumers, administration of higher doses of caffeine may be necessary to achieve substantial increases in dopamine (Varani et al. 1999). This notion is supported by more recent work examining the influence of caffeine on visual attention in both low and high consumers (Brunyé et al. 2010a). Brunyé and colleagues (2010a) investigated the effects of a range of caffeine doses (0 mg, 100 mg, 200 mg, and 400 mg) on visual attention using the Attention Network Task (ANT). The ANT is designed to test Posner's three visual attention network functions: alerting, orienting, and executive control (Posner 2004). Executive control is examined using a flanker task which requires a participant to resolve a conflict among potential responses to a presented stimulus (reactive inhibition). In the first study, thirty-six male and female low consumers or nonconsumers (> 100 mg/day; mean 42.5 mg/day) completed four test sessions, one with each level of dose, in a counterbalanced, double-blind design. For each test session, participants consumed one of the four doses and then twenty minutes later completed the ANT. Results showed that caffeine improved reactive inhibition starting at 200 mg. Reactive inhibition was also improved with 400 mg, but this improvement was not greater than at 200 mg. In a second study, Brunyé and colleagues replicated the first study with habitual high caffeine consumers (< 300 mg/day; mean 592.3 mg/day). Again, thirty-six male and female volunteers completed four test sessions, one with each level of dose, in a counterbalanced, double-blind design. The authors hypothesized that as with Tieges et al.'s (2009) work, 200 mg may not be sufficient to modulate higher order cognitive processes related to the control of visual attention in habitual consumers, and thus a 400 mg dose may be necessary to produce significant effects. This study did show that, in habitual consumers, caffeine enhanced reactive inhibition, but only at the highest administered (400 mg) dose. Thus, the work by Brunyé and colleagues supports the notion that caffeine may selectively enhance reactive inhibition, but the effects are dependent on individual consumption profile and dose.

CAFFEINE EFFECTS ON PHYSIOLOGY AND IMPLICATIONS FOR AGGRESSIVE AND RISKY BEHAVIOR

The mechanisms by which caffeine may modulate (in either direction) aggression and impulsive or risky decision making are unknown. Biochemically, caffeine consumption results in increased dopamine and serotonin, which have been linked to the enhancement of processes that require executive control and enhanced mood, respectively (Abrams et al. 2005; Berkowitz and Spector 1971; Fernstrom et al. 1984; Ferré et al. 1997; Hadfield and Milio 1989; Haleem et al. 1995; Shi et al. 1993). In addition, caffeine consumption has been shown to increase cortisol (al'Absi and Lovallo 2004). This section will briefly review the literature on neurotransmitters and neurohormones, including serotonin, dopamine, and cortisol, as they relate to aggressive and risk-based decision making.

CAFFEINE, SEROTONIN, AND AGGRESSION

Caffeine is a nonselective competitive adenosine receptor antagonist which exerts its effects primarily through adenosine A_1 and A_{2A} receptors (Ferré 2010). Normally, endogenous adenosine influences ascending neurotransmitter systems, including the cholinergic, noradrenergic, histaminergic, orexinergic, dopaminergic, and serotonergic systems. Caffeine's affinity for A_1 receptors in the basal forebrain and A_{2A} receptors in the hypothalamus is thought to involve the cholinergic, noradrenergic, histaminergic, orexinergic pathways, which are involved in its arousing properties (Ferré 2010). Its affinity for the A_1 and A_{2A} receptors in the striatum is thought to be responsible for its influence on the dopaminergic pathways, which are involved in reward and motivation, as well as the serotonergic system, which may play a role in modulating aggressive behavior (Ferré 2008).

The literature to date suggests that caffeine consumption affects the serotonergic system, and this system may play a role in modulating aggressive behavior. However the mechanism by which serotonin exerts its effects on aggression is not completely understood. This section will briefly review the relationships between caffeine and serotonin and between serotonin and aggression, and how caffeine consumption or caffeine withdrawal may influence aggressive behavior.

EFFECTS OF CAFFEINE ON SEROTONIN

Animal studies suggest that caffeine increases serotonin (5-hydroxytryptamine, or 5-HT) concentration in the central nervous system (Berkowitz and Spector 1971) and upregulates serotonin receptors (Shi et al. 1993). Doses ranging from 50 mg/kg/day to 200 mg/kg/day, administered either acutely or chronically, have been shown to enhance extracellular serotonin concentrations in rats (Abrams et al. 2005; Berkowitz and Spector 1971; Fernstrom et al. 1984; Haleem et al. 1995) and mice (Hadfield and Milio 1989).

Serotonin–Aggression Relationship

The literature to date suggests that the serotonergic system intimately modulates aggressive behavior, yet the mechanism by which serotonin exerts its effects on aggression is not completely understood. In general, results have found an inverse relationship between serotonergic function and aggressive behavior (Cleare and Bond 1997; de Boer and Koolhaas 2005; Miczek et al. 2004; for review, see Seo et al. 2008). However, there is a growing amount of work that suggests that serotonin does not influence aggression in the same manner in all individuals, but rather is modulated by a number of other factors such as gender, testosterone and other hormones, types of aggression, and trait aggression profile (Berman et al. 2009; Kuepper et al. 2010; for review, see Carillo et al. 2009). For example, recent work suggests that raising or lowering serotonin levels does not affect general aggressive behavior, but rather it can attenuate or magnify, respectively, the aggressive response to provocation (Berman et al. 2009). In one study, Berman and colleagues (2009) examined the effects of paroxetine hypochloride, a selective serotonin reuptake inhibitor (SSRI) used to enhance presynaptic 5-HT activity, in subjects with or without a history of aggression. Interestingly, they found that in aggressive subjects, paroxetine reduced the aggressive response to provocation but also attenuated hormonal response relative to nonaggressive subjects. The researchers hypothesize that any increases in the bioavailability of 5-HT at the synapse reduces response to provocation in aggressive individuals. Increased 5-HT at the synapse enhances activity in the orbital and ventral medial prefrontal cortices, areas implicated in behavioral control. In addition, other data suggest that serotonin function alone cannot explain differences in aggressive behavior, but that testosterone (T) and 5-HT together interact to modulate aggression; specifically, increased T level and decreased 5-HT availability seem to augment aggressive behavior, but only in males (Kuepper et al. 2010). In terms of aggression type, enhanced serotonergic function seems to reduce predator aggression to a greater extent than it reduces offensive aggression and does not at all affect defensive aggression (Carillo et al. 2009).

Caffeine Withdrawal and Measures of Serotonin and Aggression

Previous research has shown that caffeine increases serotonin and tryptophan (TRP) levels as well as cortical 5-HT_{1A} and 5-HT_{1B} receptor densities (Haleem et al. 1995). To determine whether these changes are constant during caffeine withdrawal, rats were given either 80 mg/kg caffeine or saline (vehicle) for five consecutive days and sacrificed twenty-four hours after the fifth day. During this withdrawal period, 5-HT and 5-H1AA levels decreased in the caffeine-treated rats relative to the saline-treated rats, but TRP level was not affected, indicating that caffeine withdrawal decreases plasma serotonin but not tryptophan levels (Haleem et al. 1995). To our knowledge, this is the only study that assessed the effects of caffeine withdrawal on the serotonergic system, and we have found no studies that examined the effects of caffeine withdrawal on aggressive behavior. However, given the extensive research on the interactions between caffeine and serotonin as well as between serotonin and aggression, it is possible to speculate on the effects of caffeine withdrawal on serotonin and aggression. Given that caffeine has been shown to increase extracellular serotonin (Berkowitz and Spector 1971; Fernstrom et al. 1984; Hadfield and Milio 1989) and upregulate serotonin 5-HT_{1A} and 5-HT_{1B} (Shi et al. 1993) and that increased 5-HT availability and 5-HT_{1B} receptor densities may reduce the aggressive response to provocation (Cleare and Bond 1997; de Boer and Koolhaas 2005; Suzuki et al. 2010), it is plausible that during caffeine withdrawal, there is insufficient brain serotonin to bind postsynaptically to the now-upregulated number of serotonin receptors, thereby inducing a serotonin deficit. Given the studies finding that reduced serotonin increases aggression (Berman et al. 2009; Kuepper et al. 2010), caffeine withdrawal may, in fact, increase aggressive behavior, especially in response to provocation.

In summary, animal studies suggest that caffeine increases serotonin concentration in the central nervous system and increases serotonergic receptor densities in behaviorally relevant areas of the prefrontal cortex. Furthermore, there appears to be a complex relationship between serotonergic function and aggressive behavior, with enhanced serotonin activity reducing aggressive behavior in the majority of situations, but also other factors modulating this relationship including gender, hormone levels, types of aggression, and trait aggression profile. Therefore, based on the limited work to date, it may be that caffeine, in fact, decreases aggressive behavior in a normal adult population, possibly due to increases in serotonin, and that a state of caffeine withdrawal in moderate or high consumers may result in increases in irritability and aggressive responses. However, more empirical work is needed to test this hypothesis.

CAFFEINE, DOPAMINE, AND BEHAVIORAL CONTROL

Caffeine influences ascending dopamine pathways by antagonizing adenosine A_1 and A_{2A} receptors in the striatum. Normally, endogenous adenosine inhibits dopaminergic neurotransmission, but caffeine blocks this inhibition, thereby increasing extracellular dopamine concentrations (Ferré et al. 1997).

Attempts to determine caffeine's effect on dopaminergic activity derives in part from clinical evidence suggesting that anti-Parkinson therapy, which increases dopaminergic neurotransmission, may increase impulsive or risky behavior (for review, see Robert et al. 2009). The evidence to date suggests that the dopamine D_2 receptor is primarily responsible for dopamine's role in risk taking and impulsivity. Multiple studies have found that increased dopamine (DA) activity increases risk-taking behavior in humans (Riba et al. 2008) and rodents (St. Onge and Floresco 2009) and impulsivity in humans (Colzato et al. 2010) and rodents (van Gaalen et al. 2006). Increases in extracellular DA via treatment with a DA reuptake inhibitor or amphetamine have been shown to increase impulsive behavior, whereas a D_2 receptor antagonist reduced this effect (for review, see Pattij and Vanderschuren 2008). Similarly, decreases in DA activity by means of a dopamine D_2 receptor antagonist decrease impulsive behavior, whereas increases via D_2 receptor agonists increase impulsive behavior (St. Onge and Floresco 2009).

In contrast, studies that have examined the relationship between behavioral control, a process associated with the ability to inhibit impulsive or risky decisions, tentatively support the notion that increased DA in the prefrontal cortex (PFC) leads to better inhibitory control and less impulsive decision making. It has been suggested that dopamine is one of the neurotransmitters involved in the modulation of behavioral inhibition (for review, see Mink 1996). For example, it has shown in rats that dopamine may be crucial for the ability to quickly and accurately inhibit a prepotent response in rats using a stop signal task (Eagle et al. 2009). Schizophrenics, characterized by an imbalance of dopamine in the basal ganglia, are slower to inhibit responses compared to healthy controls (Enticott et al. 2008). In addition, patients who suffer from Parkinson's disease, characterized by a loss of dopaminergic neurons in the basal ganglia, take longer to inhibit a response on a stop signal task and exhibit decrements in the ability to suppress conflicting and irrelevant information (Gauggel et al. 2004; Wylie et al. 2009).

Additional evidence to suggest a role for dopamine in impulsive and risky decision making comes from work with cocaine users. Habitual cocaine users typically have fewer DA D_2 receptors in the striatum (Volkow et al. 1997). Work by Colzato and colleagues (2007) showed that cocaine users took longer to inhibit a response during a stop signal task compared to healthy controls. Taken together, these studies suggest that dopamine may play a role in inhibiting inappropriate responses and controlling behavior. However, because data are derived mainly from patient and drug studies, it is unclear how these data translate to healthy populations. Indeed, it has been suggested that the beneficial effects of dopaminergic stimulants on response inhibition may be limited to subjects whose inhibitory efficiency is suboptimal to begin with (e.g., De Wit et al. 2002). In response to this issue, a recent study examined how variations in dopamine levels influence efficiency inhibiting a response in healthy participants (Colzato et al. 2009). The investigators used spontaneous eyeblink rate (an index for dopamine production in the striatum; Blin et al. 1990; Karson 1983; Shukla 1985; Taylor et al. 1999) as a marker of central dopaminergic functioning and a stop signal task to measure the ability to inhibit unwanted actions. Their data revealed an inverted-U relationship such that average eyeblink rates were associated with a greater ability to inhibit unwanted actions compared to low or high eyeblink rates. The authors suggest that the notion of an inverted U-shaped relationship between DA levels and the ability to inhibit a response should be investigated further in psychopharmacological studies that account for individual baseline levels of DA as these may account for differential sensitivity to the effects of dopaminergic drugs.

In summary, caffeine increases extracellular dopamine concentrations; however, it is unclear what role dopamine plays in impulsive or risk-taking behavior. Some work suggests that increased levels of DA may augment impulsive or risky behavior (Robert et al. 2009), while other work suggests that increased levels of DA may increase the efficiency of inhibiting unwanted actions. It may be that DA levels affect behavioral control in an inverted-U function, and therefore some of the inconsistency in the data is due to variations in baseline levels of DA. Individuals characterized with low levels of DA may benefit (i.e., reduction in risky behaviors) from dopaminergic stimulants, whereas those characterized with normal or high levels of DA may show decrements in behavioral control following use of dopaminergic stimulants.

CAFFEINE AND CORTISOL

Caffeine consumption leads to increased cortisol release (al'Absi and Lovallo 2004). Cortisol may influence risk-taking behavior independent of affective state by modulating dopamine activity within the nucleus accumbens, a region implicated in reward and punishment (Putman et al. 2010). However, the relationship between cortisol levels and risk-based decision making appears to depend on both gender and individual differences in cortisol reactivity. Although multiple studies have found a positive correlation between cortisol levels and risk taking or gambling behaviors in men (Meyer et al. 2000, 2004; Putman et al. 2010), recent studies that include female populations indicate that the relationship is more complex (for review, see Paris et al. 2010). Research investigating the relationship between cortisol and risk-taking behavior has used one of two methods: either (1) examining the relationship between salivary cortisol levels before and after a risk-based gambling task, or (2) assessing the effect of exogenous cortisol treatment on subsequent risk-based decision making. The latter method, which is most relevant to caffeine-induced changes in cortisol, has shown that compared to placebo, in men, cortisol (experimenters used 40 mg hydrocortisone administered orally) does not affect state anxiety but does increase the number of risky experimental gambles in which participants have a high possible gain at the expense of a high probability of losing (Putman et al. 2010).

Another study assessed the effects of stress-induced cortisol changes, induced by the Trier Social Stress Test (TSST) on risk-taking behavior using the Iowa Gambling Task (IGT). Participants were divided into two groups: an experimental group exposed to the stressor and a control group that was not exposed to stress (they listened to music or read a magazine). The authors then examined the relationship between cortisol reactivity and gender (Van den Bos et al. 2009). Cortisol reactivity was determined using a median split analysis comparing the difference in cortisol between baseline and after the gambling task (which took place approximately fifteen minutes after the stress induction). For men, the median value of cortisol was 4.2 nmol/l, and for women it was 0.8 nmol/l. In men, stress significantly increased cortisol levels in high cortisol responders, and slightly increased levels in low responders. In contrast, cortisol levels decreased in controls who were not exposed to stress (likely due to natural circadian rhythm). In women, stress significantly increased cortisol levels in high responders but decreased levels in low responders. Cortisol decreased in the control women. All subjects, regardless of gender, learned the IGT, specifically which decks were disadvantageous. However, whereas high-responder men learned significantly slower than low-responder and control men, high-responder women chose slightly fewer cards from the disadvantageous decks than either the low-responder or control women. These results suggest that stress and stress-induced cortisol response induce risk-taking behavior in males and risk-aversive behavior in females. Furthermore, while there were no differences in time to complete the task in men, high- and low-responder women completed the task more quickly than control women. Importantly, the increase in speed did not result in a speed–accuracy trade-off. In fact, high-responder women won more money than low-responder and control women. The authors attribute the sex-specific results to sex differences in the right PFC, which is involved in both risk-taking behavior and stress regulation. Indeed, the authors note that previous research has shown increased activation in the right compared to left PFC during the IGT in men but not women (Bolla et al. 2004), as well as impaired performance on the IGT following damage to the right but not left PFC, again in men but not women (Tranel et al. 2005). Thus, increased cortisol activity in the right PFC may partially explain the differential effect of stress on risk-based decision making between the genders.

In summary, caffeine consumption leads to increased release of cortisol which in turn may upregulate dopamine activity (Putman et al. 2010). Increased dopamine has been shown to increase the rewarding properties and decrease perceived punishment of risk-based decision making (Riba et al. 2008; St. Onge and Floresco 2009; for review, see Schulz et al. 2010).

CONCLUSION AND FUTURE DIRECTIONS

Much of the work examining the relationship between caffeine consumption and impulsivity, sensation seeking, and aggression consists of either correlational work or work using linear regression analyses with typical consumption and either self-report measures in college students or professional clinical observations in patient populations. The lack of a pharmacological manipulation in these studies renders it impossible to ascertain a causal relationship between caffeine consumption and aggressive behavior. In fact, it may be that individuals who score high on sensation-seeking scales or trait aggression seek out ways to increase their arousal level and thus are more likely to be high caffeine consumers. The conclusion, drawn from survey and observational data, that caffeine consumption increases risky and aggressive behavior is not supported by the limited empirical work. In fact, results from the limited empirical work in this area suggest that caffeine either has no effect on risk taking following periods of sleep deprivation or has been shown to reduce incidences of aggressive responding. In addition, work that examines the relationship between inhibitory control, a component of executive function involved in risky decision making, and caffeine consumption tentatively suggests that caffeine consumption may improve some aspects of inhibitory control. If caffeine does impact aggressive behavior and risky decision making, the mechanism by which it influences these behaviors is unknown. There is some evidence that increased dopamine and upregulation of the right hemisphere, which are both associated with caffeine consumption, may improve executive function and inhibitory control. However, other work has shown that increased dopamine is associated with a greater number of risky decisions. In terms of aggression, much of the work to date suggests that serotonin, which increases with caffeine consumption, leads to better mood and less aggressive behavior. Future work should examine how caffeine in a range of doses influences aggression or risk-taking behavior in both male and female volunteers who are well rested and have varying consumption profiles. In addition, future work should attempt to determine if any observed changes in aggressive or risk-taking behavior are due to the beneficial effects of caffeine consumption or to the negative effects of caffeine withdrawal in individuals who normally consume caffeine.

REFERENCES

Abrams, J. K., Johnson, P. L., Hay-Schmidt, A., Mikkelsen, J. D., Shekhar, A., and Lowry, C. A. 2005. Serotonergic systems associated with arousal and vigilance behaviors following administration of anxiogenic drugs. *Neuroscience* 33 (4): 983–97.

al'Absi, M., and Lovallo, W. R. 2004. Caffeine effects on the human stress axis. In *Coffee, Tea, Chocolate and the Brain*, ed. A. Nehlig, pp. 113–31. Boca Raton, FL: CRC Press.

Anderson, C. A., Deuser, W. E., and DeNeve, K. M. 1995. Hot temperatures, hostile affect, hostile cognition and arousal: tests of a general model of affective aggression. *Pers Soc Psychol Bull* 21 (5): 434–48.

Arciero, P. J., Gardner, A. W., Benowitz, N. J., and Poehlman, E. T. 1998. Relationship of blood pressure, heart rate and behavioral mood state to norepinephrine kinetics in younger and older men following caffeine ingestion. *Eur J Clin Nutr*. 52 (11): 805–12.

Barry, R. J., Stuart, J., Johnstone, A. R., et al. 2007. Caffeine effects on ERPs and performance in an auditory go/nogo task. *Clin Neurophysiol* 118: 2692–99.

Berkowitz, B. A., and Spector, S. 1971. The effect of caffeine and theophylline on the disposition of brain serotonin in the rat. *European J Pharmacol* 16: 322.

Berman, M. E., McCloskey, M. S., Fanning, J. R., Schumacher, J. A., and Coccaro, E. F. 2009. Serotonin augmentation reduces response to attack in aggressive individuals. *Psychol Sci* 20 (6): 714–20.

Blin, O., Masson, G., Azulay, J. P., Fondarai, J., and Serratrice, G. 1990. Apomorphine-induced blinking and yawning in healthy volunteers. *Br J Clin Pharmacol* 30: 769–73.

Bolla, K. I., Eldreth, D. A., Matochik, J. A., and Cadet, J. L. 2004. Sex related differences in a gambling task and its neurological correlates. *Cereb Cortex* 14: 1226–32.

Brunyé, T. T., Mahoney, C. R., Lieberman, H. R., and Taylor, H. A. 2010a. Caffeine modulates attention network function. *Brain Cogn.* 72: 181–8.

Brunyé, T. T., Mahoney, C. R., Giles, G. E., Lieberman, H. R., and Taylor, H. A. 2010b. Acute caffeine consumption enhances the executive control of visual attention in habitual consumers. *Brain Cogn* 74: 186–92.

Cappell, H., and Latane, B. 1969. Effects of alcohol and caffeine on the social and emotional behavior of the rat. *Q J Stud Alcohol* 30 (2): 345–56.

Cherek, D. R. 1981. Effects of smoking different doses of nicotine on human aggressive behavior. *Psychopharmacology (Berl)* 75 (4): 339–45.

Cherek, D. R., Steinberg, J. L., and Brauchi, J. T. 1983. Effects of caffeine on human aggressive behavior. *Psychiatry Res* 8 (2): 137–45.

Cherek, D. R., Steinberg, J. L., and Vines, R. V. 1984. Low doses of alcohol affect human aggressive responses. *Biol Psychiatry* 19 (2): 263–7.

Cleare, A. J., and Bond, A. J. 1997. Does central serotonergic function correlate inversely with aggression? A study using D-fenfluramine in healthy subjects. *Psychiatry Res* 69 (2–3): 89–95.

Colzato, L. S., van den Wildenberg, W. P., and Hommel, B. 2007. Impaired inhibitory control in recreational cocaine users. *PLoS One* 2 (11): e1143.

Colzato, L. S., van den Wildenberg, W. P. M., van der Does, A. J. W., and Hommel, B. 2010. Genetic markers of striatal dopamine predict individual differences in dysfunctional, but not functional impulsivity. *Neuroscience* 170 (3): 782–8.

Colzato, L. S, van den Wildenberg, W. P., van Wouwe, N. C., Pannebakker, M. M., and Hommel, B. 2009. Dopamine and inhibitory action control: evidence from spontaneous eye blink rates. *Exp Brain Res* 196 (3): 467–74.

de Boer, S. F., and Koolhaas, J. M. 2005. HT_{1A} and 5-HT_{1B} receptor agonists and aggression: a pharmacological challenge of the serotonin deficiency hypothesis. *Eur J Pharmacol* 526 (1–3): 125–39.

De Freitas, B., and Schwartz, G. 1979. Effects of caffeine in chronic psychiatric patients. *Am J Psychiatry* 136 (10): 1337–8.

de Wit, H., Enggasser, J. L., and Richards, J. B. 2002. Acute administration of d-amphetamine decreases impulsivity in healthy volunteers. *Neuropsychopharmacology* 27 (5): 813–25.

Eagle, D. M., Lehmann, O., Theobald, D. E., et al. 2009. Serotonin depletion impairs waiting but not stop-signal reaction time in rats: implications for theories of the role of 5-HT in behavioral inhibition. *Neuropsychopharmacology* 34 (5): 1311–21.

Eichelman, B., Orenberg, E., Hackley, P., and Barchas, J. 1978. Methylxanthine-facilitated shock-induced aggression in the rat. *Psychopharmacology* 56 (3): 305–8.

Emley, G. S., and Hutchinson, R. R. 1983. Unique influences of ten drugs upon post-shock biting attack and pre-shock manual responding. *Pharmacol Biochem Behav* 19 (1): 5–12.

Enticott, P. G., Ogloff, J. R., and Bradshaw, J. L. 2008. Response inhibition and impulsivity in schizophrenia. *Psychiatry Res* 157 (1–3): 251–4.

Fernstrom, M. H., Bazil, C. W., and Fernstrom, J. D. 1984. Caffeine injection raises brain tryptophan level, but does not stimulate the rate of serotonin synthesis in rat brain. *Life Sci* 5 (12): 1241–7.

Ferré, S. 2010. Role of central ascending neurotransmitter systems in the psychostimulant effects of caffeine. *Journal Alzheimers Dis* 20:S35–49.

Ferré, S., Fredholm, B. B., Morelli, M., Popoli, P., and Fuxe, K. 1997. Adenosine-dopamine receptor-receptor interactions as an integrative mechanism in the basal ganglia. *Trends Neurosci* 20: 482–487.

Ferré, S. 2008. An update on the mechanisms of the psychostimulant effects of caffeine. *J Neurochem* 105: 1067–79.

Fillmore, M. T., and Rush, C. R. 2002. Impaired inhibitory control of behavior in chronic cocaine users. *Drug Alcohol Depend* 66: 265–73.

Foreman, N., Barraclough, S., Moore, C., Mehta, A., and Madon, M. 1989. High doses of caffeine impair performance of a numerical version of the Stroop task in men. *Pharmacol Biochem Behav* 32 (2): 399–403.

Furlong, F. W. 1975. Possible psychiatric significance of excessive coffee consumption. *Can Psychiatr Assoc J* 20 (8): 577–83.

Gauggel, S., Rieger, M., and Feghoff, T. A. 2004. Inhibition of ongoing responses in patients with Parkinson's disease. *J Neurol Neurosurg Psychiatry* 75 (4): 539–44.

Gilbert R. M. 1984. Caffeine consumption. *Prog Clin Biol Res* 158: 185–213.

Goldstein, A., Kaizer, S., and Whitby, O. 1969. Psychotropic effects of caffeine in man. IV. Quantitative and qualitative differences associated with habituation to coffee. *Clin Pharmacol Ther* 10 (4): 489–97.

Hadfield, M. G., and Milio, C. 1989. Caffeine and regional brain monoamine utilization in mice. *Life Sci* 45 (26): 2637–44.

Haleem, D. J., Yasmeen, A., Haleem, M. A., and Zafar, A. 1995. 4h withdrawal following repeated administration of caffeine attenuates brain serotonin but not tryptophan in rat brain: implications for caffeine-induced depression. *Life Sci* 57 (19): PL285–92.

Hasenfratz, M., & Bättig, K. 1992. Action profiles of smoking and caffeine: Stroop effect, EEG, and peripheral physiology. *Pharmacology, Biochemistry, and Behavior* 42 (1): 155–61.

Hasenfratz, M., and Battig, K. 1994. Acute dose-effect relationships of caffeine and mental performance, EEG, cardiovascular and subjective parameters. *Psychopharmacology* 114 (2): 281–7.

Jones, H. A., and Lejuez, C. W. 2005. Personality correlates of caffeine dependence: the role of sensation seeking, impulsivity, and risk taking. *Experiment Clin Psychopharm* 13 (3): 259–66.

Karson, C. N. 1983. Spontaneous eye-blink rates and dopaminergic systems. *Brain* 106: 643–53.

Kenemans, J. L., and Verbaten, M. N. 1998. Caffeine and visuo-spatial attention. *Psychopharmacology (Berl)* 135 (4): 353–60.

Kenemans, J. L., Wieleman, J. S., Zeegers, M., and Verbaten, M. N. 1999. Caffeine and stroop interference. *Pharmacol Biochem Behav* 63 (4): 89–98.

Killgore, W. D. S., Grugle, N. L., Killgore, D. B., et al. 2008. Restoration of risk-propensity during sleep deprivation: caffeine, dextroamphetamine, and modafinil. *Aviat Space Environ Med* 79: 867–74.

Killgore, W. D. S., Lipizzi, E. L., Kamimori, G. H., and Balkin, T. J. 2007. Caffeine effects on risky decision making after 75 hours of sleep deprivation. *Aviat Space Environ Med* 78: 957–62.

Kuepper, Y., Alexander N., Osinsky R., et al. 2010. Aggression interactions of serotonin and testosterone in healthy men and women. *Behav Brain Res* 206 (2010): 93–100.

Leon, M. R. 2000. Effects of caffeine on cognitive, psychomotor, and affective performance of children with attention-deficit/hyperactivity disorder. *J Att Disorders* 4 (1): 27–47.

Lieberman, H. R. (2001). The effects of ginseng, ephedrine and caffeine on cognitive performance, mood and energy. *Nutr Rev* 59 (4): 91–102.

Lieberman, H. R., Tharion, W. J., Shukitt-Hale, B, Speckman, K. L., and Tulley, R. 2002. Effects of caffeine, sleep loss, and stress on cognitive performance and mood during U.S. Navy SEAL training. *Psychopharmacology* 164: 250–61.

Lieberman, H. R., Wurtman, R. J., Emde, G. G., Roberts, C., and Coviella, I. L. 1987. The effects of low doses of caffeine on human performance and mood. *Psychopharmacology (Berl)* 92 (3): 308–12.

Lorist, M. M., Snel, J., Kok, A., and Mulder, G.1994. Influence of caffeine on selective attention in well-rested and fatigued subjects. *Psychophysiology* 31 (6): 525–34.

Lorist, M. M., Snel, J., Kok, A., and Mulder, G. 1996. Acute effects of caffeine on selective attention and visual search processes. *Psychophysiology* 33 (4): 354–61.

Mahalik, J. R., Locke, B. D., Ludow, L. H. et al. 2003. Development of the conformity to masculine norms inventory. *Psychol Men Masculinity* 4: 3–25.

eyer, G., Gauffa, B. P., Schedlowski, M., Pawlak, C., Stadler, M. A., and Exton M. S. 2000. Casino gambling increases heart rate and salivary cortisol in regular gamblers. *Biol Psychiatry* 48: 948–53.

Meyer, G., Schwertfeger, J., Exton, M. S., et al. 2004. Neuroendocrine response to casino gambling in problem gamblers. *Psychoneurendocrinology* 29: 1272–80.

Miczek, K. A., Faccidomo, S., De Almeida, R. M., Bannai, M., Fish, E. W., and Debold, J. F. 2004. Escalated aggressive behavior: new pharmacotherapeutic approaches and opportunities. *Ann N Y Acad Sci* 1036: 336–55.

Miller, K. E. 2008. Wired: energy drinks, jock identity, masculine norms, and risk taking. *J Am Coll Health* 56 (5): 481–9.

Mink, J. W. 1996. The basal ganglia: focused selection and inhibition of competing motor programs. *Prog Neurobiol* 50 (4): 381–425.

Nehlig, A., Deval, J. L., and Debry, G. 1992. Caffeine and the central nervous system: mechanisms of action, biochemical, metabolic and psychostimulant effects. *Brain Res Rev* 17 (2): 139–170.

O'Brien, M. C., McCoy, T. P., Rhodes, S. D., Wagoner, A., and Wolfson, M. 2008. Caffeinated cocktails: energy drink consumption, high risk drinking, and alcohol related consequences among college students. *Acad Emerg Med* 15: 1–8.

Paris, J. J., Franco, C., Sodano, R., et al. 2010. Sex difference in salivary cortisol in response to acute stressors among healthy participants, in recreational and pathological gamblers, and in those with posttraumatic stress disorder. *Horm Behav* 57: 35–45.

Podboy, J., and Malloy, W. 1977. Caffeine reduction and behavior changes in the severely retarded. *Ment Retard* 15: 40.

Posner, M. I. 2004. *Cognitive neuroscience of attention*. New York: Guilford Press.

Putman, P., Antya, K., Crysovergi, P., and van der Does, W. J. A. 2010. Exogenous cortisol acutely influences motivated decision making in healthy young men. *Psychopharmacology* 208: 257–63.

Reissig, R. J., Strain, E. C., and Griffiths, R. R. 2009. Caffeinated energy drinks: a growing problem. *Drug Alcohol Depen* 99: 1–10.

Riba, J., Kramer, U. M., Heldmann, M., Richter, S., and Munte, T. F. 2008. Dopamine agonist increases risk taking but blunts reward-related brain activity. *PLoS ONE* 3 (6): e2479.

Robert, G., Drapier, D., Verin, M., Millet, B., Azulay, J-P., and Blin, O. 2009. Cognitive impulsivity in Parkinson's disease patients: assessment and pathophysiology. *Movement Disord* 24 (16): 2316–27.

Seo, D., Patrick, C. J., and Kennealy, P. J. 2008. Role of serotonin and dopamine system interactions in the neurobiology of impulsive aggression and its comorbidity with other clinical disorders. *Aggress Violent Behav* 13 (5): 383–95.

Shi, D., Nikodijevic, O., Jacobson, K. A., and Daly, J. W. 1993. Chronic caffeine alters the density of adenosine, adrenergic, cholinergic, GABA, and serotonin receptors and calcium channels in mouse brain. *Cell Mol Neurobiol* 13: 247.

Shukla, D. 1985. Blink rate as clinical indicator. *Neurology*, 35: 286.

Smith, A. 2002. Effects of caffeine on human behavior. *Food Chem Toxicol* 40: 1243–55.

Smith, B. D., Tola, K., and Mann M. 1999. Caffeine and arousal: a biobehavioral theory of physiological, behavioral, and emotional effects In *Caffeine and behavior: current views and research trends*, ed. B. S. Gupta and U. Gupta, 87–136. New York: CRC Press.

Stephenson, P. E. 1977. Physiologic and psychotropic effects of caffeine on man: a review. *J Am Diet Assoc* 71 (3): 240–7.

St. Onge, J. R., and Floresco, S. B. 2009. Dopaminergic modulation of risk-based decision-making. *Neuropsychopharmacology* 34: 681–97.

Suzuki, H., Han, S. D., and Lucas, L. R. 2010. Increased 5-HT1B receptor density in the basolateral amygdala of passive observer rats exposed to aggression. *Brain Res Bull* 83 (1–2): 38–43.

Taylor, J. R., Elsworth, J. D., Lawrence, M. S., Sladek, J. R., Roth, R. H. et al. 1999. Spontaneous blink rates correlate with dopamine levels in the caudate nucleus of MPTP-treated monkeys. *Exp Neurol* 158: 214–20.

Tieges, Z., Snel, J., Kok, A., and Ridderinkhof, R. 2009. Caffeine does not modulate inhibitory control. *Brain Cogn* 69 (2): 316–27.

Tranel, D., Damasio, H., Denburg, N. L., and Bechara, A. 2005. Does gender play a role in functional asymmetry of ventromedial prefrontal cortex? *Brain* 128: 2872—81.

Van den Bos, R., Harteveld, M., and Stoop, H. 2009. Stress and decision-making in humans: performance is related to cortisol reactivity, albeit differently in men and women. *Psychoneuroendochrinology* 34: 1449–58.

van Gaalen, M. M., van Koten, R., Schoffelmeer, A. N. M., and Vanderschuren, L. J. M. J. 2006. Critical involvement of dopaminergic neurotransmission in impulsive decision making. *Nat Neurosci* 11 (8): 1450–7.

Varani, K., Portaluppi, F., Merighi, S., Ongini, E., Belardinelli, L., and Borea, P. A. 1999. Caffeine alters A2A adenosine receptors and their function in human platelets. *Circulation* (19): 2499–502.

Volkow, N. D., Wang, G. J., and Fowler, J. S. 1997. Imaging studies of cocaine in the human brain and studies of the cocaine addict. *Ann N Y Acad Sci* 820: 41–54.

Wesensten, N. J., Killgore, W. D. S., and Balkin, T. J. 2005. Performance and alertness effects of caffeine, dextroamphetamine, and modafinil during sleep deprivation. *J Sleep Res* 14: 255–66.

Wilson, J. F., Nugent, N. R., Baltes, J. E., et al. 2000. Effects of low doses of caffeine on aggressive behavior of male rats. *Psychol Rep* 86 (3): 941–6.

Wylie, S. A., van den Wildenberg, W. P., Ridderinkhof, K. R., et al. 2009. The effect of speed-accuracy strategy on response interference control in Parkinson's disease. *Neuropsychologia* 47 (8–9): 1844–53.

Index

A

Active versus reactive inhibition control, 299–301
Acute versus chronic pain, 20–21
Adenosine, 300
  caffeine and, 272
  creatine and, 216–219, 222–223, 225–226
  theanine and, 241
Adolescents
  caffeine use by, 285–286
  cognitive function in obese, 137–139
Adults. *See also* Older adults
  breakfast consumption by, 55–58
  obesity and brain function in, 134–137
Aggressive behavior and caffeine, 295–297, 301–303
Aging process
  caffeine and, 284
  fasting and, 39–40
Alertness
  caffeine and, 277–279, 281
  common cold and, 281
  theanine effect on cognition and, 247–254
Alpha(α)-amino-3-hydroxy-5-methylisoxazole-4-propionic acid (AMPA), 243–244
Alpha-lipoic acid, 23–24
Alzheimer's disease, 136–137, 189–190
  creatine and, 220
  folate deficiency and, 199–200
  homocysteine and, 197–198
  theanine and, 243, 257
  vitamin B deficiencies and, 193, 196–197, 203–207
American College of Sports Medicine (ACSM), 74
Amino acids, 27–29. *See also* Creatine; Theanine
Amyloid beta (Aß), 242
Amyotrophic lateral sclerosis (ALS), 220–221
Analgesia. *See also* Pain
  amino acids and, 27–29
  carbohydrates and, 25–26
  common foods and, 33–35
  dietary herbs and supplements and, 35–39
  fats and, 21–24
  fish oils and, 23
  flowers and, 38
  fruit and, 38–39
  future prospects, 40
  leaves and, 37
  macronutrients and, 21–27
  micronutrients and, 27–32
  polyamines and, 31–32
  protein and, 24–25
  seeds and, 39
  soy and, 25, 33–35
  tart cherries and, 33
  tree bark and, 37
  vitamins and, 29–31
Ananas comosus, 38–39
Animals, obesity and brain function in, 139–141
Anorexia nervosa (AN), 120–122
Anti-inflammatory role of cholinergic pathways, 24
Antioxidants, 30, 220–221, 260
Anxiety, 245–247, 256, 284–285
Appetite, 86–87
Arachidonic acid, 22, 25
Ascorbic acid, 30
Ataxia, 194
Attention Network Task (ANT), 300

B

B, vitamin, 29–30, 207–208
  6, 194–195, 202–204
  12, 193–194, 200–202
  intervention studies, 204–207
  neurological effects of deficiencies in, 192–195
  subacute combined degeneration (SACD) and, 193–194
Balloon Analog Risk Task, 298
Binge-eating disorders (BED), 116–118
Bitter orange, 159, 163
Black teas, 238–239, 272
Blood-brain barrier, 240–241
Blood pressure and theanine, 254–256
Body Mass Index (BMI), 104, 110, 115, 154
Boston Veterans Affairs Normative Aging Study, 202
Boswellia, 37
Brain, the
  creatine and disorders of, 218–226
  dehydration and, 9–13
  hydration and the aging, 13
  ischemia of, 242
  obesity and, 134–146
  theanine and, 241–245, 242, 256–258

Brain-derived neurotrophic factor (BDNF)
deficiency, 114, 144–145, 193
Brain function and obesity. *See also* Cognitive function
in adults, 134–137
brain-derived neurotrophic factor (BDNF) and, 144–145
brain structure and, 142–143
in children and adolescents, 137–139
dementia and, 136–137
distribution of body fat and, 135–136
in experimental animals, 139–141
future research on, 145–146
glucose regulation and, 143–144
hippocampus, 140, 142–143
lipids and, 145
oxidative stress and inflammation and, 144
potential mechanisms underlying the interactions among behavior, 141–145
sex differences in, 136, 141
Breakfast consumption
by adults, 55–58
benefits of, 54, 61–62
by children, 58–59
cognitive function and, 55–59
definitions of, 53
glycemic index and load affected by, 62
mood and, 57, 59
nutritional recommendations and, 54–55
obesity and, 54–55
physical health and, 55
underlying mechanisms, 61–63
well-being and, 59–61
Brief Sensation Seeking Scale, 298
Bristol Dietary Caffeine and Health Survey, 245, 248
Bulbs and roots, 35–37
Bulimia nervosa (BN), 118–120

C

C, vitamin, 30
Caffeine, 1, 2. *See also* Theanine
action on the body, 272
aggressive behavior and, 295–297
alertness and cognition effects of, 247–254
average intake, 273
behavioral effects of acute ingestion of, 273–275
beneficial effects of habitual consumption of, 281–285, 293–294
circadian troughs and, 279–280
in coffees and colas, 260
common cold and, 281
content of selected beverages, *3*
cortisol and, 305–306
defined, 272
dopamine, behavioral control and, 303–305
driving and, 283–284
effect on fatigue, 2–4, 278
effect on mood and cognitive perfomance, 2–4, 273–274, 279
effect on physical performance, 274–275
effects on physiology and implications for aggressive and risky behavior, 301–303
external factors affecting metabolism of, 273
health effects of, 286
inhibitory control and risk taking affected by consumption of, 298–299, 298–301
mechanisms underlying behavioral effects of, 275–277
mental health and, 284–285
outside of work, 282
personalized nutrition and, 4
pharmacokinetics and mechanisms of action of, 245
relaxation and, 245–247
research on, 294–295, 306–307
reversal of impairments due to low levels of arousal, 277–281
risky behavior and, 297–301
sleep deprivation and, 277–279
social behavior and, 282–283
sources of, 237–238, 272
sustained military operations and, 280–281
sustained work and, 280
use by children and adolescents, 285–286
withdrawal, 253–254, 276–277, 302–303
at work, 282
Calcitriol levels, 158
Calcium, 32, 160
Caloric restriction and pain, 39–40
Camellia sinensis, 160, 164
Cannabinoid (CB) system, 40
Carbohydrates, 25–27
analgesia mechanisms, 26–27
metabolism modulators, 157
Carnitine, 28–29
Catabolism of catecholamines, 193
Catecholamines catabolism, 193
Catechol-O-methyltransferase (COMT), 193
Chemotherapy and theanine, 259
Cherries, tart, 33
Children
breakfast consumption by, 58–59
caffeine use by, 285–286
cognitive function in obese, 137–139
Chitosan, 160, 164
CHO balance, 90–93
Cholecystokinin (CCK), 24, 106

Cholesterol
breakfast consumption and, 55
cognitive functioning and, 145
theanine and, 255
Cholinergic pathways, 24
Chromium picolinate, 158–159, 163
Chronic versus acute pain, 20–21
Circadian troughs, 279–280
Citrus aurantium, 159, 163
CLOCK mutant mouse, 115
Clostridia, 55
Coffee, 260, 272. *See also* Caffeine
aggressive behavior and, 296
Cognitive function. *See also* Brain function and obesity
assessments, 139–140
breakfast consumption and, 55–59, 63
caffeine effects on, 2–4, 273–274, 279, 284
creatine and, 226–227
diet composition and, 140–141
distribution of body fat and, 135–136
folate deficiency and, 192–193, 199–200, 205, 206–207
homocysteine and, 190–192, 196–198, 206
hydration and, 7–8, 12
impairment and decline, theanine consumption and, 254–259
in obese adults, 134–137
in obese children and adolescents, 137–139
one-carbon metabolism and, 190–192
safety and, 63
theanine and, 247–254
vitamin B deficiencies and, 192–195, 200–208
Cola, 260, 272. *See also* Caffeine
Common cold, the, 55, 281
Compensation, biological and behavioral
to dietary restriction and physical activity, 78
temporal profile and stability of, 84–85
Complex carbohydrates, 25–26
Complex regional pain syndrome type 1 (CRPS-1), 30
Conjugated linoleic acid, 159–160, 163
Cortisol, 13
caffeine and, 305–306
Cravings
in healthy pregnancy, 173–174
in mild gestational diabetes mellitus, 179–182
C-reactive protein, 25–26
Creatine
Alzheimer's disease and, 220
amyotrophic lateral sclerosis (ALS) and, 220–221
benefits of, 215–216
brain-related disorders and, 218–226
cognitive behavior and, 226–227
deficiency, 218
depression and, 224–226
future research, 227–228
Huntington's disease and, 219–220
intake, synthesis, and metabolism, 216–218
Parkinson's disease and, 221–222
psychological stress and, 223–224
schizophrenia and, 222–223
traumatic brain injury (TBI) and, 226
Curcuma longa, 36–37
Cytokines, 24, 25

D

D, vitamin, 30–31
D-amphetamine, 107
Dehydration
fluid regulation and, 8–9
in infants and children, 11–12
mental performance and, 7–8, 10–13
in older individuals, 13
potential mechanisms, 13–14
stress hormones and, 13–14
brain and, 9–13
voluntary, 12
in young adults, 10–11
Dementia and obesity, 136–137. *See also* Alzheimer's disease
Dependence, food, 106
Depression and creatine, 224–226
Devil's claw, 35–36
Diabetes, 260
Diclofenac, 29
Dietary restriction. *See* Restriction, dietary
Dietary supplements for weight loss (DSFWLs)
calcium, 160
chitosan, 160, 164
chromium picolinate, 158–159, 163
citrus aurantium, 159, 163
conjugated linoleic acid, 159–160, 163
defined, 154
efficacy, 158–162
ephedrine, 159, 163
garcinia extracts, 160–161, 164
glucomannan, 160, 164
green tea catechins, 160, 164
guar gum, 158, 163
hydroxymethylbutyrate, 161, 164
mechanism of action, 156–158
pausinytalia yohimbe, 161–162, 164
prevalence and sales figures, 154–156
psyllium, 161
pyruvate, 161, 165

reasons for use, 156
safety, 162–165
DNA methylation, 112–113, 193
Docosahexaneoic acid (DHA), 22
Dopamine (DA), 108–111
active transporter (DAT) response, 112–113, 119–121
in binge-eating disorders, 116–118
in bulimia nervosa, 119–121
caffeine and, 303–305
central modulators of, 113–114
peripheral signals influencing hedonic aspects of food intake and, 114–115
theanine and, 242
D-phenylalanine, 27
DRD2 gene
in anorexia nervosa, 120–121
in bulimia nervosa, 119
Driving and caffeine, 283–284
Duration of feeding and obesity in animals, 141

E

Early morning work and caffeine, 279
Eating disorders
anorexia nervosa (AN), 120–122
binge-, 116–118
bulimia nervosa (BN), 118–120
reward deficiency hypothesis in, 116–122
Eicosapentaenoic acid (EPA), 22
Electroencephalographic (EEG) alpha effects of theanine, 249–254
Endocrine and metabolic profiles in healthy pregnancy and GDM, 171–172
Energy
balance, impact of dietary restriction and physical activity on, 75–77
drinks, 1, 294
expenditure and dietary supplements, 156
intake and substrate metabolism, 89–93
intake effect of physical activity, 80–82
nonexercise activity thermogenesis and, 82–84
Ephedrine, 159, 163
Epidemiology of obesity, 103–104
Epileptiform convulsions, 194
Ethical Naturals, 239
Ethylamine, 238
European Commission (EC), 239
European Food Safety Authority (EFSA), 239
Evaluation of Risks Scale, 298
Evening primrose oil, 39
Evening work and caffeine, 279
Exercise-induced energy expenditure (ExEE), 76
Experimental animals, obesity and brain function in, 139–141

F

Fasting and pain, 39–40
Fat, dietary, 21–24. *See also* Fatty acids
absorption, 158
postprandial trafficking of, 87–89
Fat, distribution of body, 135–136
Fatigue, caffeine effect on, 2–4, 278
Fat mass and obesity-associated (FTO) locus, 112
Fatty acids, 22–24, 39
oxidation (FAO), 89–90, 157
synthesis, 158
Feverfew, 38
Fiber, dietary, 26, 55
Fish oils, 22–23
5-hydroxyindoleacetic acid (5-HIAA), 242
Flowers, 38
Fluid regulation, 8–9
aging and, 13
Flu virus, 259
Folate deficiency, 192–193, 199–200, 205, 206–207
Food
as a dependence-inducing substance, 106
intake, peripheral signals influencing hedonic aspects of, 114–115
reward, 105–106
Food and Drug Administration (FDA), 154
Fruit, 38–39

G

Gamma-aminobutryic acid (GABA), 193, 195, 224, 241
caffeine effects on, 275–277
neurotransmitters, 14
theanine effects on, 242, 256–257
Garcinia extracts, 160–161, 164
Gender and cognitive function with obesity, 136, 141
Genetic evidence of reward deficiency in obesity, 112
German Federal Institute for Risk Assessment, 239
Gestational diabetes mellitus (GDM)
defined, 170
endocrine and metabolic profiles in healthy pregnancy and, 171–172
future research needs, 182–183
implications for clinical practice, 183–184
liking for sweet-fat dairy drinks in mild, 175–178
liking for sweet taste and glucose intolerance in, 175
medical management of, 172
risk and consequences of, 170–171

role for gestational hormones in sweet taste changes in mild, 178–179
sweet crasvings and dietary intake in mild, 179–182
sweet taste changes and appetite in diabetes type 2 and, 174–182
sweet taste changes and diet in, 172–173
Ghrelin, 115
Ginger, 36
Glucomannan, 160, 164
Glucose intolerance and liking for sweet taste, 175
Glucose metabolism
gestational diabetes mellitus (GDM) and, 170–171
obesity and, 143–144
Glutamate neurotransmitters, 14
theanine and, 243–245, 256
Glutamic acid, 238
Glycemic index and load, 62
Glycine, 241, 242
Glycogenostatic model, 87
GRAS Monograph for Suntheanine®, 239
Green tea catechins (GTC), 160, 164, 238–239, 272. *See also* Theanine
Guar gum, 158, 163

H

Harpagophytum procumbens, 35–36
Health
beneficial effects of habitual consumption of caffeine on, 281–285, 293–294
breakfast consumption and, 59–61
Healthy pregnancy, 171–172
changes in sweet taste and cravings in, 173–174
Herbs and supplements, 35–39
Hippocampus, 140, 142–143
Homocysteine, 190–192, 196–198, 206
Homovanillic acid (HVA), 119
5-Hour Energy™, 1
Huntington's disease, 219–220
Hydration. *See* Dehydration
(-)-hydroxycitrate (HCA), 89–90
Hydroxycitric acid, 160–161, 164
Hydroxycut, 155–156
Hydroxymethylbutyrate, 161, 164
Hypericum perforatum, 158, 161
Hyperlipidemia, 258
Hypernatremia, 9
Hyponatremia, 9

I

Immune function and theanine, 259
Inattention, 251–252, 276
Infants and children, dehydration in, 11–12
Inflammation
and anti-inflammatory role of cholinergic pathways, 24
obesity, oxidative stress and, 144
Influenza, 259
Inhibitory control and risk taking, 298–299, 298–301
Insulin resistance, 143–144
Internalized inattention, 251–252
International Life Sciences Institute (ILSI), 2
Iovate Health Sciences, 155
Iowa Gambling Task (IGT), 305–306

J

Japanese Food Additive Association, 239
Jitteriness, 245–247

L

Lactoferrin, 25
L-arginine, 28
Leaves, 37
Leptin, 106, 114
gestational diabetes mellitus (GDM) and, 171–172
Lewy bodies, 221
Lipids and cognitive functioning, 145

M

Macrocytic anemia, 200
Macronutrients, 21–27
Magnesium, 32
Maine-Syracuse Longitudinal Study, 203
Maintenance of Wakefulness Test (MWT), 277
McArthur Studies of Successful Aging, 203
Melanocortin-concentrating hormone (MCH), 113–114
Memory
breakfast consumption and, 57, 58
caffeine and, 274
theanine and, 256–258
Mental energy, 1–2
caffeine effect on, 2–4
methods to assess, 4–5
Mental health and caffeine, 284–285
Mental performance and dehydration, 10–13
Metabolic compensation
dietary restriction and, 78–79
physical activity and, 79–80
Metabolism
creatine, 216–218
glucose, 143–144
Metabolism, substrate, 86–87, 93–94

as a physiological driver of eating behavior, 89–93
postprandial trafficking of dietary fat and, 87–89
Methyl-phenyl-tetrahydropyridine (MPTP), 222
Micronutrients, 27–32
Milk, 54
mild gestational diabetes mellitus and liking for sweet-fat drinks with, 175–178
Monoamines and theanine, 241–242
Monster®, 1
Mood
breakfast consumption and, 57, 59
caffeine effect on, 2–4, 273–274, 279
Morris water maze, 139–140, 145
MRNA and obesity, 109, *111*
Multiple Sleep Latency Test (MSLT), 277
Mu-opioid receptor (OPRM1) gene, 118

N

National Health and Nutrition Examination Surveys (NHANES), 54, 201–202
Neuro-protective effects of theanine, 256–258
Neurotransmitter synthesis, 195
Night work and caffeine, 279
NMDA receptors, 32
caloric restriction and, 39–40
N-methyl *D*-aspartate (NMDA), 227, 243–244
Nonexercise activity thermogenesis (NEAT), 82–84
Noradrenaline, 241
Nutritional recommendations and breakfast, 54–55

O

Obesity and overweight. *See also* Weight gain
brain-derived neurotrophic factor (BDNF) and, 144–145
and brain function in experimental animals, 139–141
and brain function in humans, 134–139
brain structure and, 142–143
central modulators of dopamine signaling that are altered in, 113–114
defined, 154
dementia and, 136–137
dopamine and, 108–111
epidemiology of, 103–104
evidence disputing the reward deficiency hypothesis in, 115
food as a dependence-inducing substance and, 106
food reward and, 105–106
genetic evidence of reward deficiency in, 112
glucose regulation and brain functioning, 143–144
neurobiological evidnece for reward deficiency in, 107–111
oxidative stress and inflammation, 144
potential mechanisms underlying the interactions among obesity, brain function, and behavior, 141–145
predisposition, 108–109
prevalence, 133–134
reward deficiency hypothesis and, 104, 107–115, 122–123
side effects of, 134
theanine and, 258–259, 261–262
transgenerational, 112–113
Oenothera biennis, 39
Older adults. *See also* Adults
breakfast consumption by, 58
caffeine and, 284
dehydration in, 13
Omega-3 polyunsaturated fatty acids, 22–23
One-carbon metabolism, 190, 216
Opioid signaling pathways, 105
in binge-eating disorders, 116–117
Orexin, 113
Overweight. *See* Obesity and overweight
Oxford Healthy Aging Project, 201
Oxidative stress, 144

P

Pain. *See also* Analgesia
acute versus chronic, 20–21
caloric restriction and, 39–40
defined, 20
Parkinson's disease (PD), 221–222, 257
Pausinytalia yohimbe, 161–162
Peripheral signals influencing hedonic aspects of food intake, 114–115
Personalized nutrition and caffeine, 4
Pharmacokinetics
of caffeine, 245
of theanine, 240–241
Phasic alpha, 250–251
Phosphocreatine, 216–218
Physical activity. *See also* Restriction, dietary; Weight management and loss
anorexia nervosa and excessive, 122
biological and behavioral compensation to, 78
caffeine effect on, 274–275
effect on energy intake, 80–82
factors in weight gain, 72–73
impact on energy balance, 75–77
interindividual variability in biological and behavioral responses, 77–78
metabolic compensation and, 79–80

nonexercise activity thermogenesis and, 83–84
temporal profile and stability of compensation with, 84–85
weight loss using, 73–75, 85–86
Phytoestrogens, 34–35
Pineapple, 38–39
Plasma sodium, 9
Polyamines, 31–32
Polyphenols, 36, 255
Postlunch dip and caffeine, 279
Postprandial trafficking of dietary fat, 87–89
Predisposition, obesity, 108–109
Pregnancy, healthy, 171–172. *See also* Gestational diabetes mellitus (GDM)
Prostaglandin E2, 25
Protein, dietary, 24–25
soy, 25, 33–35
Psyllium, 161
Pyridoxal-5'-phosphate (PLP), 202–204
Pyridoxine, 29
Pyruvate, 161, 165

R

Reaction times and breakfast consumption, 56
Red Bull®, 1
Relaxation and theanine, 245–247
Resting metabolic rate (RMR), 85, 86
Restriction, dietary. *See also* Obesity and overweight; Physical activity; Weight management and loss
biological and behavioral compensation to, 78
factors in weight gain, 72–73
gestational diabetes mellitus (GDM) and, 172
impact on energy balance, 75–77
metabolic compensation and, 78–79
nonexercise activity thermogenesis and, 83
temporal profile and stability of compensation with, 84–85
weight loss using, 73–75, 85–86
Reward deficiency syndrome, 104, 122–123
in anorexia nervosa (AN), 120–122
in binge-eating disorders (BED), 116–118
in bulimia nervosa (BN), 118–120
central modulators of dopamine signaling that are altered in obesity with, 113–114
D-amphetamine and, 107
in eating disorders, 116–122
evidence disputing, 115
genetic evidence of, 112
neurobiological evidence for obesity-linked, 107–111
peripheral signals influencing hedonic aspects of food intake in, 114–115
transgenerational obesity and, 112–113
Riboflavin, 29–30
Risk taking and inhibitory control, 298–299
Risky behavior and caffeine, 297–301, 301–303
Roots and bulbs, 35–37

S

Sacramento Area Latino Study on Aging (SALSA), 197–198, 199, 201
S-adenosylhomocysteine (SAH), 190–192
folate deficiency and, 192–193, 199
subacute combined degeneration (SACD) and, 194
S-adenosylmethionine (SAM), 190–192
creatine synthesis and, 216
folate deficiency and, 192–193, 199
subacute combined degeneration (SACD) and, 194
Safety
cognitive function and, 63
dietary supplements for weight loss, 162–165
Satiety, increased, 157
Satiety Cascade, 86–87
Schizophrenia, 222–223
Seasonal Variation of Blood Cholesterol Study (SEASONS), 54
Seeds, 39
Serotonin, 241–242, 255–256, 301–303
Simple carbohydrates, 25–26
Sleep
caffeine and, 277–279
theanine and, 259–260
Social behavior and caffeine, 282–283
Sonchus oleraceus, 37
Sound hypersensitivity, 194
Sow thistle, 37
Soy, 25, 33–35
St. John's wort, 158, 161
Stress
creatine and psychological, 223–224
hormones and hydration, 13–14
theanine and, 245–247
Subacute combined degeneration (SACD), 193–194
Substrate metabolism, 86–87, 93–94
as a physiological driver of eating behavior, 89–93
postprandial trafficking of dietary fat and, 87–89
Sugars, 25–26
Supplements, dietary
calcium, 160
chitosan, 160, 164
chromium picolinate, 158–159, 163
citrus aurantium, 159, 163
conjugated linoleic acid, 159–160, 163

defined, 154
efficacy, 158–162
ephedrine, 159, 163
garcinia extracts, 160–161, 164
glucomannan, 160, 164
green tea catechins, 160, 164
guar gum, 158, 163
hydroxymethylbutyrate, 161, 164
mechanism of action, 156–158
pausinytalia yohimbe, 161–162, 164
prevalence and sales figures, 154–156
psyllium, 161
pyruvate, 161, 165
reasons for use, 156
safety, 162–165
Sustained military operations, 280–281
Sweet-fat dairy drinks, 175–178
Sweet taste
analgesia and, 26–27
change and diet in type 2 diabetes, 172–173
changes and appetite in gestational diabetes mellitus (GDM), 174–182
changes in healthy pregnancy, 173–174
glucose intolerance and liking for, 175
role for gestational hormones in changes in, 178–179
Switching-off mechanism, 252

T

Taiyo International, 239
Tanacetum parthenium, 38
Tart cherries, 33
Taste perception and preference in healthy pregnancy, 174
Taurine, 28
Tea. *See also* Theanine
green tea catechins (GTC), 160, 164
Temporal profile and stability of compensation, 84–85
Theanine. *See also* Caffeine
alertness and cognition, 247–254
blood pressure lowering effects of, 254–256
chemotherapy and, 259
consumption and risk of cognitive impairment and decline, 254–259
effects on monoamines, 241–242
effects on the GABA system, 242
effects on the glutamate system, 243–245
effects on the glycine system, 242
electroencephalographic (EEG) alpha effects of, 249–254
immune function and, 259
intake, 238–240
major sources of, 237–238
neuro- and memory-protective effects of, 256–258
neuropharmacological actions of, 241–245
pharmacokinetics of, 240–241
pharmacology, 240
public health implications of, 260–262
relaxation and, 245–247
sleep quality and, 259–260
stability, 239
weight control and, 258–259
Thirst, 8–9
Tonic alpha, 250
Total energy expenditure (TEE), 78–79
Trafficking, nutrient, 87–89
Transgenerational obesity and reward deficiency, 112–113
Traumatic brain injury (TBI) and creatine, 226
Tree bark, 37
Trier Social Stress Test (TSST), 305
Tryptophan, 242
Turmeric, 36–37
2005 Dietary Guidelines for Americans, 54
Type 2 diabetes, 260
Tyrosine hydroxylase (TH), 109

U

U.S. Army Medical Research and Materiel Command (USAMRMC), 5

V

Vasopressin, 14
Ventral tegmental area (VTA), 113
Vesicular monoamine transporter (VMAT2), 109
Vitamins, 29–31
breakfast consumption and, 54
Voluntary dehydration, 12

W

Washington Heights-Inwood Columbia Aging Project, 204
Water. *See also* Dehydration
elimination, 158
intake regulation, 8–9
intoxication, 9
Weight gain. *See also* Obesity and overweight
breakfast consumption and, 54–55
postprandial trafficking of dietary fat in, 87–89
role of diet and physical activity in, 72–73
substrate metabolism as a physiological driver of eating beahvior in, 89–93

Weight management and loss. *See also* Dietary supplements for weight loss (DSFWLs); Physical activity; Restriction, dietary
breakfast consumption and, 54–55
dietary restriction and, 73–74
diet-induced and physical activity-induced, 74–75
impact of dietary restriction and physical activity on energy balance and, 75–77
physical activity and, 74
theanine and, 258–259, 261–262
Well-being and breakfast consumption, 59–61
Withdrawal, caffeine, 253–254, 276–277, 302–303
Work and caffeine, 282
early morning, 279
evening, 279
night, 279
postlunch dip and, 279
sustained, 280
sustained military operations, 280–281

X

Xerocomus badius, 238

Y

Yohimbe, 161–162, 164
Young adults, dehydration in, 10–11

Z

Zingiber officinale, 36